On Not Dying

On Not Dying

SECULAR IMMORTALITY IN THE AGE OF TECHNOSCIENCE

Abou Farman

UNIVERSITY OF MINNESOTA PRESS
Minneapolis
London

Parts of chapter 3 were originally published as "Death and the Infinite: Cryonics as Secular Eschatology," in *The Wiley Companion to the Anthropology of Death*, ed. ACGM Robben (New York: Wiley, 2018). Portions of chapter 4 were originally published as "Speculative Matter: Secular Bodies, Minds, and Persons," *Cultural Anthropology* 28, no. 4 (2013): 737–59. Parts of chapter 5 were originally published as "Re-Enchantment Cosmologies: Mastery and Obsolescence in an Intelligent Universe," *Anthropological Quarterly,* special issue on Extreme: Humans at Home in the Cosmos, ed. D. Battaglia, D. Valentine, and V. Olson, 85, no. 4 (2012): 1069–88, and as "Mind out of Place: Transhuman Spirituality," *Journal of the American Academy of Religion* 87, no. 1 (March 2019): 57–80.

Published by the University of Minnesota Press
111 Third Avenue South, Suite 290
Minneapolis, MN 55401-2520
http://www.upress.umn.edu

Printed in the United States of America on acid-free paper

The University of Minnesota is an equal-opportunity educator and employer.

Library of Congress Cataloging-in-Publication Data
Names: Farman, Abou, author.
Title: On not dying : secular immortality in the age of technoscience / Abou Farman.
Description: Minneapolis : University of Minnesota Press, [2020] | Includes bibliographical
    references and index.
Identifiers: LCCN 2019027138 (print) | ISBN 978-1-5179-0809-6 (hc) |
    ISBN 978-1-5179-0810-2 (pb)
Subjects: LCSH: Secularism. | Immortality. | Ethnology. | Religion and science.
Classification: LCC BL2747.8 .F37 2020 (print) | DDC 306.9—dc23
LC record available at https://lccn.loc.gov/2019027138

To the fullness of her afterlife

Contents

Preface

REALM OF THE POSSIBLE

I knew I wanted to work on death and secularism. I had read the older ethnographies on death—ranging from Bloch on Madagascar to Goody on the Lodagaa and Battaglia on the Sabarl—and was just exploring some of the new work on brain death, organ transplant, physician-assisted suicide, and persistent vegetative states in the United States when, down in the basement computer lab of the library of the City University of New York's Graduate Center, I stumbled upon a footnote about something called cryonics.

Sometimes a footnote can change your life. And maybe your death.

From the little I gleaned as I first followed the trail, it appeared to me that some people with a strongly scientific and avowedly atheist orientation were arguing that death was a medico-legal fiction. They were saying that humans could be preserved in a nondead state and revived later in the future. They were saying this could happen even after they had legally died.

Quickly, the footnote led me down a rabbit hole.

Although it all felt counterintuitive and somewhat disorienting, this first part sounded at least a little familiar from the work on brain death and organ transplant that had become staples of the anthropological literature. It also echoed discussions in the work on personhood and new reproductive technologies. In both, the line separating person and nonperson, life and death, consciousness and its lack, was at stake. Who could decide where the line was? On what criteria? With what consequences? That work, originally done in medical anthropology, was increasingly being taken up in the growing zone of encounter with science and technology studies and the anthropology of science and technology. Detailed and critical ethnographies by mostly feminist scholars such as Margaret Lock, Sharon Kaufman, Lynn Morgan, Rayna Rapp, and Susan Franklin inspired me, but, having started to think and work with the anthropologist Talal Asad, I kept sensing a lack of engagement with issues raised by Asad and others following his now well-known challenge to the social sciences: What might an anthropology of the secular look like?

At times religion was an object of analysis, in part because many of the challenges to orthodox medical and scientific regimes in the United States were coming from religious groups invoking religious values at the two ends of life: for example, in cases around abortion, stem-cell research, or brain death. Religion was where the conflict was seen to be (also see Feldman 2005). Modernity, too, was an object of critique in many key works, as it was said to have a particular orientation to disease and death and biology, a desire and need to master and measure and control, as well as a catastrophically Cartesian view of the mind and the body that surely had negative effects, especially in medicine. However, the secular and secularism appeared as background assumptions or side comments, if at all. Even though the scholarly work and the bioethical debates were replete with references to secular humanist values, I could find little analysis about what secular assumptions or rules were shaping these humanist debates and practices.

Cryonics seemed to be challenging the predominant medical, legal, and scientific views of death. Yet cryonicists were themselves reductive materialists, working with the same assumptions as most medical professionals and researchers. They were suggesting that if reductive materialism is true—that is, if your mind or self or individual identity is reducible to your brain states—then logically you ought to be able to preserve that brain state if you had the proper technology to do so. One way to preserve that brain state, they further reasoned, was to cryopreserve it, just as other biological matter—even if only simple forms of it—was being "frozen" and then reused.

There is much more to what cryonicists were saying and doing, but what I had figured out up to then was that in a couple of buildings across the United States a few mostly white, mostly Western, mostly male, mostly atheist folks were suspended in liquid nitrogen, not dead but not alive, waiting to be reanimated in a distant future. They had been legally declared dead, with death certificates issued by the state's medical examiner, but they were not considered dead by their custodians, the cryonicists who were storing and taking care of their frozen bodies in Detroit and Phoenix. This smelled like anthropological territory for sure. Aside from a small book by a sociologist who had looked at their organizational structures and disputes in the 1970s (Sheskin 1979), nothing with academic heft had been written about this phenomenon. But before I committed to the topic I felt I needed to at least meet a cryonicist and get a sense of the ethnographic viability of the project, beyond articles and websites. Maybe I was making too much of all this. Maybe there was nothing to see, no there there. Online I came across someone from California, Christine Peterson, who was about to give a talk at a business conference at Princeton. Her name led me to the website of the Foresight Institute, an organization having to do with something called nanotechnology. She had set up the institute with her husband (at the time), Eric Drexler. At the time, all this felt beside the point to me really. I didn't realize how nanotechnology had any role to play in the game of death in the secular; neither did I realize that Drexler was recognized as the father of nanotechnology. All that mattered was that Peterson was a cryonicist. I emailed her to see if she would agree to meet up with me.

I was on the New Jersey Transit to Princeton Junction the next evening. My field notes on the train innocently referred to the upcoming Malinowskian ethnographic moment of first contact with a real live cryonicist.

Peterson was a quiet, smart, and gentle white woman, a little distant and wary too, as anyone should be when meeting an anthropologist. We sat for dinner at her hotel restaurant, and I watched as she removed a white paper napkin from her purse and quietly unfolded it. She arranged the dozen or more pills inside in an array laid out flat on the napkin. It reminded me of a Damien Hirst installation. I asked, and she said they were supplements. Each one had some purpose, some function, some benefit that would help her remain optimized and healthy for longer. I didn't know it then, but I would find out that supplements are an important part of longevity regimens, made most famous perhaps by the celebrity inventor Ray Kurzweil, who in big magazine features about him would say that his dozens of supplements a day kept him younger than his biological age. What's more, a supplements manufacturer, the Life Extension Foundation, is a major backer of cryonics and other life-extension ventures.

I remember being surprised that Peterson ordered a burger, which she ate delicately with knife and fork, while pacing her supplement intake. I began to ask my first questions about cryonics. I asked about death and suspension and what it meant that people were not dead and not alive. How did families cope? Was there a closure issue? How did cryonicists prepare? How could it be that cryonicists believed that someone in that state, frozen and declared dead, would not be dead, would still have a chance at a future life? How can someone who has died of cancer or dementia at old age be brought back at all? The dead will be reawakened and then die right away again, won't they?

She patiently explained the logic of cryonics to me more fully. People in a state of cryopreservation were not considered dead; they had deanimated. They were in a state of suspension. And they would not be reanimated—that's the term, not reawakened—unless it was clear that science was able to cure the problem that had caused their legal death to begin with. People whose brains were in bad condition would probably have less of a chance of being reanimated. The key was a good cryopreservation that would preserve as much of the brain structure as possible. People needed to make careful arrangements before they died, fill in all the relevant paperwork that would allow the cryonics organizations to intervene legally and efficiently to prevent further decay of the brain and its cells at the point of legal death. But as long as there was the brain matter in relatively good shape, not overly deteriorated or mushed up by an autopsy, then there was a nonzero chance: it had to be accepted in the realm of the possible.

Strange realm, the realm of the possible, with its internal territory of accepted imaginaries and the impossible dragons beyond. Is the possible a realm that is open, or one that is closed off? The possible is beyond the actual, the present. But the word can both open up the permutations of reality (nothing is impossible) and restrict the wilderness of the imagination (stick to what's possible). Right then, the possible felt like an opening. I thought to myself, wow, this is some radical extension of materialism—pushed to its very limits. I was a little dizzy by the implications of taking on this view. We had gone somewhat beyond supplements and healthy lives and the secular ritual of preserving bodies.

Peterson asked me if I understood.

I had a strange flashback at that point—strange in that way in which your past pushes itself into your consciousness from somewhere else when you are fresh in the field, because after all that's all you have, for at the beginning the field has not yet given you any memories of its own. I remembered an accident from a long time ago. I was walking to the tram stop in Geneva, Switzerland, with my teammates from the high school soccer team. It was fall and getting dark earlier every day; it was foggy and rainy and medieval. We turned the corner from an alley shortcut into the main street only to witness a truck turning and hitting a moped. We ran toward the scene. The moped was bent out of shape on one side, and its rider was splayed on the asphalt on the other. His head had obviously been crushed by the large wheels of the truck. The skull was cracked like a walnut, and bits of brain were splattered all over the street. We watched the rain wash blood and tissue down the gutter under the cold streetlight. We were in shock, unable to do anything. There was no hope there, no chance of rescue, of medical intervention. A series of contingent events—the rain, a moped slipping in the rain at a certain angle, a truck being just at that corner at that moment in time, turning, with enough speed and not enough visibility—and from one second to the next a life was gone, a mind extinguished. The body lying there was the young man's, but the young man was not in it. The truck driver was banging his guilt-filled head against the steel of a lamppost. A passerby threw her jacket over the rider's crushed head. It was not clear who was being protected in that gesture: the mutilated dead man from viewers or the viewers from the dead man. Or both, for dignity in the end inheres in the relationship, not in the singular body. One thing was clear: that was the absolute end of that life. Nothing to recover. What better image of absolute, irreversible death could have struck me at dinner with Christine Peterson?

I replied, "Yes, I get it. Essentially you are saying that theoretically the brain's information is recoverable from the brain so that theoretically you don't die unless, say, an eighteen-wheeler runs over your skull and splatters your brain to bits."

She paused, the last of the hamburger in her hands.

"Well," she said, "it depends."

Depends?! Now what? A brain crushed by an eighteen-wheeler and death is still not absolute?

"Depends. Did you make a backup of yourself last night? Was your information preserved somewhere?"

Ethnographic encounter indeed.[1] I now had to make the connection between the reductive materialist view, dying, brain states, nanotechnological brain repair, mind uploading . . . So all that was not beside the point; it was the point. This wasn't just about freezing life as a way of preserving a body. In a sense, biological life per se seemed incidental to the whole affair, whereas the mind–brain nexus seemed central. It became obvious that this project wasn't going to be just about death in the humanist way I had imagined it, as the abstract, universal backdrop readily invoked in the social sciences and the humanities, that horizon "we" all face, the "human experience of death" as the core of a "universal code," to quote Johannes Fabian (1973), as the "supreme mediator of those oppositions and contradictions by means of which the human mind constructs its universes." Fabian had called for death universalism to argue against the secular primitivization of death in the anthropological encounter with other peoples and practices; he was writing against folklorized accounts in which, "usually without critical examination, . . . 'primitive' reactions to death are placed in the domain of religion (which in turn is taken to constitute a self-contained aspect of human activity) and this makes it possible to replace the general evolutionist perspective with a view of 'secularization,' religious devolution, as an intrinsic component of modernity" (1973, 52) At the same time, Fabian argued, the folklorization of death and death-related behavior placed the matter at a "safe distance from the core of one's own society," that is, the West. So his was just the sort of position that had initially interested me. But in fact, Fabian's re-universalization of death would come at the cost of its reification into an existential horizon or at best a baseline assumption about the psychology

of death in the panhuman "we," without troubling the assumptions—
ontological and epistemic, materialist and rationalist—that underlay the
secularization process that articulated matters of life, mind, and personhood to each other in very particular ways.

Accounts of existence I heard from immortalists (such as information-
theoretic death, explained in the introduction) showed the extent to which
a secular materialist and rationalist matrix could be, perhaps had to be,
stretched out. That people were working on projects hoping that not
just life but mind could be stopped and started, deanimated, suspended,
reanimated; that mind and life could be transposed to other media, running on other nonbiological, non-carbon-based substrates; that the universe was made of something called information that was concentrated
in a highly complex pattern in the form of human brains but that would
keep on increasing in complexity in other material formations, and that
the convergence of certain sciences (nanotechnology, biotechnology,
information technology, and cognitive science, known collectively by the
acronym NBIC) would help the development of this complexity not just
on earth but in the whole universe—all that blew my mind.

It was like discovering a parallel cosmos of materialists whose cosmologists were now called transhumanists, immortalists, Singularitarians,
extropians, and even cosmists. Except it was different from everything
I had thought of under the rubric of materialism. It didn't seem like the
kind of materialism most materialists I knew could jibe with, and indeed,
as it turned out, it wasn't. It seemed like more than materialism. It was
precisely here at this site where materialism bubbled up in excess of itself
that it revealed itself: rather than being a literal-minded description of
what is out there, it indicated its own conventions, its own openings and
closures, tensions and contradictions, especially as an ontological view
espoused in a world where there was also that strange thing called mind,
which, somehow apart from the world, apprehended the world around
it, had an experience of it, changed it deliberately, and was aware that it
was apprehending, experiencing, changing, and being aware.

All this talk of deanimation, suspension, and reanimation, of information and convergence, was landing me squarely in the emerging
anthropology and science studies debates about disenchantment, the

ontological turn, and posthumanism. Instead of being fueled by postcolonial Amazonian animism and contra-modern Latourianism, these blurrings were being propelled by West Coast futurists through a hypermodern, rather than antimodern or postmodern, view in which humans, through technology and reason, could move beyond themselves and give rise to a whole new stage of the cosmos where biology and technology would merge seamlessly and thus infuse the whole matter of the universe such that the cosmos would be *culturally* constituted in what some people were calling a "cosmoculture" (Dick and Lupisella 2010). What seemed to be missing from the medical anthropology debates on death and technology was also missing in much of the discussions in science and technology studies, posthumanism, and the ontological turn: an analysis of secular assumptions and limits.

When I contacted Christine Peterson those many years ago, it was harder to find people open about their cryonics affiliations or interest in immortality research, especially in the scientific realm. It was too kooky and ostracized. Today, fueled by the giddy successes of informatic platforms (i.e., digital cognitive technologies) and easy and powerful gene-editing technologies, defeating death and aging and pursuing digital or biological immortality are active projects all over the United States, from Silicon Valley to university gerontology labs. The projects have gained power as well as legitimacy. Of course there were many causes, including the rise and concentration of the tech sector, the circulation of biocapital and biovalue (Sunder Rajan 2006; Waldby and Mitchell 2006) through the hype and hope of molecular biology (N. Brown 2003), and the proliferation of libertarianism as a political ideology anchoring the spread of neoliberalism especially in the tech sector and in transhumanist and immortalist circles. Although I do delve into a series of overlapping, sometimes surprising histories, this book is not a causal history of that transformation. I am, rather, exploring the cultural logic and the techno-scientific imaginary (Marcus 1995) through which immortality projects have found their niches and taken particular forms in the nexus of medicine, science, law, and other secular institutional practices and discourses in the United States. As mentioned above, one important part of this has to do with materialism and the discontinuity of mind and matter, of human and the universe, which reflects the secular discontinuity of life and death. Another part has to do with discontinuity writ large, that is, with the possibility of the annihilation of the species, of the collapse of civilizations, and other forms of secular futurelessness (Lifton 1987), or what transhumanists have come to call "existential risk." That is, immortality as an imaginary becomes more salient when social and political futures are at risk, when, for example, continuity in the American empire and Western civilization seems threatened. In the midst of that there is the great hype and excitement of technological progress improving everything, turning humans into better humans and harnessing all the information in the universe. The two actually are coiled together, an inevitable end and a bright future, especially in the United States, where the national (white) American narrative has toggled between apocalypse and redemption, "bad endings" (Harding and Stewart 1999) and new beginnings. It's a modern secular drama as much as an American one.

Acknowledgments

In trying to understand and narrate this drama, I have benefited directly and indirectly from the intellectual generosity and support of many people, including Talal Asad, Mayanthi Fernando, Debbora Battaglia, Hussein Agrama, David Valentine, Murphy Halliburton, Rayna Rapp, Patricia Clough, Paul Boghossian, Ida Susser, Shirley Lindenbaum, Miriam Ticktin, the anthropology department at the New School for Social Research, my cohort and all my teachers at the Graduate Center CUNY (still one of the most amazing and radical educational institutions), and the anonymous reviewers of this press and of many journals, who are our colleagues, our selves, doing the work without remuneration or official recognition. At various stages, I was helped in preparing the manuscript by Sara Soudavar, Sarah Chant, and Agnes Eshak. I am indebted to many in the field who sat with me and let me into their homes and organizations and events; I would especially like to thank Regina Pancake, Todd Huffman, and Mike Perry. I am sure there will be much disagreement on their part with what I am saying in these pages, but our political differences were apparent from the beginning: I was often called "that hippie." For all errors of fact, method, and reason, I only have some bad algorithms to blame.

Introduction

THE PROBLEMS OF DISCONTINUITY AND INDETERMINACY IN A SECULAR WORLD

Over the last decade, a small, marginalized, and often discredited assembly of groups that has had as its goal the elimination of biological death by technoscientific means has grown to gain major influence in Silicon Valley and become accepted as a bona fide part of the world of biotechnology and scientific research more generally. From motel room meetings, DIY operating rooms, and garage labs a decade ago, technoscientific immortality projects have found their way into major initiatives at companies like Google and highly capitalized ventures such as Life Biosciences; research on the possibilities of rejuvenation has found its way into important research labs all over the country, such as that of George Church at Harvard University; and the front pages of important popular and scientific publications and media outlets now lend fame, rather than infamy, to the early advocates of technoscientific immortality—people such as Ray Kurzweil, Aubrey de Grey, and Martine Rothblatt.

Even though the movement is composed of a diverse range of groups, with different approaches, practices, and agendas, and while in the new limelight they almost all shy away from the word "immortality," they all nevertheless have a common goal: extending personal identity beyond its current *biological* limitations, beyond what is seen as its current and contingent instantiation in a defective biological platform that leads to death. Many turn to bioengineering or informatic strategies, considering, for example, artificial intelligence (AI) as a potential method of capturing and extending personal identity through mind uploading. Not all immortalists want to jettison the cellular carapace that is our human body, the wetware of our being, and reconstitute themselves as machine or as code. But even those who don't want to be transubstantiated as computer algorithms nevertheless want to get past biology's betrayal of the body, its degenerative descent into death; that is, even the more biocentric of the immortalists favor at least a hybrid body, subjected to computational interventions or nanotechnological reconstructions of biological mechanisms. In a sense, then, they all want to extend being beyond life.

I use the term "immortalism" to denote the movement as a whole, since the core groups have been tightly interconnected, with members and main actors supporting and advocating for each other regardless of their particular focus, appearing on each other's boards and conference panels and audiences, advocating for transhumanism, Singularitarianism, space colonization, and a range of other futurist visions. Members are mostly atheist, white, male, and often libertarian futurists who may be scientists, venture capitalists, organizers, writers, computer engineers, or scientifically oriented laypeople fascinated by the prospect of defeating death and advancing technological futures more generally. Some groups and members advocate rejuvenation therapies through molecular biology and engineering, others through the translation of identity to artificial neural networks, while others place their bets on cryonic preservation and innovations in nanotechnology that could lead to reanimation or bodily repair on an atomic scale. Cryonics, which requires the preservation of body and brain in liquid nitrogen upon the legal declaration of death, is the oldest of the immortalist strategies, but today, for many, it has become the backup option, not the preferred one. If in the next few decades molecular biologists don't solve the problem of aging, or if neuroscientists and AI researchers do not manage to create conscious-enough digital replicas, then at least cryonics will allow dying people to be suspended until such time as science will have advanced far enough to do all those things.

Two common premises underlie the strategies. First, they do not hold a *necro-ontological* view: death is not an inevitable fact about the way existence is constituted in the universe. Since it is only a biological fact, a contingency of evolution's, and not an ontological necessity (Marcuse

1959), it may be defeated. Second, and less predictably, immortalists hold not only a *neuro*centric but also an *informatic* view of human persons (and often of the universe); in immortalism, the precise pattern of atoms in the brain (the information) essentially constitutes the self, including its memories and abilities. The Brains "R" Us proposition was mapped out in the nineteenth century by materialists who did not abide any notion of a transcendental soul as the locus of personhood, and thus sought qualities of the person such as consciousness, intentionality, and identity in *matter* itself. The matter at hand was the brain, and its mapping has been literal at least ever since Paul Broca in France and other atheist anthropologists and phrenologists began to cut open the skull and plot the brain's regions onto capacities and behaviors, beginning what, in her well-known study of the Society of Mutual Autopsy, Jennifer Hecht (2003, 254) has called their "deconsecrating projects." Poking around in the skull, they were happy to declare that there was no soul found in there; instead, they found a clear link between aspects of personality, such as memory or language, and particular structures and regions of the brain. As a basic ontology along with a set of strategies, neurological correlation and causality continue today in neuroscience, in forms such as neurochemical reductionism and brain mapping, flattening out the "space between person and organ" (N. Rose 2007, 198).

That we are our brains entails that if the structures and functions of the brain were known or preserved or replicated at a detailed enough scale, then the self could also be theoretically known or preserved or replicated. For technoscientific immortality projects, this has deep consequences: theoretically, if those patterns—the exact connections and states of your dendrites, axons, nerve endings, and other features of neurons and their synaptic connections and functions—are known or preserved somewhere somehow, then they can in some theoretical future be rebuilt or reanimated again on a biological or nonbiological substrate. This notion of death is what is known in those circles as *information-theoretic death*.

Its original formulation came from Georgia Tech computer scientist Ralph Merkle, who worked with Dr. Robert Freitas, a nanotechnologist who in turn worked closely with the Foresight Institute, an organization set up by Christine Peterson (see Preface) and Eric Drexler. As chapter 2 will describe in more detail, nanotechnology was all about thinking mechanically at the smallest scale, thinking of the very building blocks of life and mind as bots of some sort. It was in that context, then, that death was turned into an information-theoretic matter, where a new kind of potential person emerges whose personhood depends on the development of future technologies and on a notion of a mind–matter continuum conceived through an informatic vision of being. At the nanoscale everything could be code plus function; the person's fleshiness becomes secondary, part of a contingent carbon substrate. Indeed, Merkle's (1992) definition of death carries no biological terminology whatsoever:

> A person is dead according to the information theoretic criterion if their memories, personality, hopes, dreams, etc. have been destroyed in the information theoretic sense. If the structures in the brain that encode memory and personality have been so disrupted that it is no longer possible in principle to recover them, then the person is dead. If they are sufficiently intact that inference of the state of memory and personality are feasible in principle, and therefore restoration to an appropriate functional state is likewise feasible in principle, then the person *is not dead*. (1992, 9)

It was a definition that took hold in the futuristic context fusing nanotechnology and cryonics and computation. It was the only definition of *absolute* death that seemed plausible for cryonics and for technoscientific immortality more generally. In this information-theoretic scenario, function and biology no longer provide sufficient criteria for the evaluation of death, so death becomes largely an information recovery and engineering problem. The implications of such a view affect not only death and dying but the rest of existence too, for it is as "information-beings," or informatic selves (Farman 2014a), that bodies and minds are to be cultivated, preserved, extended, or even edited. Biological life and death would be transformed unrecognizably, and to be immortal you would first have to be nonlife, assume a posthuman form.

Immortality in the information age, then, is a very particular manifestation of not dying, requiring particular assumptions about life and mind and personhood, engendering particular technological desires and futures, constructing particular nonbiological notions of human and posthuman. Many immortalist and transhumanist projects (the memberships overlap but are not identical), such as those run by the trans transhumanist Martine Rothblatt, are geared toward having your "mind" or your "information" uploaded on to a digital platform, becoming an avatar, a robot, or some other nonbiological form of superintelligence that would still be functionally you. The ubiquitous transhumanist Elon Musk, the founder of the private space venture Space X, has been promoting chip implants in the brain that could interface seamlessly with vast computational networks, merging human minds with AI to make the species superhuman and help it survive. In the immortalist and transhumanist communities, all this is cause for rejoicing. For humanists like Harvard professor Stephen Greenblatt, this spells doom for humans and humanity. Writing in response to Musk's plans,[1] Greenblatt argued that humanity is in part defined by biology and especially "biological flaws," without which "we lose an essential part of who we are."[2]

Immortalists I worked with always expressed surprise at how little traction immortality, or life-extension research, had gained among the wider populace. After all, avoiding death and striving for healthy longevity seemed to them an uncontroversial human desire. Was it not clear to everyone that death was a bad thing, something people already tried to avoid on a daily basis, that humans had been striving to overcome death forever? Now, after millennia of cooking up illusory techniques like mummification and concepts like the soul or the afterlife, here was a real scientific possibility of conquering death, and no one was coming on board?! They often pegged the blame on the ongoing influence of religious doctrines. Even those who weren't churchgoers were still under the sway of fake news about death and the afterlife. At the same time, I myself was struck by the extent to which secularists, scientists, and humanists such as Greenblatt, rather than religious folks, were the most vocal critics of the prospect of technoscientific immortality. Physicians, hospice workers, physicists, cryobiologists, lawyers, philosophers, psychologists, evolutionary biologists, my social science colleagues, and atheist friends consistently berated immortalists and transhumanists, viscerally reacting against their ideas and, in turn, calling *the immortalists* religious, often in public debates, often in secular and scientific publications.

That tension points to the core problematic of this book. What happens when a category like immortality goes from being a project given over to religion to a project adopted by technoscience? What can that tell us about secular assumptions? How are death and not dying related to secular notions and ways of being human and of moving beyond the human?

Addressing these questions entails tracing the ways in which technoscientific immortality constructed a posthuman vision of being, a whole cosmology even, in order to advance its immortalist agenda, thereby roiling those who have adopted a hard-fought secular ideal of what human is, what history is, and how the future must progress. That is so because part of what the secular humanist ideal has entailed has been a specific orientation toward death, the acceptance of its inevitability, its finality, and its meaningfulness in human life and civilization. The goal of this book is to examine contemporary "immortality" not just as a set of technoscientific techniques that are building a posthuman future but as a historical, rather than universal, concept within the West, infused with assumptions, problems, fears, tensions, and concerns that arose as the secularization process took hold of the minds of men [*sic*] and the institutions of law, medicine, and science.

PROMISE AND PROHIBITION

Mortality and immortality tend to lend themselves to universal narratives, to sweeping claims about humanity and civilization and the metaphysics of being and not being. Often, these grand narratives go something like this: Members of *Homo sapiens sapiens,* the doubly knowing species, are said to be aware of their impending demise, of the possibility of their own ending, and that makes of humans a strange and poetic species in the universe, for we humans know not only that we exist but that we will also cease to exist; we are aware not only of being but of the cessation of being. Indeed, this "death awareness" is said to define us as a species. What's more, it is said that it's precisely this relationship to death that lends meaning to life; it is through mortality that humans find purpose in life; it is the deepest font of motivation and creativity. Humans are moved—dare we say animated—by death. Without death, the narratives say, we would not be human.

Because of our awareness of death, and because death is at the same time the negation of awareness, we humans are said to be also afraid of death; and so we are the species that consciously seeks a way out of dying. Unlike other species, we have tried a wide range of inventions to transcend this biological limit, to defeat it, deny it. Out of the prospect of mortality, humans have imagined the possibility of immortality, conjuring up futures in which we might continue to exist *despite* the reality of death. Refusing this mechanical inevitability, this biological condition, is also what makes us human, if tragically so, as the humanist philosopher Miguel de Unamuno (1972) pointed out. From the first ritualized burials where we laid out some food and left our dead their goods to take with them to the other world, through the mummies and monuments of big civilizations to art forms and quests for the fountain of youth and backroom deals with gods and devils to today's biotechtopian ventures, humans have pursued immortality in all sorts of tragic, ironic, comic, absurd, and very human ways. When archaeologists identify a formal burial site, they immediately know it is human because, the story goes, only humans have attended to their dead, have created rituals around their dead, memorialized them and continued their existence in this way (T. Taylor 2004). To die is human. To not die is also human (or at least to die trying not to die). One is biology, the other is culture. One is given, inevitable; the other is an overcoming and a fantasy at the same time. The human condition, in this narrative, is the classic interplay of nature and culture, of biological mortality and symbolic immortality.

That interplay is imagined as an underlying force in the motivations of human beings—captured by E. M. Forster's well-known statement that "death destroys a man but the idea of it saves him"—and the telos of human history. Many, like Sigmund Freud ([1930] 1962) and Ernest Becker (1973), have gone as far as to suggest that without this quest to overcome death not only would our lives feel meaningless but also we would have no religion and no great civilizations. The psychologist Robert Lifton also proposed as much: "Much of history can be understood as a struggle to achieve, maintain and reaffirm a collective sense of immortality" (1987, 283). A recent, popular iteration of this narrative puts it this way: "A will to immortality is the underlying driver of civilization" (Cave 2012, 14; also see Gollner 2013). This is a story that not only seems intuitively correct but is widely accepted given the supposed historical evidence, those monuments connected to large empires and priestly cults, erected to satiate the human yearning to last beyond the limits of physical finitude.

Far from being a clear idea, desire, or category through which a history of humanity may be neatly organized and teleologically narrated, immortality has been a contested and shifting category, changing meaning and carrying different charges in different domains at different times in history. Those presumptuous stories about immortality and history, about the human relation to death and not dying, do not describe an actual history of human development; rather, they are describing Western secular humanism's entanglements with these issues. The secular as a social formation (Asad 2003)—as a set of epistemological rules, metaphysical premises, scientific practices, political laws, cultural assumptions, and colonial histories—was shaped from its beginnings and in important respects through the emphatic stance secularists assumed in relation to death and death's finality, projecting finite limits and definitive endings not just for individuals but for species, civilizations, the planet, even the cosmos. Immortality was said to belong to the realm of religious delusions, and so it has been one of the markers through which distinctions between key categories of the secular and the religious, the modern and nonmodern, person and nonperson, life and nonlife, human and nonhuman, were constructed and managed, though not very cleanly, since immortality overspills the boundaries of those secular binaries, in law as much as in medicine and public life.

"Secular" and "immortality"—two key words in the subtitle of this book—sit together tensely. If immortality is said to belong to the realm of religious delusions, which the secular tries to keep at bay, the promise of immortality today is nevertheless coming through science, one of the pillars of secular authority. That has been immortality's history in the modern era, where on the one hand it has been pitched as an illusion by secular humanists (e.g., Lamont [1935] 1990), who associate it with some kind of soul that survives the materiality of the body, and on the other, it has also been a real category within scientific inquiry, where its fate ranges from viable research concept to bona fide goal to anathema (e.g., Hayflick 2000). As a figure, then, the secular immortal is suspended between tense polarities—immortality is illusion, immortality is possible; accept death, extend life; life is all there is, death is all there is; the future is open, the future is finite; life has meaning, the universe is meaningless.

Using research on immortalism, this book is more broadly about those tensions and how they shape each of the concepts (secular, immortal), how they are carried forth and reproduced in relation to death and continuity, mortality and immortality, and materiality and the immaterial through institutions of science, law, and medicine. The tensions I refer to arose through a historical process of secularization in the West, which I have grouped in three domains, each taking on the problem of continuity from a different angle: the elimination of the soul and hence the mind–body problem (dualism), the elimination of the afterlife and hence the problem of time and finitude, and the elimination of a cosmological order and hence the problem of telos and purpose. Each process of secularization has produced a set of gaps or discontinuities around which grow zones of indeterminacy (Kaufman and Morgan 2005) where science, medicine, and law have at best offered answers that have been controversial and at worst have declared that no proper answer can be furnished.

First, the rise of materialism and the elimination of the soul, among the key processes of secularization (Aries 1975; Lacquer 2015; Makari 2015; R. Martin and Barresi 2004; N. Rose 1996; Sloane 1991), produced important new practices and dispositions, emphasizing the materialist determination and self-understanding of the human animal as body and biology—for example, as gene, as brain, as code, as cell or organism, as evolutionarily hardwired, and so on.

Equally and alongside this came the elimination of the afterlife, which ushered in a sense of the finality of death and the nothingness that follows it, a feeling that, in the words of Unamuno, was "more terrifying than hell" (1972, 48). Whatever the affective reactions, the scientific elimination of the soul and the afterlife led to a particular view of the finitude of life. This second shift in the secular frame occurred, then, when certain temporal orders were rendered illegitimate (Kosselleck 2002; C. Taylor 2007) and secular scientific worldviews discarded such "illusions" as eternity, resurrection, and cyclical time (Leach 1961), instead generating their own secular eschatologies.

The third is the convoluted and tense secular scientific relationship to the cosmos and to telos, according to which cultural, physical, and biological evolution are nondirectional and without a predefined purpose, even though a strong notion of progress animates all secular social and scientific projects. Many immortalists, in the meantime, seem quite happy positing a kind of purpose, based on their notions of information and a "waking universe" (e.g., Kurzweil 2007; Tegmark 2017). According to theories about the technological Singularity, once the accelerating pace of technological developments leads to an intelligence explosion in computation and machines become conscious enough to reproduce themselves and have an interest in doing so, the future will no longer be predictable. Thus the Singularity suggests the collapse of the mind–matter divide. Indeed, as I will argue further, immortalism frequently challenges the consensus views around all three zones, pushing the boundaries of secular norms from within.

Aspects of these problems may also exist in some version in other cultures and ages. In some sense, dualism, for example, seems to be common cross-culturally (Porath 2007), and the thought that something like the soul (or many souls) and body were separate predates Descartes in what is commonly called the Western tradition. Equally, it is wrong to think that the Western materialist tradition is the only one to truly think of individual death as final. Certain hunter-gatherer groups such as the Hazda regard death in ways that would strike a materialist as perfectly reasonable (Woodburn 1982). Death is considered the end of the individual, and that's simply that. They don't even conduct any elaborate ceremonies. *Plus materialiste que les materialistes!*

But these questions get raised and tackled in the Western secular and scientific context in particular ways. The tension that charges dualism in the modern Western context, for example, is a tension between two historical strains claimed by secular thought, the materialist and the rationalist, which maps roughly onto the body and the mind (also see Jonas

[1966] 2001). Rationalism (or idealism) and materialism, "which in the seventeenth century had been sitting peacefully side by side" (Collingwood 1960, 12), turned into rivals in the eighteenth as a result of secularization. One might say that once Descartes's God was removed from the equation there was nothing left to mediate the relationship between mind and body, rationalism and materialism. This internal secular "rivalry" changed the nature of dualism such that "mind–body" went from describing a relationship, a continuum, to describing a problem, a discontinuity. Ongoing contemporary struggles over the determinations of life and death in the biopolitical sphere are refractions of these dualist problems: as chapter 4 will explore, for the establishment, continuity, or end of personhood, secular courts and medical institutions require physical correlates for concepts and categories such as intentionality, autonomy, and consciousness that are social, metaphysical (rational), or undetermined.

Sometimes these gaps and tensions serve to reproduce the secular-religious binary by drawing strong lines of demarcation, as though there were somehow two stable (and preexisting) realms—as when a secular judge is trying to adjudicate whether to grant a religious exception for a practice and must somehow determine what counts as religion or religious, and why (see Agrama 2010 and W. Sullivan 2007 on the arbitrariness of this). At other times the tensions may be internal, as between materialism and rationalism (Tambiah [1990] 2000).

While materialism searches explicitly to eliminate or overcome the boundary, rationalism since Immanuel Kant has issued a waiver,[3] saying that certain matters are not in the domain of empirical and scientific determination. Since determining ultimate meaning, value, and purpose are generally said to be outside the finite purview of scientific knowledge, some secularists make room for religiosity in the provision of values and accounts of ultimate meaning. Secularists such as Clifford Geertz, Jürgen Habermas, and many others say religion has served the function of assuaging existential anxieties, providing comfort regarding the unanswerable metaphysical and cosmological questions about human origin and purpose. Materialism, in the meantime, moves forward with its totalizing project, abiding no gaps, a stance exemplified, for example, by the quest in physics for "a theory of everything," or sociobiology's bid to explain values, morality, and consciousness, or the quest in neuroscience to materialize the mind, to close the gap between mind and matter. Materialist or rationalist or both, a secular subject might well think there must be a larger meaning to existence, but she knows that it's not something she can assert in a secularly viable way; or rather, that asserting it would mean that secular norms would categorize her as religious, spiritual, primitive, deluded, schizophrenic, or worse, what one interlocutor called "whackadoo."

Once it produced a void in answer to the question "Is there nothing after this?" the secular spawned its own temporalities of anticipation, its own eschatological dispositions, and fears of finitude. In chapter 3, I will argue that the final ending of the individual stands in contrast to the ongoing evolution of the world and especially to progress, in which the future is always supposed to get better. So it may be that in other cultures there is an obvious sense in which you end and the world goes on without you; but time does not go on in the same way if your notion of continuity is eternal return, or reincarnation; time does not go on in the same way if you assume there is a better continuity elsewhere in the land of ancestors or in God's heaven. What's more, shifts in the eschatological horizon brought about by the elimination of the afterlife along with the development of scientific knowledge about evolution have produced a particular relationship not just to individual death but to other kinds of larger endings and continuities such as extinction, obsolescence, survival, and succession.

The secularizing West may not have been the first to ask these questions, of course, but, at least from Blaise Pascal onward it may have been the first to be terrified by its own particular answers. What seems to mark the secular is not only the finality of individual death, the expiration of individual consciousness, but the further claim that there is no purpose whatsoever in existence, that since matter itself is not meaning making (no mind in matter), then there is a discontinuity between this absurdly meaning-seeking animal called human and the meaningless unfolding of the universe. Indeed, positing a purpose for the meaningless processes of the universe, as many religious and nonmodern cosmologies do, would be to commit a fallacy. Thus, the dogma has held that meaningful and teleological cosmologies are not to be considered in any valid theory. The secular human is cosmically alienated.

Stitching these ideas together, I would say that the mind–body problem is part of the problem of cosmic alienation or secular disenchantment, and throughout the book I approach the secular as a set of tensions around these larger dualities, including the relationship of mind to the universe, how the former (the mind and its intentions) can or cannot be a part of the latter (the universe of purposeless matter). Part of the reason the mind–matter relationship is so salient is because it is a key problem for science, and science is one of the pillars of secular authority, through which the boundaries of religion get established. The psychologist Steven Pinker, for example, has grandly declared the hard divide between matter and consciousness to be the only real chasm left in science. For the liberal secular humanist George Kateb, the mind–matter problem is a fundamental issue, and the secular becomes the exclusive frame for the investigation of its unresolved problem. "Secular investigation," he wrote, "must begin with the postulate that mind developed out of matter" but that there is yet no proper scientific account of it and neither can there be a religious one: "This is a mystery that religion does nothing to penetrate, and that only scientific reason could ever dissolve, provided it did not replace religious presumptuousness with its own" (2009, 1015).

Technoscientific immortality projects become an exemplary site of these anxious demarcations if only because they work on the continuity of mind and person beyond death and across matter. After all, what is one to make of figures such as "the patient" in cryonic suspension, who is considered legally dead by secular institutions of medicine and law but potentially alive by cryonics criteria because his or her brain information is intact? Or digital avatars of persons whose informatic selves are recognized as partial agents and beamed out by satellite into the universe? Or the Singularitarian claim that the cognitive, computational purpose of the universe is equally the purpose of humans and so we must develop computational superintelligence? What I saw and heard in immortalist circles required a nondualistic vision of the world, but it was not the nondualism of the New Age nor exactly the reductive materialism of science; for example, in immortalist views the person or the self or the mind *could be* disembodied and separated from the original biological body, but it was not, at the same time, posited as a soul or spirit or some other *unaccounted-for* substance.

So much about immortality seems to turn on the mind (or consciousness) and its relationship to matter, to the body, at least as these have been conceived and understood in the West for the last few centuries: the way the mind or consciousness seems to exceed matter and yet has always also to be contained by it; the way the universe has been cut up between the two, matter being inanimate and purposeless, mind being the opposite. Reductive materialism has prescribed a kind of monism in which the two were not conceivable independently, not without betraying materialism and bringing that thing called religion (or rationalism!) into the mix, as Cartesians or transcendentalists or spiritualists had done, by invoking a god or a nonmaterial world of spirits or Platonic ideals or a universal consciousness. That is why some philosophers, like Daniel Dennett and Paul and Patricia Churchland, fight hard to depict consciousness as irrelevant, a temporary illusion that neuroscience and AI will eventually dispel, just as spirits, demons, and homunculi were dispatched in a previous era. If you are a materialist, what can you do with consciousness but dismiss it or materialize it? Or else you end up a dualist and get charged with being at best a mysterian, at worst a closet mystic, and maybe if you are lucky a rationalist, a metaphysician.

No surprise, then, that immortalist forays become part of the contested metaphysics of the secular–religious divide. What was more surprising was the extent to which immortalists embraced, played with, and altered spirituality, religion, and religious-like tropes. Rather than managing that boundary as per traditional secular institutions, immortalists and their epistemic and material entities were bringing out the tension and contributing to the blurring and dissolving of that boundary—even as they also kept pulling back in order to maintain some distance from the religious, supernatural, and whackadoo. A good example of this appears in a recent publication by a transhumanist philosopher who argues that digital entities carrying your information also carry forward some aspect of you, which he calls a ghost: "Any possible future digital ghost carries information about your life. Since it carries that information, you remain, after death, in your ghost." But he is very quick to point out that "of course, digital ghosts are not like the ghosts described by spiritualists or occultists. Digital ghosts are entirely physical patterns of energy in material computing machines. They are not supernatural" (Steinhart 2014, 2). Immortalism depends both on the establishment of the boundary between science and religion (to remain legitimate) and on its breach (to reach its destination). Secular and scientific immortality projects need to distinguish themselves from religious ones, yet they themselves are often designated as a religiously inflected pseudoscience precisely because they pursue immortality. After all, to claim nonsymbolic immortality, to deny the finality of endings like death, is to enter the zone of the supernatural, the whackadoo. At the same time, immortality is the very promise embedded in modern secular futures, not only because the biopolitical or life-saving logic of modern medicine slips down that slope with its promise to stave off death and debility, but because immortality is the redemptive end of progress and technoscientific advance. In this way, in the secular, immortality is forbidden but promised. To be secular is to emphasize endings and their finality in order to assert one's own nonreligiousness, but the secular also needs to transcend endings in order to make sense of its own order, its own triumph, to redeem immanence beyond the absurdity of transient finalities. A salvation from the absurdity of materialist death and other final endings, immortality figures at the end point of the secular promise of the good life, for how can the good life be good in any nonabsurd sense with death looming as the absurd end of everything, of me, you, the universe?

LIMITS OF HUMANISM AND EMERGING COSMOLOGIES

I hope that in exploring these zones of tension and indeterminacy I will be able to say something more general about a secular subject whose life is embroiled in a double coil of hope and anxiety, of endless possibility and absolute limits, the futures of progress and of an absurd death that ends everything always prematurely. I am also hoping to explore ways in which those dualities and tensions are shifting as secular humanism gives way to forms of posthumanism, including transhumanism,[4] and as secular scientific imaginaries struggle with their cosmologies, trying to imagine existence beyond the contingency of a random birth, beyond the splendid isolation or alienation of the conscious figure of modern humanism embedded in an unconscious cosmic context.

Before I continue, it is worth saying a few more words about the concepts of secularism and the secular, which have benefited from a recent abundance of analysis. Secularism—the more familiar designation—is a doctrine about political and institutional arrangements, insisting on the separation of religious authority from political or governmental authority. Such a separation is also about the contest over *epistemic* authority, about what knowledge from which institutions counts as valid for the conduct of projects mainly in law, medicine, science, and education. The arrangement called separation, then, is organized around less visible background concepts, foundational assumptions, validity rules, and dispositions that define a proper relationship to reality and may together be called *the secular*. As a general condition for the emergence, legitimation, and management of secularism, *secular formations* (Asad 2003) may be considered to be underlying secularism's normative politics. What's at stake, then, is not just the shallow and supposedly neutral separation of institutions but "the unceasing material and moral transformation" of citizens (their bodies, affective dispositions, and minds) "regardless of their diverse 'religious' allegiances" (Asad 2003, 191).

Charles Taylor, with a different agenda from Asad's, trains his lens on "experience" in a *secular age,* or rather how the secular age has produced "the possibility or impossibility of certain kinds of experience" (2007, 11). For Taylor, an envelope of secular assumptions covers the majority of people's experiences and actions in today's West in general[5]—even for people who are "religious." In other words, religious or nonreligious, a person in the West lives in a context in which the existence of God is not taken for granted, in which the world is largely made sense of in terms of this-worldly causality, and in which the tension between belief and unbelief becomes in itself constitutive of experience on a wide scale. It is interesting to note, however, that Taylor's massive tome scarcely mentions the topic of immortality, even though its main thrust is to describe the cultural setting, or the conditions of possibility, that enabled the rise and spread of a largely secular point of view, what he calls the "immanent frame." Yet, surely, the elimination, or rather the reconfiguration, of immortality or an afterlife constitutes one of the key experiential conditions of modern secular living—and dying.

A common position in recent years has been to note the extent to which secularism is imbued with a repressed and hidden religiosity, specifically a set of cloaked Christian and, more specifically, Protestant concepts and dispositions whose past is veiled and which get posited as universal secular positions. I rather lean toward Hans Blumenberg's position (1983), which suggests that at some point modernity and reason must be understood as having attained their own analytic concepts and social forms rather than be taken as recastings of religious pasts. This is not to legitimate secular concepts and dispositions as those that allow for the best purchase on the real (the story secularists tell). Rather, my point is to assert that the secular has its own logic and tradition by now, that there are rules that try to define and redefine proper secular attitudes and logics—these are related to atheism, materialism, modernity, medicine, and liberal ideas of a self. If secularism as a form of governmentality has separated certain domains as real and illusory, the secular is what suggests what should be permitted in which, what attitude is appropriate where, what hopes and fears are valid, where the boundaries lie: the conviction that humans not spirits will make the world better; that entropy, not Christian Armageddon, is to be feared; that anxiety about the nothingness at the end of life is the right affect, not terror at the punishment meted out by the Lord. A secular subject might believe in God or wild spirits privately but also know that in a secular court of law she cannot say "The green fairy made me do it" in order to justify an action, unless she wants to claim incompetence to stand trial; she knows that in a science paper she cannot say "I know my eyes can see because God made sure my retina had ganglion cells" or even that in a multiverse there is "an eternal time beyond this one" or "I know what is right because my ancestor spirit told me." If any of these is accorded in a secular court of law or a university classroom it is either through insanity pleas or religious exceptions (even if secular worlds such as medicine are full of uncanny but mainly repressed presences [see Whitmarsh and Roberts 2016]).

But, as I have argued, there are indeterminacies in fundamental parts of the apparent norms and consensus, and thus the secular carries with it its own distinct conflicts in zones of indeterminacy. "Indeterminacy" is also a term used by Hussein Agrama (2010) to describe the ways in which secularism, or the secular state, maintains its full sovereignty through the courts by constantly blurring the public–private boundary on which secularism depends. For to designate a realm as religious and thus subject to religious authority is to cede a chunk of sovereignty. Agrama shows how in the law and in family court cases in Egypt, the family is at once associated with religion while at the same time placed at the foundation of public order, defined explicitly as a secular concept. Secularism in general folds religious freedom and privacy into invocations of public order, a cunning strategy (Fernando 2014b) that allows the secular state dominion over what it itself designates as a religious domain. I am influenced by Agrama's notion of indeterminacy, especially in relation to the legal relationship to corpses (chapter 4). However, I am not convinced that the blurring is part of a managed attempt at sovereignty; rather it betrays deep-seated problems in secular conceptualizations of the world—what I call the gaps. Thus, the public–private distinction of secularism is less salient here compared to the existential issues, which secularism also deploys to draw the secular–religious boundary. Rather than seeing indeterminacy as a deliberate exercise of power in secularism, I understand those indeterminacies and their ensuing tensions as constituting the secular and forming secular subjects. In my rendering, the secular is less secure and more anxious.

I am not proposing the secular as some homogeneous formation to be found everywhere secularization has taken hold; the same disclaimer holds for the domain I keep calling science. But I do see commonalities across the differences. One of the conventional ways to think about differences in secularism has been to suggest that in Europe, secularism took an explicitly anticlerical form that instituted a rigid separation between religion and the state, while in the United States the impulse was more about accommodating pluralism, which gave shape to the First Amendment (e.g., Casanova 1994; for a more detailed account of shifts and negotiations in this history see Feldman 2002). This may be true in part for the legal structure, but it is hardly the whole story. Outside the courts, American secularists and freethinkers were quite outward and firm in their rejection of religious tenets, authority, faith, and forms of knowledge. As Noah Feldman (2005) and Susan Jacoby (2004) recount, early American scientists and writers were "strong secularists" adamant about replacing religion with reason. In nineteenth-century creeds by the likes of John William Draper (1876), separation was understood not only as institutional neutrality but as an opposition between reason and traditional authority, otherwise framed as a conflict between science and religion. The same was true for psychologists, who replaced the "science of the soul" with the science of the mind (see chapters 1 and 4). Today the secularist position is reflected not only in the very vocal declamations of the new atheists (who have ardent fans in immortalist circles) but also in middle-of-the-road secular humanists like Kateb, whom I mentioned above.

Even as we can acknowledge that science, the secular, and secularism arise and function differently in, say, Turkey or France than they do in China or the United States, it seems to me that both have the capacity to cross geographic and national boundaries in cultural networks of their own. Thus, for example, a scientific paper on immortal jellyfish in Japan or on planaria worms in Germany and related claims about immortality in nature will echo loudly in biological circles—not to mention public media—in many separate locations globally. So despite the differences, many of the secular conceptions that animate the particularities of the American secular approach to immortality are shared across secular scientific and Western cultures and institutions. When I invoke secular science I am not referring to specific practices or methods. In part, science as a social project is secular insofar as it maintains that the transcendent or supernatural can add nothing to proper explanations of reality, that the afterlife is illusory, that the universe has no goal or meaning, and that there are no spirits in things, or mind in nature. Science becomes a clear part of secularism in those instances when, by virtue of these secular negations, it is enrolled in enforcing the secular–religious divide. But I am also referring to something more specific: the secular in science denotes an aspect of the scientific project that is caught between its Kantian/rationalist abnegation on matters of meaning, mind, and value as concerns beyond the purview of science proper, and its convictions that the explanation of everything, including matters of meaning, mind, and value, are to be found in a naturalist or materialist model.

In the case of death in the United States, whether you are a Buddhist or Christian or a neopagan, your dying and later dead body will be processed through secular institutions of medicine and law, according to rules that adhere to materialist determinations of the body's condition (rather than an assessment of the state of the soul), using specific material technologies, and applying the authority of secular legal institutions (the state medical examiner, state laws regulating the determination of death). Even if most people in the United States believe in the soul and its survival after death,[6] most people must face this secular regime and its assessments. If for some reason or other one were to object to some part of this series of procedures—say, to autopsy—then usually if there is any accommodation it is only granted when there are religious reasons for the objection. It is granted as a "religious exception," thereby only confirming the secular grounds of the regime as a whole. And yet, consistently, facts of the matter about personhood, consciousness, and the line between life and death get unsettled by science and technology themselves. Thus, political conflicts in American secularism have not just been about whether this particular activity, say education, is religious or not; they have been over epistemic and metaphysical content. They have been about questions like What is a person? What is consciousness? When does a person begin or end? How do we know?

I call the set of rules, institutions, and authorities that legitimize the cultural changes of status entailed in the passage from life to what follows life an "immortality regime." As I'll describe in more detail in chapter 4, the role of secular immortality regimes is to try and maintain stability in the face of indeterminacy when it comes to these and other existential issues. In doing so, they also point to zones that lie beyond epistemic legitimacy, zones in which science and secular knowledge have not fixed the facts of the matter, and where these cannot be authoritatively fixed either, which is why the regime is concerned with their stabilization. These gaps are produced by secular modes of knowledge—that is, they are gaps because, while remaining important concerns within secular society and central to secular accounts of the world, the relevant modes of knowledge seem unable to establish any truth of the matter regarding them. In those gaps, key categories like life and death, animate and inanimate, secular and religious, may collapse or merge; and in those gaps immortality thrives.

The secular indeterminacies regarding animation, deanimation, and reanimation are most familiar in disputes at the extremes of life, where contests over the authority to define and manage life and death are rendered more spectacular because they pitch religious groups against science on such matters as abortion, embryonic stem-cell research, or end-of-life decision making. Anthropologically, I am interested in how these tensions and unresolved domains are understood, negotiated, managed, and occupied *within* the secular albeit by people with different claims and interests; I am interested in the new, possibly postsecular forms that are generated by technoscientific projects like immortalism and its visions of continuity.

A secular problematic, continuity has had to be constantly reimagined and reinscribed in other terms, requiring, in the words of Benedict Anderson ([1983] 2003), "a secular transformation of fatality into continuity, contingency into meaning." For Anderson, the idea of the nation was one vehicle for this: "If nation-states are widely conceded to be 'new' and 'historical,' the nations to which they give political expression always loom out of an immemorial past, and, still more important, glide into a limitless future" ([1983] 2003, 11).[7] Today, as collective futures are undermined and as biotechnology gains power beyond the boundaries of nations, a particular chronobiopolitics (Freeman 2007) of progress and infinitude is emerging. The immortalist imaginary's[8] extended future, a form of continuity writ large, stands against the projected future of destruction and demise, of climate collapse, nuclear holocaust, bioterror; it would have to be the redemption of modernity's failed teleologies, of empires that need to continue their dominion, of technologies and forms of knowledge and rationality that had promised to fix everything, especially the apparent human inability to make good rational decisions, which the correct Bayesian algorithm run on the proper supercomputer promises to resolve. If the immortalist imaginary has gained traction in the secular technoscientific culture at large, it is in part because it takes on not only individual death but also collective endings and the humanist absurd.

This book is driven by the conviction that social imaginaries and philosophical discourse and many of the conundrums they address play a role in determining the texture, the shape, the experiences of the secular (or any) lifeworld. I don't consider this an idealist approach, and I don't lean on beliefs as such (accusations that Wendy Brown [2007, 2010] has leveled against Charles Taylor's Hegelian approach to secularism). Concepts and conceptual issues cannot be written off as removed epistemological matters or merely the ideological superstructure of material and institutional processes of power, especially in the speculatively driven world of futurisms. Concepts are what Michel Foucault called "dispositifs";[9] they are constitutive aspects of knowledge structures, legal apparatuses, institutional networks, and administrative mechanisms deployed in the exercise and perpetuation of power. Concepts also have dispositional and affective states associated with them, particularly in secular contests where identities turn on concepts with epistemological content (e.g., atheist, believer) and where technoscientific projects unsettle our existential securities.

The immortalist project today is part of a wider technoscientific shift from biopolitics and a culture of life to a cosmology of mind and information (appearing often under the label of convergence) that is increasingly favoring nonlife, or parsing biology in terms of the nonbiological or digital. In part, this shift is a reflection of the ubiquity of computation and the growing functionalities of AI; in part it's about a growing awareness of the inadequacy of secular views on mind and matter. Among the more critical accounts is the book *Mind and Cosmos* (2012) by the philosopher Thomas Nagel, in which he explicitly argues against the "secular consensus" on matters regarding teleology, meaning, and the presence of mind or subjectivity in the universe. In doing so he does not fall back on religion but rather takes note of the dogmatic rules or limits of the secular. He states bluntly that "almost everyone in our secular culture has been browbeaten into regarding the reductive research program as sacrosanct, on the ground that anything else would not be science" (2012, 7). Nagel's sober goal is to seek "an alternative secular conception . . . that acknowledge[s] mind and all that it implies, not as the expression of divine intention but as a fundamental principle of nature along with physical law" (21). It is not clear to me that such a conception would continue to be a secular one, but it does jibe with the visions of some emerging—I would say postsecular—cosmologies that are seeking to put (or imagine) mind back into matter. How that might be possible is, of course, "the hard problem."

What the philosopher David Chalmers (2002b) made famous as the "hard problem" of consciousness holds that neuroscience's increasing understanding of the brain is functional and mechanical and that, while it can explain many complex cognitive phenomena, including the integration of vast sensory data into unified percepts, it falls short of providing any basis for the range of first-person experiences and feelings that fall under the rubric of consciousness or mind. The third-person, objective mechanics and lawfulness of materialism seem to be insufficient to explain first-person subjectivity—in brief, the problem of the dualistic gap. To bridge the gap, Chalmers took the audacious step of suggesting a new physics. Consciousness, he proposed, should be considered one of the irreducible fundaments of physics, such that alongside physical laws we would develop psychophysical laws. Absent any data on what constitutes a psychophysical substance, it is not clear what might be meant by psychophysical at all, though Chalmers tries to get around this via the vague notion of information as a way to bridge the physical and psychological. It is not coincidental that Chalmers has become involved with transhumanist groups and conferences, as his proposal folds well with immortalists and transhumanists who, before him, had adopted an informatic view of life, death, and the universe.

The notion of "information," as I will explain, is key to immortalist conceptions of personal identity and continuity. It has also become the basis of a cosmology, for the possibilities of immortality in a purposeful universe are constructed through ideas about information as the very stuff that both people and the universe are made of, that constitutes humans and their minds, and that will constitute posthuman intelligent forms. It is in this respect that I consider the immortalist project part of a broader trend in the Western secular imaginary that is trying to occupy what I have called the secular gaps, to eke out new conceptions of the universe and our place in it, not through spiritualism or appeals to transcendence (although that also happens) but through variations on secular scientific positions and disciplines that try to reintegrate mind and purpose into their accounts, recalibrating the mind–matter and human–cosmos relationships. In *Reinventing the Sacred* (2008), for instance, the biologist Stuart Kauffman speaks exuberantly of "a new, emerging scientific worldview" that could bring it all together, while others have echoed the same scientistic optimism, declaring the "emergence of a novel scientific worldview that places life and intelligence at the center of the vast, seemingly impersonal physical processes of the cosmos" (Gardner 2010, 379).

EXISTENTIAL TECHNOLOGY AND THE QUESTION OF CONTINUITY

So far I have proposed (a) the secular as the site of persistent tensions — epistemic, metaphysical, and political. In other words, my notion of the secular is not that which, like reductive materialism, essentially takes the side of matter over mind, or chooses the side of the machine by eliminating the ghost from the machine; rather, I see the secular as the site of a face-off between the ghost and the machine, finitude and infinitude, soul and cell, mind and matter, materialism and rationalism. These discontinuities require each other as well as the boundary that separates them, a boundary that is also part of the very boundary that is invoked by secular institutions to distinguish, however sloppily, the secular from the religious (i.e., secular*ism*). I have also suggested (b) that immortality is one of the key terrains on which these tensions are played out, because immortality is just where the boundary of discontinuities gets blurred and some version of continuity between life and death, mind and matter, meaning and the cosmos is demanded. By pushing and unsettling the normative discourses, language games, and practices of secular and scientific consensus around these boundaries, technoscientific immortality exposes these tensions though it is also caught up in them. And so, I also argued that (c) by mobilizing the concept of information to bridge those gaps (or discontinuities), immortalism calls for the convergence of the ghost and the machine, of mind and matter, of life and cosmos, and yet constantly reproduces their history, their difference, and their distance.

These ideas and arguments came largely out of the field as I found myself surprised to hear the range of reasons immortalists themselves proffered for wanting to extend their lives into the distant future. Early in my fieldwork I noted that immortalist motives didn't necessarily have to do with individual survival. It's not much reported on, but one of the strongest activists in the immortalist firmament, Aubrey de Grey, frequently has said (and not just to me) that he is not interested in his personal survival—which may explain why he hasn't cut back on his famous and prodigious beer intake. Others may not be beer guzzlers, but even among the health-conscious I did hear many say things like "It's not death as such . . ." or "I'm not interested in my personal survival . . ." Something more seemed to be at stake.

Yes, there was talk about calorie restriction, rabbit kidney preservation at liquid nitrogen temperatures, developments in neuroscience, or the most recent biotech innovation to treat macular degeneration, and even moral talk of *saving* lives, but one of the recurring topics is what in one of our conversations Ian Marks called "all the unanswered questions."[10] Marks, well-versed in Buddhism, had recently dropped out of his university program in molecular biology, with a year left to finish, in order to join a number of other white, male dropout biologists in a project they thought would accelerate genetic studies of disease and aging. They were one of a few DIY biology ventures, set up garage-style in the environs of Phoenix, Arizona. They were looking for funding. And funding would soon come, because on the other side, venture capital was looking to build biocapital. They would first move their whole team farther out into a large suburban house where they lived and researched and worked and also partied collectively, then eventually on to Silicon Valley, like almost everyone else I got to know back then in Phoenix. The company would finally fold because of several missteps, and Marks would get caught up in the fall. But right then he was shuttling between places, without living quarters of his own, often sleeping on other people's couches and for a period in a tent in a friend's backyard.

He wanted to focus on "the big choices," he said, which meant "not focusing on a CV" but rather "dropping out of school to do what I think is right." What is right is a future in which medicine will have developed "the potential to save almost everyone's life or at least extend everyone's life on the planet" and we will have "learn[ed] enough about biology so that we can truly defeat all diseases."

As it turned out, his own relationship to the logic that "more life is better" ended up being both shakier but also more ample than the first rhetorical layer. When I inquired about his motivations, he struggled between his convictions and his ambivalences:

> IM: Living is good. Being alive is good. So more of it is good.
>
> AF: What's good about it?
>
> IM: Love. [Laughs] Yeah love, that's it. [Pause] Well, maybe life is not good. Maybe that's what the Buddha was trying to say. Maybe I should just stop right now and go to a Zen Buddhist retreat and try to extinguish myself without suicide.
>
> AF: Well, logically one would have to say that life is not good, which is why we are trying to improve it.
>
> IM: Yeah, yeah life absolutely sucks. No disagreement with that. Life sucks. But as long as there's tomorrow, tomorrow has the potential to be better than today, and even if life got suckier and continued to suck for a long long long time, as long as I thought there was light at the end of the tunnel, I'd want to keep going. But if there was ever a point where I thought there was no light at the end of the tunnel, and things are just going to keep sucking and get worse and worse, then would I still be an immortalist? Maybe, I don't know.

Later, he described his motivation for seeking extended life: "Curiosity. About everything. How the universe really works. All the unanswered questions. . . . Why are we here? I guess I could find out by dying if there's anything to me besides this. I'm not betting on it."

This, in fact, ended up being a common refrain among immortalists I spoke to. A prominent transhumanist, cryonicist, and molecular biologist, who also moved from Phoenix to the Bay Area to get his own start-up off the ground, stated it to me in these terms:

> I am interested in cryonics primarily out of a sense of curiosity and adventure, not because I think this ensures my sense of survival. I don't

have a particular problem with the issue of death, because when I die, I simply cease to exist. It's not like I go to hell and spend eternity suffering. If I cease to exist it is a totally neutral state for the universe, for me. . . . To start off there's a great sense of adventure in terms of "the greatest mystery is the future." There are all these decisions that I make and society makes where we have no idea how it's going to end up in the future. From the perspective of all the things that are going on today, it would be fulfilling to be able to personally witness at some point in the future the consequences of all the things we're trying to do. So I consider cryonics worth the effort if I could regain consciousness, even for a twenty-four-hour period, one hundred years from now, just to satisfy the curiosity that surrounds the future.

The motivations vary across the field, but these more existential or cosmic narratives caught my ethnographic attention. The recurrence of these themes in immortalist zones echoed a larger secular scientific promise: having gotten rid of religion and emptied the universe of (enchanted) meaning, it had promised its own total explanation of it, which was yet to come. Perpetually yet to come. Thus suspension, the key temporality of cryonics, is in fact a central technoscientific temporality. Despite the powerful local achievements of science and technology, the "big picture" has remained obscure; yet it has always been part of the promise of science, suspended above the horizon, held dangling beyond the limits of death.

Not only do these larger questions seem inseparable from the quest for immortality, they also shape it. Another immortalist told me, "There's the daily things: going out, seeing friends, having a drink. But there's bigger things there: Why are things the way they are? And the only way I'm going to understand the universe is if I live long enough, be able to explore and expand my own intelligence." The quest for extended life, extended intelligence, and extended time seem to take shape in these zones of metaphysical indeterminacy, what one transhumanist has called the "long-term perennial existential questions: Why do we exist? Why do we exist here and in this way? What are we? Where does everything come from? Where is everything going?" (Van Nedervelde 2008)

I will open this up further through a conference I attended at Lincoln

Center in New York in June 2013. The organizer of "Global Future 2045," Dmitry Itskov, is a young multimillionnaire who made his fortune in the internet business in Russia but has now left his company to pursue all matters regarding immortality (for more on Itskov see Bernstein 2015, 2019). Although Itskov's main focus is on robotics, AI, and mind uploading, his goal for the conference was to bring the so-called spiritual and technological together in that same zone of indeterminacy, to show their compatibility rather than their contradictions, their continuity over their discontinuity. In an interview with the *New York Times* prior to the conference, Itskov said, "We need to show that we're actually here to save lives. . . . To help the disabled, to cure diseases, to create technology that will allow us in the future to answer some existential questions. Like what is the brain, what is life, what is consciousness and, finally, what is the universe?"[11]

The quick shift from technology that saves lives and cures disease to technology that will answer questions about the nature of the universe used to strike me as a quantum leap between two unrelated states. But it began to make more sense when I realized that this attempt to address aging, disease, and death *technically* has taken shape as a reaction to those secular tensions and indeterminacies that have a specific relationship not just to the problem of individual death but of endings, purpose, and continuities in general. I am not suggesting that, say, the gene-therapy technique of delivering telomerase through a viral vector is a secular technique or is the direct result of trying to resolve the mind–body problem or meaninglessness in the cosmos—though certainly, as we'll see in subsequent chapters, some of these questions influence directly some immortalist techniques and practices, from brain imaging to beaming mindfiles by satellite into outer space. But if immortality is about breaking from or going beyond the limits of existence as these are felt and understood at any given moment, then surely what matters is to explore what those limits are *conceived* to be, what those confinements and those bounds are *felt* to be, and how they are experienced, rationalized, and naturalized.

FUTURE MAKING AND PASTS

Another part of my argument should by now be obvious: immortality in general, and immortalism in particular, should be analyzed as social and historical projects of future making, and not as natural or timeless bids for individual survival or preservation. Secular relations to death and the dead are commemorative; they are about preserving the past, because there is no after-life time. In contrast, immortality in general and immortalism in particular are future-oriented. Immortality is a category under which particular people at particular times organize the continuities that matter to them in particular ways. Thus, to analyze immortality is to grasp the kinds of futures as well as the kinds of endings that are at stake and to consider what lives, what minds, what units of survival matter to a given social and cultural moment. As a form of future making, and as with many other technoscience futures, technoscientific immortality collapses prediction and promise; therein lies its power. The prediction holds that immortality will come because the pace of scientific breakthroughs is unstoppable, because animal lives have been extended, because molecular biology can extend telomere length, because neural nets are working, IBM's Watson beat the best carbon-based intelligence in *Jeopardy!*, and Google's AI just beat a Go master; the promise says it will come and it will be a good thing, a monumental achievement, it will cure everything, it will come and fulfill an ancient human desire: to not be subject to arbitrary death. If immortality is a form of future making rather than the perpetuation of a past, if what is proposed in immortalist projects is not just extended lives but particular futures, I ask, What future is being imagined or promised when technoscientific immortality is being imagined or promised? What future forms of life are immortal life-forms embedded in? Who is saving whom and from what kind of finitude? What futures are being displaced in the process?

The secular has strongly conditioned technoscientific imaginaries of possible futures, of limits and endings today. The elimination of the soul and of the afterlife, the invalidation of cyclical views of time and being, as resources for imagining forms of identity and creating practices of continuity, produce and prohibit specific experiences, affects, and challenges in relation to the mind–body problem (what is to survive?) as well as to temporality (to survive what? for how long?) and purpose (why survive?). What is it that is going to be continued? Life? The organism? The individual? Which is to beg the question, What is life? What exactly is the individual? Is it the biological body? Mind? The legal or social person? Online avatar? Mental algorithms? Mere organism? Or is what is to be continued a collective entity, the social group? The ethnic or religious group? The population? The ancestors? The culture or a specific tradition or practice? A civilization? In all of this, is the individual, if anything, a contributor to or the key unit of continuity? Or are all of these—individuals and groups—social and psychological epiphenomena of an evolutionary process, vessels for the perpetuation of genes or *bits* or life itself through time? Or unwitting agents in the development of other forms of complexity or intelligence? Or local transient manifestations of a wider unfolding of consciousness, of the noosphere, or of the Hegelian universe coming into self-awareness? The specificities of the problem of continuity in the secular and scientific world—the ways in which that extension is imagined, prohibited, discussed, pursued for individuals or larger formations and forces—are inseparable from the ways the secular, imbricated with its colonial history, inflects the three gaps, namely, the problems of mind–body dualism and personal identity, the temporality of finitude, and the cosmological problem of meaning and the teleology of progress.

Immortalist futures are overdetermined by the technological landscape in the United States, by the tools and ideas emerging out of the centers of tech, and by the social structures and informatic premises that undergird that landscape. It should come as no surprise, then, that immortalism is a white and largely male movement. This is not a random formation but part of a specific and repeating history. As chapter 1 will describe, scientific research into immortality has consistently been tied to eugenics. While many liberal immortalists and transhumanists today disavow the racist eugenics of the past and espouse a democratizing immortalism—no death for all!—the immortalist and transhumanist movements have often given room and voice to those (such as Michael Anissimov and Peter Thiel) who espouse illiberal, right-wing views on such matters as disability and genetic, cultural, and mental inferiority.

There is a colonial history underlying the secular modern episteme more generally. Relying on views about matter and mind, the division between modernism and animism was one of the key sites of colonial politics, where some humans called "primitives" and "savages" would be sorted from the secular moderns, those with the realist view of death, personhood, and matter. To be secular and modern was to face up to the reality that there was nothing after death, that the person was only the body (as there was no soul), and, crucially, that matter was not infused with spirit (or soul or mind). These imperatives sanctioned and monitored the separation between, on the one hand, "religious" believers or "primitives" (deluded and enchanted) and, on the other, those "moderns" with a proper grasp of reality (not deluded but disenchanted). Even more so, the anti-animist divide between subject and object also helped sort the world into the two legal categories of person and property—those with reason and consciousness falling into the former category, the others falling into the latter. As is well known, and historically enshrined in the 1857 U.S. Supreme Court's decision in *Dred Scott v. Sanford,* black people in the United States were desouled and given over to the category of property, that is, legally owned and enslaved. That division, then, was part of the racialized trajectory of *secular* modernity, as it was part of the development of colonialism: desouling matter and disenchanting the universe were and are a means of asserting a secular modern identity (Flanagan 2002; Keane 2013; also see Casanova 2011), of *feeling* secular and modern, where the epistemic stance also becomes the affective disposition; they were also part of modernity's secular and colonial imperatives, the premise for brutal exercises of power. These sorts of epistemic designations and affects, through which identities were established, were and continue to be designations crucial to colonialism and empire, justifying and enabling forms of governmentality (Fernando 2014; Mahmood 2016), allowing a certain formation of power to impose its dominion on parts of the world and suppress what it presented as other—irrational, delusional—social formations. Colonial regimes always outlawed animist practices and claims. These are crucial aspects of the ideological conditions of possibility of what Wendy Brown (2007) calls "secularism as an instrument of empire"; equally, they are the grounds from which the secular immortal rises toward its transhuman horizon.

Importantly, immortalism and transhumanism have embraced a libertarian ethos that enshrines the individual and serves to hide the structural reproduction of the past in the future under the banner of technological and biopolitical universalism. As I will argue in chapter 6, this means that matters of race, ethnicity, class, and to some extent gender are ignored as important shapers of the culture of immortalism and of the futures it claims. Even more, under that universalism what is deemed important to move toward in the future is the specific telos of Western civilization, whose progress is indexed to technological innovation and the eventual emergence of transhumanism. Technoteleology erases the structural history of inequality and power that has shaped and sustained those "civilizations" and of the racialized, colonial, and geopolitical legacies of the ideology of progress. As Afrofuturists (A. Nelson 2002) have pointed out, the future, as progress or as science fiction, was always a domain claimed by whiteness, for its own perpetuation, with the exclusion of those whose lives were deemed to be stuck in the past or not worthy of continuation, forms of life unfit for futurity.

A critique of immortality and immortalist discourses and practices exposes core problems and tensions not in order that Western epistemology and power should resolve them—to produce a better secularism! Rather, the tensions and problems are important to point out because they are veiled or erased aspects of what Nagel (2012) has called a secular consensus. That consensus is not simply epistemological, as per Nagel's arguments, but also has political consequences. Designations such as "primitive" and "animist" may not be Nagel's concerns, but they are part of the social and historical picture regarding the relationship between mind and matter as they designate *what* Westerners assumed other people believed (cf. E. B. Tylor) regarding the relationship between mind and matter, and *how* Westerners assumed other people knew what they thought they knew. To say that the body is all there is, to assert the finality of death, and to deny that nature may have purpose was to assert one's secularity, one's modernity, one's enlightened Western self. As Michael Taussig (1986) has been at pains to point out over the last few decades, the "primitive" or "animist" is invented as a figure, a figment, the ghost of an eliminated form of life at the nexus of a violent epistemic and imperial encounter that in fact comes back over and over to haunt not the so-called primitive but the colonizer—or in my terms, *the modern secular subject*.[12] I would say that "the primitive" as a figure provided the "ghosts and souls" that, having been abandoned in secular modern reality, kept haunting that reality, and thus had to be displaced elsewhere. The reason why they keep returning, rather than remaining buried, is because the range of replacements have not been fully compelling: mind, self, person, psyche, consciousness can be in themselves enspirited and ghostly, their relation to matter fraught, emphasizing the discontinuity between humans and their universe.

Current intellectual movements such as new materialism, the ontological turn, perspectivism, and the new animism have arisen in their own way as a challenge to some of the consequences of subject–object duality. Importantly, given the backdrop of the Anthropocene, much of that literature has taken the division between life and nonlife as the key separation to address. What has been at stake has mostly been agency—what might be considered an agent, what can produce an effect, what has the power to interact and respond independently? However, animism is not only a view about life and agency—rather, it is a view about mind or soul, and about what can count as consciousness. Exemplified by E. B. Tylor's nineteenth-century account (1871), the original conundrum of animism for the modern episteme was about what kinds of entities can be attributed qualities that are mind-like—thought, feeling, subjectivity, intentionality, purpose, meaning. That, rather than what counts as "life" or "agency," continues to be the key division. The separation of mind from matter, the evacuation of mind from the universe in the disenchanted order, is a constitutive part of law and politics, of orders of visibility and legibility (who/what is allowed to speak? to be heard?) that form structures of power and governance.

Even as the return of new animism and panpsychism and ontological turns signals the rethinking of the desouling project in certain academic quarters, for most theorists the attribution of mind to nonhuman matter is one step too far.[13] The haunting of secular modernity is now taking place not through the specter of animism, the return of Christian souls, or primitivism, but through their opposite, the horror of imagining modern humans as nothing but emptied shells, code-filled bodies, algorithmically generated beings, zombies who have fooled themselves into thinking they have selves at all, that something special like a mind animates them into unique conscious beings. The horror of the machine is the horror of discovering that humans are nothing but machines. Having desouled the rest of the world, the secular modern West finds itself facing its own materialist vision of finitude. If many everyday humanists and materialists express horror at the extremes of the secular immortal, at its incessant production of nonhuman forms of humanity, of frozen corpses and algorithmic avatars, it is because desouling or deanimation haunts the secular modern today.

I see the secular immortal as the limit figure at the outer boundaries of the secular human, sitting at the border of machinic or zombie existence, always at risk of falling over to one side or the other. Although it constantly aims to transcend the limits of finitude, often despite itself, the secular immortal relies on a reductive materialist version of human life and mind; the secular immortal must transcend biology but also constitute itself through reductive materialist processes, that is to say, representable, replicable, predictable material processes (that nevertheless may be emergent as opposed to strictly mechanical) focused on the unified self of the modern individual. Caught in the ping-pong between animism and deanimation, the secular immortal is a figure shot through with secular tensions and colonial histories.

THE *-ISM* IN IMMORTALISM

Today, most immortalists don't want to be labeled by what many call the "I" word, because, they will tell you, it is not accurate. The strict definition of an immortal is a being not subject to death at all, as per *Merriam-Webster*'s concise and apt wording: "exempt from death." By some cosmic oversight, we were dropped from the list of the exempted! And the search for immortality is an attempt to gain or regain the state of exemption. Immortalists say what they are after is to stave off death and extend healthy life indefinitely. Oh, yes, and also transubstantiate onto digital platforms. But *not,* they say, to gain Zeus- or Jesus-like immortality or an afterlife in heaven. In actuality, why and how the term was dropped shows a much more complicated set of reasons (a history discussed in chapter 2), having to do with the term's proximity to religion, which discredited their efforts in the eyes of mainstream science and investors.

I choose to use the terms "immortalism" and "immortalist" to denote the movement and its adherents for several reasons. First, it is the adoption of an earlier emic appellation. The first two founding texts of cryonics use the term in their title: Evan Cooper's *Immortality: Physically, Scientifically, Now* and Robert Ettinger's *The Prospect of Immortality,* both dating back to 1962. Aside from the fact that some have continued to hold on to it—"joyful immortality" is the term Martine Rothblatt happily uses to describe the mission of her group, Terasem—the term "immortality" is generally understandable and capacious enough to be a working label. In common usage the term carries a somewhat broader meaning that does not entail an absolute exemption from death. Indeed, death is often a necessary precondition of immortality modalities. The survival of a name or reputation across generations (Wordsworth has achieved immortality), principally for good deeds or achievements, is counted as a sort of immortality, especially in secular contexts. In another vein, Karen Armstrong conceptualizes immortality in ways that echo contemporary immortalist concepts: a "liberation from the constraints of time and space, and from the limitations of our narrow horizons" (2006). Here immortality becomes a relationship between finitude and infinitude, limits and limitlessness, endings and their absolute horizons.

Hannah Arendt ([1958] 1998, 20) points out that the condition of transcendent deathlessness should actually be glossed by the term "eternal" rather than "immortality," which at least in Greek discourse referred to earthly existence. The earthly included the Greek gods, who were not transcendent figures like the Judeo-Christian god but were of a similar nature to the human. Eternity, being temporally and spatially removed, "has no correspondence with and cannot be transformed into any activity whatsoever"; by contrast, immortality refers to an *action or deed* that lasts, and it has an origin and a consequence within the world, in earthly time. Arendt's is itself a secular rendition of history, presenting the only viable survival as a survival through "deeds," the very basic unit of action in the public realm, to do and make such things that would seem permanent. Similarly, in Hindu cosmologies, there are eight immortal figures, very long-lived (literally called that) deities who manage to maintain their powers over time. Unlike Arendt's heroic view, the Hindu example points to immortality as a form of power and the perpetuation of an order of existence—a view I espouse and will discuss in a different vein in the final chapter.

In a weaker sense, any significant extension of the life span is also loosely interpreted to fit the category: from the point of view of our eighty-year life expectancy, a five-hundred-year human life is tantamount to immortality. When you have immortalists predicting thousand-year-long lives in the foreseeable future, immortality does not appear as such an inappropriate term. Aubrey de Grey once used figures like a thousand or five hundred years as achievable estimates of life span; he also is one of those who advocates against the use of the term "immortality" because it seems unscientific. In fact, scientists have used it and continue to do so, if in specific ways. Since the German evolutionary biologist August Weismann (1891) began to speak of biological immortality, it has denoted not "death exemption" but the "contingency of death" in the natural world, its nonnecessity. Often it relates to cells or organisms that do not seem to have an internally organized or mandated trajectory toward senescence and eventually death. These days, death is no longer considered an inevitable part of biology, and more specifically of "natural" human life. Indeed, lives in various species have been extended radically, talk of immortal cells is common, and some species have been observed to not have a "natural" process of senescence and death (the Japanese biologist Shin Kubota's now famous immortal jellyfish, for example [Kubota 2011]). The biological contingency of death is a well-accepted possibility in science and the broader culture informed by science. And frequently, cells or organisms that do not die but may be killed have been called immortal.

In sum, "immortality" is one of a cluster of related terms whose meanings and conditions of signification vary. We use "immortality" in many ways, some of which actually *imply* death and others which simply denote an unimaginably long life, irrespective of the potential for death. James Frazer wrote, more than a century ago, that "faith in the survival of personality after death may for the sake of brevity be called a faith in immortality" (1913, 25). If what is at stake in many transhumanist and immortalist projects is the survival of personality beyond biological death but on a different substrate, then why hesitate to call this immortality?

The main way I use the term is sociological, insofar as the term names a social project, a movement under which I gather related and overlapping groups. For some such groups, the term "radical life extension," aside from being cumbersome, would be even less accurate, the whole point of their endeavor being to get out of the defective carbon-based substrate out of which this thing called life has emerged and back into which it keeps crumbling.

Finally, immortalism helps mark a distinction with the more general health obsession and longevity concerns in the United States, an \$88 billion business of Botox injections, hormone replacement therapies, extracts, and supplements (Weintraub 2010). Of course, there are continuities and entanglements between health obsessions and immortalism. Some immortalists do use cosmetic alterations, creams, vitamins, supplements, and fitness regimens that tend to make up the bulk of the youth and health culture. More important, the two founders and directors of the supplement company Life Extension Foundation are ardent, veteran cryonicists, contributing large sums of money to organizations like Alcor and Suspended Animation, with the latter founded and highly influenced by them. On another level, some basic health and longevity practices such as calorie restriction may be seen as almost self-sufficient interventions into the evolution of aging mechanisms (the theory is that calorie restriction works to extend healthy life because it triggers an evolutionarily developed crisis mechanism that tells the body to wait a while longer before it enters full-on aging mode). Immortalists have a range of views and practices regarding these less radical interventions; when they use strategies that seem broadly to fit these categories—for example, dietary regimens and supplements—they are mainly considered to be backup strategies until such a time as science and technology might finally break through with the long-awaited innovations. The continuities and contradictions were there when at cryonics gatherings I would see one calorie restrictionist carefully weighing his avocado slices and counting every unit of protein and calorie entering his body, while slabs of ribs and Idaho potatoes were sizzling away on the barbecue outside; or when a couple on a calorie-restricted diet were running marathons while at the same time spending their time and money improving algorithms that could lead to real AI and mind uploads so that they could leave their bodies behind.

While one could posit a continuum of life-extension practices,[14] it is important to mark the point where the practices shift categorically. Most health-obsessed people—including botox users *as well as* most mainstream neuroscientists or computational biologists, who even might think of the world theoretically in terms of information—don't consider *themselves* information-beings, and so don't see the point of having their heads stored in liquid nitrogen or their "mindfiles" beamed by satellite into outer space in order that some future superintelligence might reconstitute their person; they do not store their medical, personal, or official data, such as genetic information or DVDs of family photographs, in highly secure vaults hundreds of feet under the salty desert of Utah; they do not attempt brain-optimization techniques or drive their cars with helmets on in order to protect their brains in case of an accident. But these are just the sorts of things that immortalists do and the rest of society, as they see it, sneers at.

Another difference may be captured in the distinction between life expectancy and life span. Life expectancy, being a matter of averages, is a social matter, a matter of general health, of biopolitics; it rises when the general health of a specific population, defined by class, nation, geography, and so forth, rises (Marmot 2004). In principle, one can break the limits of a species' life span in an individual of that species without greatly affecting a population's life expectancy. Life span, on the other hand, is considered to be biologically limited. It is the genetic hand dealt to the species, or perhaps the evolutionarily optimal developmental trajectory for an organism (given its mass and structure) in the service of the perpetuation of the species form. Immortalists are interested in the kind of work it takes to change inherited life spans, to upend phylogeny, more than they are interested in changing life expectancy or even extending life by a few years. Although immortalists frame extended life as a natural desire, they are interested in changing nature's ways more than our own—or rather to change nature's way in order, they say, to change our own.

A volunteer who worked with one of the immortalist projects put the dilemma in stark terms when she told me about her outreach strategy:

"The way I pitch it is: 'Learn about health and wellness.' If I call and say I want to freeze your head, that'd be whackadoo to some people." So, yes, crossing over into the realm of "whackadoo"—the whackadoo barrier—is what is required, though it is also an acknowledgment that whackadoo is nevertheless part of the continuum, that what separates immortalism and American health obsessions is a semipermeable barrier, or people would be neither convinced nor anxious. Most of this book, like most of immortalism, will require moving back and forth across that barrier.

A conversation I had with Ian Marks, whom I introduced earlier, underscores the point:

AF: What's an immortalist?

IM: Someone who wants to live as long as they want to.

AF: Do you identify as an immortalist?

IM: Maybe I wouldn't do that publicly. Maybe that word has to die.

AF: Seems to have stuck.

IM: Yeah. Yeah. If you ask Tim he'll give you the definition of a "strong immortalist": someone who is doing everything they possibly can to achieve that.

AF: As opposed to just taking a couple of supplements?

IM: Yeah. That's sort of a weaker immortalist: someone who doesn't take responsibility for making the future happen. People who figure someone else is going to cure aging, someone else is going to make cryonics work, I don't have to be responsible for it. It's how much responsibility you take to bring about that outcome.

"Immortalism" as a commitment to certain views and practices oriented toward making a certain "future happen" entails a categorical shift in the understanding of the future, of what it is to be human, of what science is destined to achieve, and what is culturally permissible or desirable.

SUSPENSION, DEANIMATION, CONVERGENCE

It may be a promise for the future, but in important ways, technoscientific immortality is already here: it is spread all over popular media; the scientific interest has grown; there are practices and labs; very powerful players in Silicon Valley have started their own versions of the project; many of the projects and ideas also overlap with American state power, in places like the United States Department of Defense's Defense Advanced Research Projects Agency (DARPA), NASA, and the military. In the next two chapters I will describe the many ways in which technoscientific immortality is here and some of the history of how it got here, starting with a selective genealogy of the term "immortality" as it winds its way through scientific discourse and practice (chapter 1), followed by a history of the main projects of immortalism (chapter 2). After that come three more ethnographically focused chapters, organized through three terms taken from the immortalist vocabulary—"suspension," "deanimation," and "convergence"—each also corresponding to one of the gaps mentioned above.

The first, suspension, tackled in chapter 3, deals with the temporality of endings. My argument is that suspension—a state of being made possible through cryogenic technologies—is the space of translation between biological and chronological time; in suspension, the frozen body *emerges as meaningful* through the broader historical and cultural relation to finitude and infinitude, to secular eschatologies that on the one hand confined time to the body and on the other expanded the time of the world in which the body finds itself thrown. The secular immortality regime culturally eliminates the possibility of an afterlife, thereby rendering final the end of the person at a certain point in the trajectory of the body. Yet secular time continues, and more important, it continues not as repetition or the eternal return of the same but via the secular notion of progress, a future ever changing, ever filled with promise of better things. Suspension is the secular space in which the promise of the unfolding future can assume material embodiment and in scientific garb negate the absurdities of individual death.

The second term, deanimation, taken on in chapter 4, deals with the general problem of dualism, especially in terms of its consequences regarding the legal, epistemological, and practical indeterminacies of death and personhood. The question here is, When the body ends, what continues to persist in, for, and around the person? And how is that persistence or continuity managed and understood technologically, legally, and socially? For cryonicists, the brain frozen contains the potential for future personhood and should be protected as such, whereas for the state's immortality regime there is no person there. So cryonics brings out the ontological insecurities underlying the secular, suggesting that secular personhood, forged in the dualistic gap, is itself shaky and ambiguous.

Convergence, the subject of chapter 5, deals with cosmology, or the "cosmic question" (Nagel 2010, 135) in the sense of the human relation to the cosmos, especially the question of purpose in a universe secularly emptied of purpose. Humans, in the secular frame, are made discontinuous with matter (if not with life) because they are and have minds that stand out, or stand outside, while taking everything in. Immortalists imagine continuity through the notion of "information," outlining an evolutionary unfolding in which human intelligence (mind) and matter not only converge in machinery and computational processing but also set out to infuse all the matter in the universe with intelligence, producing a vision of an integrated universe. Indeed, that is posited not as a merely human project but as the very telos of the cosmos of which the human is but a phase, a position rather opposed to secular scientific eschatologies in which not only is there no telos but everything in the universe comes to an end in what might be called the entropic cosmology. The cosmic grandeur of the informatic cosmology belies the sense of cosmic insignificance, among the central and most anxious components of secular subjectivity.

In the final chapter I will return to think about the assumptions of biopolitics and the reemergence of a strong civilizational discourse. Why should life be extended? Whose life will be extended? Is extending life and defeating death a matter of universal human or even humanitarian concern, as many immortalists argue, or is it a matter of power and a new configuration of biopolitics, of racial and geopolitical reconsolidation via technological rather than social arrangements? How are biopolitics and secularity related in their focus on immanence and the body? And how does progress, as a secular temporality and a civilizational vision, relate biopolitics to the future and to technological futures more specifically? I present immortalists as among the new mythmakers of secular progress, themselves necessary to the wider view of civilizational progress that upholds Silicon Valley as a whole. On the one hand, immortalism and transhumanism seem to me to break down the liberal fictions of the unified self and the modern Western figure of the human in interesting and potentially powerful ways; on the other, they reproduce continuity as a Western, mostly American, and mostly white project of extreme individuation, where immortalism and transhumanism become the form whiteness takes as it makes new claims on the future.

1

After Life

VARIETIES OF IMMORTALITY IN
THE SECULAR WORLD

In 2008 the Immortality Institute initiated an online forum to discuss the possibility of a name change. The dilemma was over the word "immortality." On the one hand, it was said, the word connected the project to an ancient human dream, lending it universal appeal. It also made a bold statement in the cautious world of science, proclaiming that this is not about tacking on a few paltry years, as gerontologists would want, but about stopping aging and conquering death. On the other hand, the term also evoked false religious promises of existence after death.

It was a perfectly secular bind: what was being sought was immortality in a physical sense, that is, within the accepted parameters of a materialist scientific understanding; yet the idea itself seemed to float outside of acceptable scientific and secular parameters, indicating claims and possibilities about the self, consciousness, personhood, matter, and humans that seemed to have become irrevocably bound up with that thing called religion.

The link between immortality and religion had moved another group to initiate a name-change vote a little earlier. In July 2006 the nonprofit Immortalist Society, a close collaborator of the Cryonics Institute, wrote to their members asking them to vote on a name change. After living with the term for over thirty years, the directors contemplated a name change because, as they explained it, "the word 'Immortalist' has too many religious connotations [a few are acceptable?] and is a public relations liability."[1]

In fact, the use of the word "immortal" has been a mild source of controversy from the beginning of cryonics in the 1960s. Upon its found-ing, Alcor decided very consciously to forgo reference to immortality and attach the phrase "Life Extension" to its official title: the Alcor Life Extension Foundation. Other groups called themselves by such names as CryoVita and Transtime, focusing on concepts of life and time. Neverthe-less, terms like "immortality research," "immortality science," and "physi-cal immortality" were commonly used, as exemplified by the Immortalist Society itself and their publication *The Immortalist*. Besides, the book credited for founding the cryonics movement, Dr. Robert Ettinger's *The Prospect of Immortality,* wore the word on its sleeve with pride. Similarly, in 1969 Alan Harrington published a popular book of polemics called *The Immortalist,* making an unabashed and defiant case for the word because, he argued, the true goal in human history was to transcend our biologi-cal limitations, the ultimate limit being death, in order to preserve and extend human consciousness. Later, a number of science fiction novels and nonfiction books featuring cryonic suspension and reanimation, chief among them *The First Immortal* by James Halperin, became successful in the cryonics and life-extension communities. Many older immortalists told me that *The First Immortal* either introduced them to the concepts behind cryonics or influenced their decision to sign up with a cryonics organization.[2]

But as public exposure grew in the new millennium, so did self-consciousness about the public terms by which the various groups would be known. Immortalist projects were beginning to make headlines in more serious places than the "weird news" columns. Public relations became a major focus as media and outside authors began to write about some of these issues in the late 1990s, frequently using the term as a sensational titling strategy (e.g., science writer Stephen Hall's *Merchants of Immor-tality* [2003]). In response, immortalists began to consciously avoid the word. In many discussions and meetings, I heard "immortality" referred to only half-jokingly as the "I" word—the word that could not be spoken to the outside world. The Immortality Institute's collection of essays, for example, was called *The Scientific Conquest of Death: Essays on Infinite Lifespans*. Aubrey de Grey started out with the Methuselah Foundation, named after the long-lived biblical figure, before eventually shifting to the more mechanical-sounding "Strategies for Engineered Negligible Senescence" (SENS).

In interviews and online testimonies, many members shared how reluctant they were in their personal and professional lives to associate themselves with the term because of a real fear of stigma and ostracism, including in scientific or secular settings. One member told me, "I don't mention the word 'immortality' around my family." Another testified, "I'm not stupid enough to use the word 'immortality' about anything I do. I prefer not to be seen as a certified wacko." These were typical accounts, and the apprehension is by and large justified, as the two following stories from email listservs show:

> When I mention the name Immortality Institute to my friends (people who are computer science or biotech majors . . . who are really interested in life extension and science) they get turned off immediately because they think it's some sort of new age group, or a cult, or a pseudoscience group.

> I have only mentioned the Institute on one occasion to my colleagues. My colleagues are well-published and recognized in the field of aging. Let's use Ass/Prof. G.L. as an example. He has published a book [on the medicine of aging] as well as having close to 100 published papers to his name. Before I could even explain what the mission was, the room was full of laughter. I won't be doing that again in a hurry.

"Immortal" was a catchall, recognizable term, but one with the disadvantage of inviting ridicule or scaring people away. Scientists and regular people were much more open to other terms, such as "negligible senescence," "longevity science," "anti-aging research," or "life extension." Foremost was "the religious problem": the word brought up all the religious connotations the very secular and atheist Immortality Institute was trying to avoid; it scared away mainstream folks, made the serious scientists snicker, and instead attracted the kinds of people that life-extension advocates regularly referred to as "the mystical types." Because immortalist communities want to claim that their project is scientific and materialist, and not dependent on imagined worlds or the "whackadoo,"[3] religion becomes the line they draw to separate their projects from that other sort of immortality project.

The Immortality Institute's online poll and discussion on a name change offered numerous alternatives—over 220, actually. In the end, though, none seemed to win the day and the members voted to retain the name Immortality Institute, but added a subtitle: Advocacy and Research for Unlimited Lifespans.

At the time the general consensus seemed to be that "immortality works fine." It defined a broad enough concern inclusive of people who were focused not only on a biological approach to immortality but also on an informational one. Equally important, the director at the time, who goes by the handle *Mind,* explained to me in an email, the sense was that Immortality Institute had already "staked out a spot in the memespace." *Mind* emphasized, "we own 'immortality' on Google."[4] Who could ask for more?

"Immortality" survived—but only for another year until the online forum changed its website name to Longecity.

As for the Immortalist Society, its earlier effort had ended in a similar outcome. The necessary two-thirds majority required for the name change did not materialize either. Instead, the directors agreed to at least change the name of their publication, *The Immortalist*. In the first issue of the magazine now called *Long Life,* the directors wrote:

> Yes, the magazine you are looking at was formerly called THE IMMOR-TALIST . . . after very extensive arguing the I.S. Directors agreed on the magazine name change.[5]

The ambiguities infusing these two episodes are indicative, on the one hand, of the ways in which immortality has served as a marker in the separation of the religious and secular realms, and on the other, of the extent to which the line between religion and the secular is itself blurry and unstable.

It is often assumed that with the rise of science, scientific material-ism, and evolutionary explanations of the origins of life, immortality was simply discarded as a scientific topic. But in fact immortality was a topic of rigorous debate and research from the 1880s through to the early part of the twentieth century *within* the sciences, resurfacing again recently. In this chapter I will review the ways in which immortality historically emerged and became constituted as a tense object in and around the sciences, including biology, the social sciences, and psychology, sometimes an object of study and fascination, sometimes an anathema.[6] I focus on the ways in which the technoscientific and wider cultural contexts informed each other, and show how this object was uncomfortably stretched between finitude and infinitude, marking the possibility of infinite openness against the maddening limits of finitude and death. This is not a comprehensive history or genealogy; rather, I take a selective approach with several emblematic periods and ideas, especially as they might relate to immortalism today. I move from the early European materialist elimination of the immortal and "primitive" soul to the rise of modern psychology in the United States in the 1880s and 1890s; I examine the biological sciences in the period from the 1880s to the mid-1930s, when immortality was indeed a viable research project, with a brief review of its decline with acceptance of the Hayflick limit, at which point decay again became understood as an inevitable part of normative biology; and I detail the banning of cryonics and immortality research from the American Society for Cryobiology as a way to explore scientific boundary work (Gieryn 1983). As science and technology studies scholars have consistently shown, part of the key work of official science and scientists has been the boundary-maintenance work of keeping vitalism, mysticism, supernaturalism, and other versions of the incredible, unpalatable, primitive, and whackadoo out of their domain. Immortality, as the space of whackadoo par excellence, where the boundary between credible and incredible, acceptable and taboo, constantly dissolves, is also the space of science's boundary-maintenance work.

DISPLACED SOULS

For secularists, the concept of immortality has served as one of the strongest markers of religiosity from the beginnings of materialism in Europe in the eighteenth century. Under the general category of immortality, secularists lump together a diverse set of concepts, entities, and practices—from ghosts, to resurrection, to rebirth, to possession, to afterlife, to research on planarian worms—whose main unifying relationship is not their relationship to each other as a class but to a materialist ontology that undergirds the space and procedures of secularity, including the teleology that leads from a unique birth to a final death. As Sharon Kaufman and Lynn Morgan have written, "The broad topics of reincarnation and resurrection, along with the particular practices of exhumation and reburial, pose a challenge to our terms beginning and end, and to the discrete, linear, Eurocentric trajectory these terms imply" (2005, 320). Even within the history of Christianity in the West, immortality has taken many different forms, as Carolyn Walker Bynum (1995) documents, from detailed debates about resurrection of the body to the immortality of the immaterial soul to their reunification. In Japan, large cross-sections of the population seem to believe in an afterlife without accepting the existence of a transcendent deity (Lock 2002); in the United States, many younger people who do not believe in the existence of God and many others who do not identify with a particular religion nevertheless believe in some sort of life after death, including a paradisiacal world beyond this one (Pew Forum on Religion and Public Life 2010; also see Gallup and Proctor 1983). Platonic arguments about the immortality of the soul were not "religious" arguments insofar as they did not depend on deities or spirits, though they did posit a world of ideals. Nor does every single set of practices and beliefs that has been categorized under the label of religion place much emphasis on afterlife belief systems. The most common examples are forms of Judaism and Confucianism (A. Segal 2004), but as mentioned earlier there are also non-monotheistic hunter-gatherers like the Hazda who tend to dispense with the usual hullabaloo over burial rites as well as beliefs in continuity of the person after death (Woodburn 1982). Even where there was a structure of judgment after death, the assessment did not necessarily involve a specific deity, as can be seen in the principles of karma in Hinduism or their equivalents in gnosticism or Orphism (Brandon 1967). In other words, the links between afterlife beliefs and theistic beliefs are neither universal nor necessary. The fact that afterlife beliefs[7] and religion have been yoked together, even though they do not harbor a necessary connection, only emphasizes the point that the category of religion itself is problematic, as it seems to be cut up in arbitrary ways to fit various assumptions or agendas (Cantwell Smith [1962] 1991; Asad 1993).

Similarly, the elimination of the doctrines of the soul and of immortality, far from being a mere side effect of the elimination of a belief in God, was in itself an independent and a crucial step on the way to establishing a materialist or "immanent frame" (C. Taylor 2007), and later a secular space devoid not only of religious dogma but of metaphysics altogether. The series of discussions in England and France that gave rise to a robust materialist conception of the world in the eighteenth century had as much to do with the existence of God as with a *separate* debate over the concept of the soul and its survival after death (R. Martin and Barresi 2006; Hecht 2003).

Materialist doctrines, broadly speaking, had their strongest foothold in the French Enlightenment and the coterie of philosophers, mathematicians, and writers revolving around Diderot and his *Encyclopédie* project, culminating in the most renowned early materialist exposition, *Systeme de la nature,* written by Baron d'Holbach ([1771] 1835), a friend of both Diderot's and Hume's.[8] What came to be known as French materialism was driven by two goals: first, to explain the world based on simple materialist premises of sense perception and cause and effect; and second, equally important and overtly political, to refute religious doctrines as such, in particular the existence of God and an immaterial soul, and by extension its immortality. In other words, it was a question not only of presenting a materialist argument about the composition of the world but also of preparing that materialism to effectively counter some of the doctrines posited by religion and "spiritualism," including of the Cartesian sort, which separated mind from its material basis.

Jennifer Michael Hecht's (2003) detailed account of the Society for Mutual Autopsy, a group of French materialists who were dedicated to dissecting brains after death, illustrates the extent to which early materialist agendas were driven by an obsession with the concept of the soul. On the one hand, the society's mission was to show that there were correlations between the physical features of a brain and that person's mental characteristics. On the other, that objective was determined by an ideological goal: to prove the nonexistence of the soul. Hecht suggests that this "deconsecrating project guided a host of endeavours at the dawn of professional science" (2003, 254). Materialist scientists and anthropologists wanted to discredit the sacred and transform it into the profane by taking on specific aspects of what was considered religious or supernatural: religious ecstasy, demonic possession, faith cures, and the soul's immateriality. The crucial criteria for scientific work included methodological rigor, repeatability, and verifiability, but they also included the production of such knowledge as would "lock out philosophy and religion" (2003, 50). Against the Catholic notion of the sanctity of the whole body, the integrity of which was crucial to resurrection, and against the noumenal soul, they publicly denigrated their bodies after death, calling it "rotting garbage" or "an assemblage of decomposing matter," positions that were also articulated as anticlerical. Because of their materialist commitments, and therefore their aversion to anything resembling a soul, atheists, freethinkers, and materialists eventually took aim at Plato, Kant, and Descartes, that is, at the metaphysicians or rationalists. Grappling with and proving the nonexistence of souls or soullike concepts—whether or not sourced in religion—has been a boundary-maintenance activity of modern secular authority. At the same time, Hecht shows, they "agonized over the end of the soul and its consequences for humanity" (2003, 2). The creation of the Society for Mutual Autopsy was an attempt to make sense of their soulless death by offering up their own bodies and brains to science and progress, giving them a chance to connect their existence to eternity, or at least posterity.

In early United States history, materialism had significant influence in some circles, around Jefferson for example (Richardson 2005),[9] but most commonly it was presented as a threat, a prelude to total moral dissolution, a French corruption that could only be countered by reemphasizing the immortality of the soul. In reaction, antimaterialist and millenarian tracts proliferated through the nineteenth century, including during what historians of American religion call the second and third great awakenings—often blaming the evil on the French. For example, in *Immortality of the Soul and Destiny of the Wicked,* Rev. N. L. Rice ([1871] 2005) attacks materialism as a French invention and the greatest evil to have been thought up. "In making man a mere material organism and denying his free agency, they furnish the best apology for all wickedness." The argument was common that sin afflicted every individual and could only be battled with help from theology and the church. The church and its discipline prepared the individual for a proper life, and immortality was a promise delivered upon the completion of a good Christian life. Without that threat and promise, humans would turn to the wickedness inherent to this world.

Among less reactionary religious groups and institutions, there was a frenzy of attempts to examine and grapple with the effects of science on doctrines of immortality. A number of symposia were organized at the turn of the century, including the "Christian Register Symposium on Science and Immortality" in 1887 and another one in Chicago in 1902 called "The Proof of Life after Death: A Twentieth Century Symposium." In the discussions of these symposia, as in a number of books and articles, it is easy to see the now familiar acrobatics of religion attempting to adjust to a scientific worldview: to either attempt to reconcile the Christian view of immortality with a scientific view of the world by suggesting that their concepts did not contradict each other, that they were merely given to different domains, that science had nothing to say about the spiritual; or else to fall back upon inner experience—including its modern, secular vocabularies—which in the case of immortality meant that a *"natural and not a miraculous"* order of existence was acknowledged, in the words of Rev. Theodore Munger, since an inner sense of immortality could nevertheless *be experienced* within the self through the self's consciousness of Christ (Munger quoted in Leuba [1916] 1921, 150–51). Liberal theologians like Friedrich Schleiermacher wanted to reconcile theology with Enlightenment critiques by authenticating religion as a personal and empirical experience (see Proudfoot 1985, xiii). A similar defensiveness plagues current theological accounts of the afterlife that tend to justify immortality doctrines with reference to the increasing authority of doctrines of secular immanence and the biological body. They argue, for example, that belief is justified until disproved (Zaleski 1996), or else they justify belief through indirect evidence, by marshaling data from scientific experiments on near-death experiments and the irreducibility of the mind to matter in order to posit the possibility, even within science, of the existence of a soul (Hick [1976] 1994; Zaleski 1996). Even for the religious, discussions about the afterlife and immortality are already entangled in secular discussions about their impossibility.

The naturalist or materialist view on death was not the only or the predominant view, but from the end of the eighteenth century onward, a growing number of people in Western Europe and North America came to live with the fear that both their lives and their selves could or would come to an absolute end with their death (Sloane 1991; Aries 1975). Matthew Arnold, the British poet and schoolmaster, said of his own teachers that they "purged my faith," and one result was a new relationship to living and dying: "What doest thou in this living tomb?" he queried of humanity, signaling a moment when living in the face of the finality of death was suddenly being experienced as a form of dying, the person—"thou"—trapped in the living tomb of the body. Similarly, Wordsworth, in one of the first secular explorations of memory, nationality, and personal identity, *The Prelude,* laments that all we're doing here on Earth is "unprofitably traveling toward the grave." Narratives of people who had "lost" their faith or been "purged" of it through secular proselytization proliferated. The possibility that there was no afterlife became, if not a certainty, at least a definite consideration that in itself began to shape a particular orientation toward life imagined now as a long death, a living toward dying.

The materialization of the soul was never properly completed. Instead, there arose new concepts that tried to capture or explain various aspects of what once had been delineated by the concept of the soul (see Flanagan 1991). And these concepts were developed through (and in turn generated) new disciplines and modes of knowledge, starting with the social sciences as they separated themselves from the humanities and theology on the basis of the secular differentiation of spheres of value and expertise (Calhoun et al. 2011, 4), followed by new studies of the "mind," mainly psychology. Descartes was the first to begin to use "mind" in this sense. Though he believed in God and his epistemology ultimately depended on God's existence, Descartes started using *mens,* "mind," in order to avoid using "soul" as his referent and still refer to some of the same capacities and attributes. Søren Kierkegaard turned spirit into self, and William James, who studied the physiology of the brain, developed the term "spiritual self" to denote "inner subjective being." James approached *religious experience* as an irrefutable if individual empirical event rather than as a doctrinal or metaphysical proposition; he thus provided religion with an authentic, real, and autonomous domain based on mental states that could be called mystical or religious (also see Coon 2000). This pragmatism did little to troubleshoot problems faced by materialism, however, and mind and matter remained irreducible to each other. One might cite later attempts to occupy the space left by the elimination of the soul: concepts of self, person, and personality in sociology, anthropology, and psychology; and the cluster of concepts glossed as consciousness (awareness, autonomy, subjectivity, capacity to feel pain, etc.) in political science, law, philosophy, critical theory, and neuroscience.

Dilemmas over the proper way to think of persons and their extensions beyond the body (and so beyond death) are caught up in the secular–religious, modern–primitive divides over beliefs, and so it is no accident that the problems of death and immortality appear as crucial pivots in the founding of anthropological theories of culture as well as psychology just when materialism was shaping its own view of a world without a prime mover, without souls, and with a final death as the central fact of life itself. Originally, in eighteenth-century France, *psychologie* was considered to denote doctrines about the soul, that aspect of being human without which one would not be human. But as psychology began to make claims as a scientific discipline, "soul" was slowly extradited to the domain of religion and there was a clear effort to steer away from such formulations (N. Rose 1996). Prior to 1880 there were no academic departments of psychology in American universities (Fuller 2001). Classes in psychology, if there were any, were usually taught in theology or philosophy courses, often as an extension of Protestant moral theology (F. Fuller 2001, 125). The brain, for its part, was entirely under the purview of physiology departments. Within a decade, however, almost every American university had established a separate department of psychology—to teach a topic now being referred to as "psychology without a soul." The history of early psychology in the United States clearly traces a shift from the study of mind as soul to a study of the mind without soul—in other words, as a subject within the purview of science. In this sense, the "mind" is a secular concept; it is one of the specifically designated nonreligious placeholders for the discarded idea of soul.

That first generation of American professional psychologists, themselves the sons of Protestant ministers, saw their main task as the separation of the study of the mind from theological concerns (Fuller 2001, 125). It is to one of them, Professor James Leuba, that we owe the first figures recording American attitudes toward an afterlife dating back to the turn of the twentieth century. A relatively influential member of the first generation of American psychologists and the son of religious parents, Leuba had anticipated a career in the ministry but in a crisis of faith realized that he could not reconcile his new scientific understanding of the world and the difficulties of modern pluralism with his traditional religious teachings (Leuba [1916] 1921). His "conversion" to psychology was a means of studying human nature more scientifically—and, in his case, to refute religious thought (Fuller 2001, 197). Taking on the soul and immortality was one way to achieve this and, indeed, a crucial one for a discipline like psychology. The nonexistence of the soul was thus not just a fact about reality for early psychology but an elimination project to be carried out.

Leuba sent out a questionnaire to a thousand scientists chosen at random from the publication *American Men of Science,* the majority of them college and university professors. About half of the respondents said they disbelieved in immortality, had doubts about it, or were agnostic. Leuba presented the same questionnaire to lists culled from the membership of the American Historical Association, the American Psychological Association, and the American Sociological Association, obtaining similar results. Interestingly, of those who said they did not believe in immortality, 45 percent said they nevertheless desired it ([1916] 1921, 261), and overall, *a larger percentage said they believed in immortality than said they believed in God,* a result that seems consistent with contemporary results.[10] Leuba's own conclusions were twofold. First was a version of secularization theory: Christianity faced a problem of belief as more and more people were abandoning the two most important tenets of the faith—the existence of God and the immortality of the soul, which, it must be noticed, were held independently. Second, he theorized sociologically in a Durkheimian vein that the condition was related to the rise of individualism and the attendant loss of institutional authority, specifically Christian institutional authority.

If psychology's genealogy runs through immortality via the secularization of a Christian soul, one might say that anthropology's runs through the secularization of the souls of others. Anthropology begins with E. B. Tylor's well-known and influential schema of cultural evolution ([1871] 1958), where the animism of "savages"—the belief that a whole range of objects had spirits or souls—is transformed through progressive stages of abstraction into supposedly more complex monotheistic concepts of an immaterial soul and doctrines of retribution. Having conceived of this schema of cultural evolution, based on a Spenserian social evolution model, Tylor sought to explain how a belief in nonphysical spirits could have arisen to begin with. As a materialist, he took it for granted that these entities do not *in reality* exist because they cannot be observed through sense perception in the same way that objects of science could. As a believer in the psychic unity of mankind, he had to explain the phenomenon by referring to reasoning and experience that could translate across time and culture. He speculated that the idea of spirit arose from dreams and memories of dead relatives—instances in which a primitive mind might take literally a vision of something that, Tylor says, does not exist outside that vision. Apparently confused by the goings-on inside their own minds, "ancient savage philosophers" (12) concluded, according to Tylor, that every person also has a "phantom" or an image. According to this view (again popular today in the cognitive anthropology of religion, not to speak of the world of selfies), the entire apparatus we call religion arose through and stands on an original cognitive error related to death and the absence of the person.

In 1911, James Frazer, whose more famous work connected "savage" doctrines to Christian ones through the category of sacrifice, turned his focus to immortality in a series of lectures at the University of St. Andrews, comprising the first volume of *The Belief in Immortality and the Worship of the Dead*. Before foraging through his usual sources of travelers, missionaries, proto-ethnologists, colonial administrators, and other Western mythmakers to catalog the range of beliefs, Frazer, who influenced one of the founders of cryonics, dedicates a few sections to speculating on the theme of immortality in the constitution of humankind as a whole. By Frazer's time, the Tylorean immortality story, with origins in a sort of "primitive" illusion refined by the Greeks and Christians and superseded by modern science, had already become a common explanation of the stubborn fact of religiosity in light of empirical facts and theories of human rationality. Frazer (1913, 29) himself admitted: "It is perhaps the commonest and most familiar that has yet been propounded." Frazer's goal was to argue that beliefs were not consistent across cultures and, more, that Christians and moderns still harbor illusions of continuity. He stated his central question thus: "How does it happen that men in all countries and at all stages of ignorance or knowledge so commonly suppose that when they die their consciousness will still persist for an indefinite time after the decay of the body?" (1913, 26). That the opposition of the illusions of the mind and putrefaction of the body should be the binary through which the question of immortality, of life, death, and the afterlife, gets raised is itself a modern way of thinking about the issue—after all, in many places putrefaction is only a part of the material trajectory of the body on its way to clean skeletal remains (Bloch 1971; Seremetakis 1991) and so is considered part of the continuity, rather than rupture, of the relations between the living and the dead, part of the process of engaging with the reality of death and the ensuing grief.

At any rate, Frazer's answer to his own question regarding immortality was like Tylor's regarding religion: it is some sort of cognitive error based on the illusion that dead people are not dead because they appear in dreams. This not only requires the assumption that people, mainly nonwhite people, confused dream with reality, but is also a way of drawing and monitoring the boundary of the unreal and the real and designating who belonged on which side, with rational white Euro-Americans having the most proper access to reality. Between Tylor and Frazer, immortality had become racialized as a key marker of primitivism for moderns—and so, again, nineteenth-century anthropologists prove themselves better as mythmakers of modernity than as chroniclers of other peoples in other places.[11]

It was Durkheim who, in his distaste for psychological explanations, first pointed out that the argument about dreams and the dead *presupposes* what it sets out to explain, that is, the mysterious idea of the spirit or deity: How did these memories and dreams get transformed into concepts like spirits and/or gods? As Durkheim asked in his critique of Tylor, couldn't they just have been viewed simply as dreams or memories? Or did the dreamers have what Durkheim called "the faculty . . . of adding something to the real" (1995, 469)? Though Durkheim did not take on immortality directly, he suggested that it was the endurance of society or social forms, preceding and surviving the individual, that materially or literally represented immortality. Society was the soul.

Durkheim's student Robert Hertz (1960) further developed this sociological hermeneutics of afterlife and immortality. While examining what he saw as the mystery of the double burial, Hertz argued that the practical reasons of hygiene are insufficient to explain a second burial. The belief that the soul is free to depart the body only once the bones, and nothing but the bones, are left would not necessitate a second burial, especially one including such pomp and cost. Death and the afterlife must be understood not in terms of beliefs but as a process dealing with the social person. The person is never simply a person, but a member of a group, formed by the group, invested in by the group. "The society of which that individual was a member formed him by means of true rites of consecration, and has put to work energies proportionate to the social status of the deceased: his destruction is tantamount to sacrilege" (1960, 77). Thus, he concluded, "when a man dies, society loses in him much more than a unit; it is stricken in the very principle of its life, in the faith it has in itself. . . . Because it believes in itself, a healthy society cannot admit that an individual who was part of its own substance, and on whom it has set its mark, shall be lost forever" (79).

Hertz suggests that a person's mortality is to be reckoned with as neither a biological nor an individual psychological event but as one that threatens the continuity of the social group. Consequently, he comes to understand burial rites as a process addressing these two problems: the gradual replacement of the person as a social member within a permanent social structure, and healing the wound in the image society has of itself as something lasting or permanent. After a death, society must "regenerate" itself. Hertz's argument, therefore, is that "in establishing a society of the dead, the society of the living recreates itself" (1960, 72).[12] He concludes: "Society imparts its own character of permanence to the individuals who compose it: because it feels itself immortal and wants to be so, it cannot normally believe that its members, above all those in whom it incarnates itself and with whom it identifies itself, should be fated to die" (77). In moving away from false belief to explain the problem of death and immortality, Durkheimian functionalism reified society as a transcendent agent (it feels, it incarnates, it believes).

As the afterlife was transformed into and theorized as social continuity, the psychology of grief (as opposed to the ontology of ghosts), the survival of one's "works" or a person's life project (Aries 1975; Walter 1997; Lifton 1975), the affective and practical ways in which contact between the dead and the living persist within secular settings came to be generally ignored or sequestered. Tony Walter (2005) has astutely outlined some of these persisting influences in the modern secular world between "the world of the living and the world of the dead." Contrary to the common assertion that "the western mourner has little to do but feel the grief and reconstruct a life without the dead" and that in the "modern West" "the dead are cut off from the living, with no traffic between the two worlds," he writes, there is "considerable traffic, and several professions make a living out of directing the traffic, or perhaps more accurately, as messengers or telephonists, bringing information from the dead to the living" (2005, 407). The traffic is managed formally and institutionally by the army of counselors, clergy, and psychologists administering to mourners after a death, as well as networks of organ procurers and recipients, coroners and medical examiners, lawyers, pathologists, and even police officers and spiritualist mediums. These specialists, Walter suggests, are mobilized to "mediate" between the living and the dead, an activity opposed to the dogmas of both the church and of secularism, for which there is nothing after death to mediate with (2005, 406). In fact, Walter argues, these specializations may have arisen precisely as a response to the structural separation of the living and the dead in a modernized, urban setting, where most dying takes place away from home and in hospitals.

The externalization of aspects of personhood onto digital media—digital personhood—has opened up new secular "afterlife" practices and analyses, even new immortality regimes, concerned with the legal and personal dilemmas of continuity. Questions raised by the maintenance of Facebook profiles after death, avatars that can continue tweets on your behalf, the management of digital estates and identities, and a gamut of other recent afterlife practices have brought these questions to the fore (e.g., Kneese 2018), producing impure domains in which continuity of the person after death cannot automatically be relegated to the dustbin of religious illusion or humanist metaphorization.

IMMORTAL BIOLOGY

Despite the range of psychological, anthropological, and sociological engagements with notions of immortality and forms of continuity, in a secular immanent world in which the biological body appears as the fundamental locus of existence—a biocentric world in which the meaning of life is reducible to the condition of the body (Wynter 2003; Anidjar 2011; Jasanoff 2018)—questions of immortality rebound inevitably back to biology, at different scales, from the cell to the body to the population (Palladino 2016). Biological immortality was a live debate in early evolutionary biology in the 1880s and emerged as a productive scientific object through the 1930s. On the one hand, the body itself, resolutely biological and finite, was filled with time via measures of cellular duration and turned into a timekeeping device through which biological time became a concept with its own measures aside from chronological time. That is, time was transformed from a notion in physics and metaphysics to a process in biology. On the other, rather than making death a condition of evolution and a necessity of biology, evolutionary theory opened up the possibility that biological death was not inevitable or natural and that biological entities could live on indefinitely. Biological finitude and infinitude thus rose in tandem from the same horizon, the limits of the former constantly conjuring hopes of the latter.

A German evolutionary biologist following fast in the footsteps of Darwin, August Weismann was among the most famous and controversial figures to posit that death was a contingent effect of natural selection, "secondarily acquired as an adaptation," and thus not an ontological necessity. Making the evolutionary argument, however, triggered "the most difficult problems in the whole range of physiology—the question of the origin of death" (1891, 20). If it was not in the nature of living things to die, then why would biological death come to exist at all? In a famous 1881 lecture called "The Duration of Life," Weismann concluded that "life is endowed with a fixed duration, not because it is contrary to its nature to be unlimited, but because the unlimited existence of individuals would be a luxury without any corresponding advantage" (1891, 25). For evidence he marshaled examples from "low organisms" that "do not die" but may be "destroyed." The crucial example came from protozoa that multiply by dividing, not reproducing. Division gives rise to new individuals but does not result in the death of any, prompting him to muse: "This process cannot be truly called death. Where is the dead body? what is it that dies? Nothing dies; the body of the animal only divides into two similar parts, possessing the same constitution" (26). In a lecture a decade later, the biologist Alfred Binet rewrote that protozoan proposition into the more felicitous formulation: "In multiplication by division there are no corpses" (1890, 22).

Weismann's "most difficult" question regarding the origins of death then arises in this manner: How did complex species lose the capacity to not die or the capacity for unlimited reproduction? Weismann's answer, based also on the work of other biologists on reproduction,[13] depended on the remarkably prescient idea that differentiation necessitated a separation between germ cells and somatic cells, such that germ plasm or reproductive cells continue the work of reproduction, whereas the others take on the work of specialization. Somatic cells lost this capacity to reproduce indefinitely in order to be able to develop into specific things, to gain distinct and complex capacities. The trade-off meant that their reproductive capacities got limited to a "fixed number." As a result, the organism's specialized parts could get damaged in the long run, and the accumulation of damage is not advantageous. "Normal death could not take place among unicellular organisms, because the individual and the reproductive cell are one and the same: on the other hand, normal death is possible, and as we see, has made its appearance, among multicellular organisms in which the somatic and reproductive cells are distinct" (1891, 29).

Weismann's theories of division and the continuity of germ plasm in time, beyond individual deaths, were controversial. Other biologists kept trying to refute or prove them, and a series of experiments on unicellular organisms and their ability to persist in time ensued from the 1880s on (see Binet 1890; Calkins 1914). There was no final agreement, with most experiments into the 1920s apparently showing that protozoa would eventually die, though after hundreds of divisions, and at least one showing that they would continue indefinitely if the nutrient fluid is cleaned and replenished regularly. Reviewing the empirical evidence, Binet assessed Weismann's theory based on a philosophical notion of identity: the concept of continuity is warranted only if "in division a very small number of elements is replaced." By this criterion, Binet concluded, Weismann's theory could neither be proved nor refuted according to the "observed facts" (1890, 37).

Nevertheless, from Weismann on, immortality as a concept and research object in biology got established in terms of the contingency or nonnecessity of death. In arguing that death was not to be explained by *internal* cellular and physiological mechanisms, that it was not inherent in biology as such, Weismann was also pointing to a crucial and ongoing problem in biology. In the proliferation and continuation of life, what unit of analysis was paramount: the population or species, the individual organism, or the germ cells? (See Palladino 2016 for an important discussion.)

Weismann's ideas on immortality resonated beyond the realm of biology, getting the attention of people such as Sigmund Freud as well as James Frazer. In his work on immortality discussed above, Frazer makes the astounding suggestion that in what he called "savage" culture, humans were all naturally immortal but were subject to sorcerous intentions or accidents that could be ascribed to such intentions. Immortality, then, was the original human idea, with the necessity of death only added on later. He disapprovingly links that mode of thought to modern evolutionary ideas. Liberally quoting Weismann and referring to Alfred Wallace's view that death is not a "natural necessity," Frazer writes wryly:

Thus it appears that two of the most eminent biologists of our time agree with savages in thinking that death is by no means a natural necessity for all living beings. They only differ from savages in this, that whereas savages look upon death as the result of a deplorable accident, our men of science regard it as a beneficent reform instituted by nature as a means of adjusting the numbers of living beings to the quantity of the food supply, and so tending to the improvement and therefore on the whole to the happiness of the species. (1913, 86)

Frazer's humanist consternation must be understood in a context where mortality and immortality were heated topics of debates and research as science was transforming ideas about the living organism, by theories like Weismann's in one vein but also in what Martin Pernick (1988) calls the "decentralization of the person." By the early twentieth century, work on organs as separate, replaceable entities was well on its way, and Jacques Loeb's conceptualization of biology through an engineering ideal (Pauly 1987) had gained major ground. The first eye transplant came as early as 1905, and by the 1920s the biologist, eugenicist, and Nobel Prize winner Alexis Carrel was preserving organs outside bodies using a perfusion pump that would become the model later used for preserving organs before transplant. The pump was designed by and coproduced with Carrel's partner at the Rockefeller Institute, Charles Lindbergh, the first man to fly across the Atlantic. It was their common interest in "immortality" (and eugenics) that provided the grounds for their collaboration.

Indeed, Carrel's most famous experiments consisted of his work with "immortal" chicken heart cells he isolated and maintained for twenty years in a flask, claiming they divided and reproduced as long as they were given the necessary nutrients. The chicken's cells thus outlived the chicken. The chicken tissue experiments were an attempt to prove Carrel's hunch that immortal cells did exist in nature and that rejuvenation was possible (Landecker 2007; Friedman 2007). In a *Time* magazine article from 1925 we read this intentionally jarring juxtaposition in the headline "Science: Physical Immortality." The article is about Professor Carrel's experiments as reported to the magazine by Professor Green returning home to Manhattan after a stay in Europe, where he examined Carrel's experiment with chicken heart tissues. Professor Green was reportedly so excited that "he could not keep talking with a ship-news reporter," exclaiming, "Dr. Carrel introduces immortality in a physical sense. It is there before your eyes, and so long as this tissue is nurtured and irrigated it will live. It cannot die."[14]

The mainstream press was effusive about the prospects of immortality. The *New York Times* headline called it a "Miracle" that "Points Way to Avert Old Age," while *The World's Work* was more symbolic: "Flesh That Is Immortal." Carrel's experiments led to an outpouring of applause from colleagues in the field of tissue culture and inspired a series of—what else—science fiction ruminations, another case of science fused with science fiction.

In this mechanization of the body what was also at stake was *time*, and Carrel's concepts were influenced by Henri Bergson's ruminations on time. Time was being linked to the body, but by making time an interior unfolding, Bergson did not intend to bring it into the realm of natural laws or of subjectivity. To the contrary, he was working explicitly against the Kantian locked-in syndrome, where the exterior world of matter was "forbidden" to the interior world of perception and mental goings-on. Duration and the processes that gave rise to it were not an effect of mind, of a closed-off perception; they were accessible by mind, by a kind of "introspection"—which Bergson termed "intellectual auscultation" ([1947] 2007), the latter being an older medical term for the action of listening to internal organs via a stethoscope. Bergson actually wanted to give time over to the realm of biology, which he seemed to think of as a special kind of matter. His biology was not the usual biology of mechanistic parts and measures, of a finite, determinable causality. His kind of biology, which would be derogated as vitalism, had as its object living forms whose main characteristic was that they were not determined, that they continually gave rise to new possibilities. This meant that the future of biological forms was unknowable, unpredictable, not contained in the prior moment or event. Biology was not a series of states that were fixed by their past or by human-derived physical laws; they were states that were "constantly becoming" and "not amenable to measure" ([1910] 1950, 231). Freedom and measure were antagonists, as were duration and time, the latter being subject to divisions that were measured, the former being an indivisible flow.

Bergson's distinction between duration and successive time, interior motions and exterior measurements, is mutually exclusive—one was fabricated, the other real; one was full of happenings, rhythmicity, and interconnected pasts, the other was "homogeneous" (in his own, rather than in Benjamin's later, words). Despite disagreements with his contemporaries, famously with Einstein, Bergson was symptomatic of his age. In an age in which time was no longer unified and relativity was just being validated in physics, Bergson's opposition between duration and time was reproducing an already-existing bifurcation, a secular disconnect between individual finite time and an ongoing universal time (a bifurcation explored fully in chapter 4).

In the multiple ways in which biology, death, and time became entangled and the debates about the naturalness of death or not dying got framed, another revealing and crucial figure was Raymond Pearl, an American friend of Carrel, a zoologist, biometrician, statistician, geneticist, and longtime professor at Johns Hopkins, where he also founded the important journal *Human Biology*. A eugenicist initially interested in heredity and population genetics, Pearl started his work with planarian worms and then later on the egg-laying capacities of chicken. But it was his work on the scientific study of population growth and statistical and experimental biology that got him noticed. In 1919 he joined the newly established School of Hygiene and Public Health at Johns Hopkins as professor of biometry and vital statistics, where he began his pioneering work on rates of life and death and the prolongation of life. He is the originator of what has come to be known as the "rate of living" hypothesis—the theory that life duration is a function of energy expenditure, or, put in Victorian social and moral terms, the faster you live, the sooner you die. Among the range of positions he held was a two-year stint (1934–36) as president of the American Association of Physical Anthropologists; he corresponded with Franz Boas, Alfred Kroeber, and Bronisław Malinowski.

For Pearl, the prolongation of life was a fundamental concern of all medical science and biology. "The fundamental purpose of learning the underlying principles of vital processes," he writes in *The Biology of*

Death, is "that it might ultimately be possible to stretch the length of each individual's life on earth to the greatest attainable degree" (1922, 17). What the "greatest attainable degree" might be was the key unanswered question. Was there an internal limit to life, to "life duration"? More precisely, was there a final upper limit to the duration of an individual body? Or was biology plastic?

Answering that was Pearl's lifework. He carried out one of the first comprehensive longitudinal studies of longevity in humans and worked on life tables, survivorship (he did the first study linking cigarettes and early death), risk, and ultimately life duration—which he preferred over the term "longevity."[15] Synthesizing Weismann's theories and other new work on germ cells, the work of his first mentor, H. S. Jenning, on "infusoria,"[16] and aspects of evolutionary theory, Pearl noted that not only had germ cells to be "immortal" but that—unlike what Weismann held—regular cells could potentially live forever too. His conclusion came out of the hypothesis that "the processes of mortality are essentially physico-chemical in nature, and follow physico-chemical laws" (1922, 51). In part these formulations were ways for Pearl the eugenicist to counter the claims of social reformists who attributed everything to social events and thus advocated for social remedies to illness (Ramsden 2002). But he also was making a philosophical point: both immortality and mortality could be conceived of within nature, according to natural laws, and need not be taken as processes and possibilities belonging to theology and animated by a mysterious vitalism or some notion of the soul, against which he would soon rail.

Although his main efforts directed him to the finitude of biology—inherent vitality, he would write, is of the "nature of a constant for the individual" (1928, 127), that is, the upper limit of the length of life for a species and an organism was relatively fixed barring external factors—Pearl did not consider death as an inherent function in nature. Drawing on Weismann, he made of life an abstraction unified over time by an overarching process:

> A break or discontinuity in its progression has never occurred since its first appearance. Discontinuity of existence appertains not to life, but only to one part of the makeup of a portion of one large class of living

things. This is certain, from the facts already presented. Natural death is a new thing which has appeared in the course of evolution, and its appearance is concomitant with, and evidently in a broad sense, caused by that relatively early evolutionary specialization which set apart and differentiated certain cells of the organism for the exclusive business of carrying on all functions of the body other than reproduction. We are able to free ourselves, once and for all, of the notion that death is a necessary attribute or inevitable consequence of life. It is nothing of the sort. Life can and does all the time go on without death. The somatic death of higher multicellular organisms is simply the price they pay for the privilege of enjoying those higher specializations of structure and function which have been added on as a side line to the main business of living things, which is to pass on in unbroken continuity the never-dimmed fire of life itself. (1922, 42)

Particular life-forms (an organism or a species) are painted in as formal manifestations and as specialized portions of life itself, whose fire has not been extinguished since its first appearance on Earth! Underlying Pearl's focus on life duration one can excavate an explicitly secular set of tensions between the materialist finitude of the organism and the possibility of breaking or denying those limits. Relying on a transcendent abstraction, life as a form, he nevertheless had to be careful to ritually ward off mysticism from his version of continuity in the undying infinitude he called life. The first few pages of *The Biology of Death* are dedicated to a fiery secular sermon against the soul and spiritualism. Humans have always tried to prolong life by natural and supernatural means, Pearl starts. But advances made by the former would always appear slight—on the order of days or years, at best, which is an excruciatingly narrow result. "When conceived in any historical sense," that is, compared to the depth of biological time and the infinite historical timeline, "man's body"—that is, the body without the soul—"plainly and palpably returns to dust, after the briefest of intervals, measured in terms of cosmic evolution" (1922, 17). No wonder, then, that humans devised another means for "infinite continuation" by conjuring up that "impalpable portion of man's being which is called the soul." This illusion "has permitted many millions of people to derive a real comfort of soul in sorrow, and a fairly abiding tranquility of mind in general from the belief that immortality is a reality." But this also had the "evil" effect of "opening the way" to "recurring mental epidemics of that intimate mixture of hyper-credulity, hyper-knavery, and mysticism" (18). Pearl felt his era, far from being a secular turning point, was suffering "the most violent and destructive epidemic of this sort which has ever occurred."

As a book, and perhaps as a science, the biology of death necessitated a salvo against the theology of the soul, because grappling with infinitude, immortality, and continuity, even on supposedly biological grounds, risked bringing science too close to the transcendence it was trying to shed. The elimination of the notion of the immortality of the soul from the minds of men, the Enlightenment's desouling project (Flanagan 2002), was a key part of the development of the scientific project; thus any Methuselah would have to be created via the internal or finite goings-on of biology. That these biologists concerned with time, aside from being secularists, were also avid eugenicists should not be a surprise (Pearl abandoned his views, but Carrel went on to work alongside Vichy France on biological superiority). Evolution had embedded within it a notion of time in its expositions on descent and generation. In the progressivist vision, changes were improvements that were linked in the European mind to race. White Europeans were considered the most advanced race and thereby had a claim on the future, being at the tip of the arrow of biological time. In other words, staking a claim on the future via biology was precisely what eugenics was about, joining the progress of history to the progress of biology. Even as science has changed, it is important to keep this in mind while thinking about immortality and the ways in which claims are made on the future and who makes those claims, a topic I will deal with in the final chapter.

Once Carrel's experiments seemed to scaffold the theoretical ideas of Weismann, biologists happily took up the assumption that cell immortality was a fact and continued attempting to produce immortal cell lines. Cell immortality remained an exciting dogma, accepted to such an extent that when cells did die in labs, as they often did, their death was attributed to mistakes in the technique by scientists themselves, not to the mortality of the cells (Hall 2003, 26). In the end, the opposite turned out to be the case. It was Carrel's cells that had remained "alive" due to experimental sloppiness: feeding the cell cultures with nutrients taken from freshly killed chicken embryos, Carrel's lab was accidentally supplying fresh new cells to the culture, creating the illusion of immortality in cells. This "sociology of mass delusion," as Hall (2003) calls it, held sway for almost half a century.

At the very end of the 1950s, the biologist Leonard Hayflick showed that cells divided a finite, not infinite, number of times. Hayflick was convinced that cell death was not a function of experimental sloppiness but intrinsic to cell biology itself, or "senescence at the cellular level." His first attempt at publishing the results of his observations was famously rejected, but eventually, after publication of his paper in 1961, it was Hayflick's conclusions that became accepted in cell biology (Palladino 2016; M. Cooper 2006)—"we overthrough [sic] that dogma," Hayflick exclaimed triumphantly later (Hayflick 1998). Hayflick showed that most normal cells—with the exception of cancer and germline cells—divide up to about sixty times, after which they stop dividing and wither away. This limit is known as the Hayflick limit. Hayflick's discovery put an end to the dream that cells were inherently immortal, and the common understanding became that immortality "was the hallmark of the pathological cell line" (M. Cooper 2006, 1), such that aging and its outcome, death, were inevitable processes of normative biology. The cell immortality myth and research into immortality turned into research into cell death and eventually the mechanisms of "molecular disorder" (Hayflick 1998) and what came to be known as apoptosis, or cell suicide, all quite contrary to the enthusiasm over immortality.

Nevertheless, the isolation and reproduction of cells outside the body and inside a laboratory indicated at one and the same time the technological ability to intervene into cells and the concomitant transformation of cells into technologies, of life into artifice and back, indicating that the difference between the two was itself an artificial one, meaning that cellular life was open, potentially subject to manipulations that could extend its functions indefinitely—a possibility that gave rise to biogerontology, which, unlike regular gerontology's soft and social approach, focused on senescence at the cellular level. As Melinda Cooper (2006) documents, tissue engineering as the key vision of biogerontology led around the turn of the twenty-first century to the reinvention of immortality through embryonic stem-cell research, making the category of potentiality, with all its undetermined and open future possibilities, the central idea of the research—as well as its capitalization. In a way, that was an echo of Weismann's idea that specialization leads to the introduction of death into the natural world, whereas the pluripotency of embryonic stem cells prior to specialization could help revolutionize biology, and especially rejuvenation techniques. The promise was that death could again be delayed indefinitely if pluripotent cells could repair the damage accrued by specialized cells. Thus, techniques that demonstrated pluripotency at the cellular level promoted the transformation of the present into an imagined or desired but not inevitable future form, resurrecting an older immortalist imaginary.

WHACKADOO SCIENCE AND BOUNDARY WORK

Immortality as a concept and a research object has had a fraught contemporary history—at times being stigmatized, at others carrying forth a central hope, at times denoting primitive error, at others carrying modern promises, at times bearing a secular promise, at others threatening it with atavism. Because of its blurry status, the boundary work of science has remained vigilant around immortality and its associated terms.

The tensions between cryonics, marginalized as a "taboo science" (Frickel et al. 2010), and cryobiology, the cold science, provides an interesting case of such boundary work. The principle that cold preserved flesh had been posited at least since Francis Bacon was suddenly struck by a bolt of inspiration while traveling with the king's physician on a snowy day. He jumped out of the coach, found a woman who sold chicken, bought and disemboweled one, and stuffed it with snow to see how it would fare in the long run. The chicken may have kept well but it's been said that the cold made Bacon come down with a bad case of pneumonia which soon thereafter killed him, a martyr to his own scientific methodology. Other seventeenth-century scientists became fascinated not by preservation for consumption but by the effect low temperatures seemed to have on the life and death of organisms. In 1670 Henry Power reported success in freezing and reviving vinegar eels, and a decade later Robert Boyle, apparently inspired by Bacon's experiment, noted similar success in experiments on frogs and fish (Thomson 1964; Parry 2004). At the time, Boyle and others noted two problems with freezing. The first was their inability to control, reproduce, or maintain low temperatures artificially. The second was that certain frozen goods, such as plants, fruit, vegetables, organs, and organisms, seemed to be damaged after thawing (Parry 2004; Leibo 2005).

The first problem would be solved in the nineteenth century with the development of refrigeration, used for storage and transportation, particularly of meats. The liquification of gases achieved by several chemists at the end of the nineteenth century would change the landscape of low-temperature experiments and usher in the era of cryogenics proper, meaning the study of the behavior of matter at ultralow temperatures. Among these chemists was a Scot, James Dewar, who liquefied hydrogen and developed a vacuum-insulated storage capsule that could contain liquefied gases without allowing them to rapidly evaporate (Parry 2004; Leibo 2005). Today's cryogenic storage devices are called dewars.

Although experiments on the effects of cooling on biology continued to take place in the early twentieth century, the research was not formalized until a Roman Catholic priest began his experiments on the survival rates of frog spermatozoa and yeast in the 1930s (Schmidt 2006; Parry 2004). Born in a Swiss mountain village, Basil Luyet immigrated to the United States in 1929 with doctorates in biology and physics from Geneva (Schmidt 2006), later taking up a position at St. Louis University and setting up the American Foundation for Biological Research, a laboratory designed to study low-temperature phenomena, where he conducted the majority of his work on cryopreservation (Leibo 2005; Schmidt 2006). As it was for some of his seventeenth-century counterparts, for Luyet cryogenics was a way into a larger existential question that arose at the intersection of his scientific and priestly duties: What is life? In the modern era that question has been scientifically, philosophically, and anthropologically as generative as it has been confusing. For Luyet, death was the way in: "To study and to know life I started to study death since death is the destruction of life," he said in an interview (Mulvenna 1974).

Tacking back and forth between the metaphysical or existential questions and the material experimentation, Luyet observed that life could indeed be restored in some cases, that animation could cease and restart, and so he introduced a new in-between concept into the study of life, "latent life," and the prospect of restoring "animation" to this state. The problem Luyet faced was the main problem of cryogenics, namely, the damage done to cells as a result of ice formation during the cooling process. This destroyed a great majority of cells (though not all), and as a result tissue and organs did not thaw viably. Luyet speculated on some solutions: rapid cooling, dehydration, and the use of certain materials as "protectants," all of which he mentioned in his comprehensive 1940 study of cryogenics, *Life and Death at Low Temperatures,* a book that established him as the "father of cryobiology." Although Luyet's proposed solutions would prove to be in principle the right ones, in practice he himself never solved the problem of freezing injury and cell damage.

An experimental mishap involving chicken semen helped solve it. Spermatozoa had become the focus of low-temperature research since the 1930s, when two biologists experimented extensively with rabbit semen. In the late 1940s the Frenchman Jean Rostand used glycerol to freeze and thaw frog sperm but did not test its subsequent fertility (Leibo 2005). The main interested parties in these techniques were the various food industries, which understood the potential for controlled fertilization and artificial insemination. In 1947 the poultry industry in the UK charged Dr. Christopher Polge with the task of developing sperm preservation and artificial insemination techniques for fowl (Leibo 2005, 357). The team Polge worked with at Mill Hill included Alan Parkes, who would coin the term "cryobiology," and Audrey Smith, who would come to be known as the "mother of cryobiology." They were later joined by James Lovelock, who is responsible for accidentally inventing the microwave oven, and for developing what is today known as the Gaia hypothesis. Polge had his chemicals shipped to him at the Mill Hill labs and went about freezing chicken semen using what he had always used in such experiments as a protectant: fructose. Previous experiments with semen had yielded very poor results. After thawing, no more than 5 percent of the cells remained viable. Yet all of a sudden, Polge and his colleagues were repeatedly obtaining a 50 percent rate of viability. Having exhausted the old supply of fructose, the team turned to a brand-new supply. When the rate immediately dropped back down to the expected 5 percent, they deduced that something must have happened to the old bottles of fructose. The last few milliliters remaining in an old bottle were sent to a lab, where analysis showed they had been filled with glycerol, not fructose. The bottles had been mislabeled. By accident glycerol became the first effective cryoprotectant.

The accident revolutionized the practice of storing and reviving cells. These could now be frozen and stored at ultracold temperatures for very extended periods before being revived. Today's tissue and sperm banks, along with practices of artificial insemination, embryo preservation, and assisted reproduction, owe their existence to these developments in cryogenic science. So aside from being the main experimental "subjects" of cryobiology, spermatozoa also have been the main beneficiaries of low-temperature research, as shown by the proliferation of human sperm banks and the extensive use of cryopreserved semen in animal husbandry.

In the 1950s, Audrey Smith, James Lovelock, and Alan Parkes turned their attention to freezing and reanimating small mammals. Using glycerol as a cryoprotectant, they cooled twenty golden hamsters down to a colonic temperature of -5°C for fifty to seventy minutes before rapidly thawing them with an apparatus made of material from a military radar and reassembled by Lovelock. Of the twenty frozen hamsters, seventeen survived the thawing process, with seven of these dying within twenty-four hours of reanimation, two within ten days, and eight living out the rest of their days close to their species' normal life span (Parry 2004). Kept at a colonic temperature of only -5°C for no longer than an hour or so, the animal was not frozen through and through. That is, not all the water in the animal's body had turned to ice—intracellular ice formation having been identified by then as the clear culprit in causing cell damage during the freezing process. In subsequent experiments in which they tried to keep animals frozen for more than seventy minutes, the animals rarely recovered. They also discovered that animals frozen at ultralow temperatures did not survive. They concluded that "suspended animation" (their term), or the long-term storage of animals and their subsequent revival, had no prospects (Parry 2004, 404)!

But these experiments were enough to convince Robert Ettinger, a high school physics teacher, that suspending human life was in theory possible, the difference being a matter of complexity and scale, not of principle. Indeed, once Ettinger circulated his first manuscript on the topic, it was enough to convince the pioneering cryobiologist Jean Rostand, and Rostand contributed an optimistic preface to Ettinger's book *The Prospect of Immortality*. "We don't have long to wait before we shall know how to freeze the human organism without injuring it," Rostand wrote. "When that happens, we shall have to replace cemeteries by dormitories, so that each of us may have the chance for immortality that the present state of knowledge seems to promise" (Rostand 1965, 9).

Rostand was an exception in his unabashed enthusiasm for cryonics, but the budding fields of cryobiology and transplant surgery had at least a warm relationship to cryonics in the early days. The new fields did yield some cryonics advisers and advocates serving well into the 1970s, one of them even claiming that he was on his way to the full freezing and reanimation of canines.[17] A few cryobiologists and transplant surgeons advised cryonics pioneers, and even served on cryonics organization boards, and at least one cryonics presentation was made at an annual Society for Cryobiology meeting. The first exchanges of letters with the Society for Cryobiology appear to have been cordial and warm. Later, some cryobiologists accepted research grants from members of cryonics organizations.[18]

Over time, however, cryobiology's official position hardened into an unreceptive stance. According to one account, by the 1970s, cryobiologists who were serving on cryonics scientific advisory boards had been "approached by one or more of their colleagues in the Society for Cryobiology and pressured to resign their positions" (M. Darwin 1991a, 7). By the 1980s, on the heels of a series of cryonics public-relations disasters, there was open hostility against cryonics at the society's annual gatherings. Mike Darwin, a leading cryonics pioneer and Alcor member, called this a "cold war" between cryobiology and cryonics, blaming the Society for Cryobiology for taking steps "to destroy cryonics."

In 1982 the Society for Cryobiology passed a new bylaw explicitly denying membership to anyone engaged in "any practice or application of freezing deceased persons in anticipation of their reanimation."[19] The original proposal for the bylaw was much more harshly worded, specifically citing cryonics and calling it a fraud.[20] But the most revealing aspect came from the preamble from the board justifying the bylaw. It sets up a familiar opposition between cryonics research and real science by placing the former into a religious domain, classifying what it calls "cadaver freezing" as "an exercise of faith, not of science." Yet based on the board's own description, it is hard to understand the difference between the research agendas of cryobiology and those of cryonics:

> The Board also recognizes that the goals of cryobiology include not only achieving an understanding of freezing injury and its avoidance but also applying this knowledge to the preservation of cells, tissues, organs, and organisms. A future achievement may well be successful mammalian cryopreservation. However complex the social consequences of such a development might be, this is no basis for discouraging research in cryobiology. The cryopreservation of biological systems remains a legitimate scientific endeavor which the Society for Cryobiology is chartered to support.
>
> . . . There is no confirmed report of successful cryopreservation of an intact animal organ. It can be stated unequivocally that mammalian cryopreservation cannot be achieved by current technology.
>
> Nonetheless, certain organizations and individuals are advocating that person be frozen subsequent to death on the premise that science may ultimately develop the capability both to reverse the injury of freezing and to revive the cadaver. The Board does not choose to involve itself in a discussion of the degree of remoteness of this possibility. The Board does, however, take the position that cadaver freezing is not science. Freezing and indefinitely storing a cadaver is not an experimental procedure from which anything can be learned. The knowledge necessary for the revival of whole animals following freezing and for reviving the dead will come not by freezing cadavers but from conscientious and patient research in cryobiology, biology, chemistry, and medicine. The sole motivation for

freezing cadavers today is the remote hope on the part of individuals that this may be a means of avoiding death.[21]

It concluded: "The Board finds human cadaver freezing to be at this time a practice devoid of scientific or social value and inconsistent with the ethical and scientific standards of the Society. The Board recommends to the Society that membership be denied to organizations or individuals actively engaged in this practice."

The board claimed that "successful mammalian cryopreservation" is a "legitimate scientific endeavor," and at the same time it asserted that since "mammalian cryopreservation cannot be achieved by current technology,"[22] research into human cryopreservation is not scientific. But logic is not what pushed the argument through. The argument was driven by several other assertions in the text outlining secular and social taboos: that cryonics is faith, not science; that cryonics is a fraudulent activity because it accepts money; and that cryonics is simply a means for individuals to avoid death. Each of these assertions appeals to a particular history. The first assertion, linking cryonics and faith and opposing them to science, plays into the history whereby immortality, death, and related fields have been largely given over to the domain of religion. The second assertion plays on cryonics' history of public-relations disasters and the mishandling of cases. The third assertion, that cryonics is an attempt at "death denial," plays into the death acceptance movement emerging at the time.

To trace the history of cryonics and immortalism in general is also to follow the ebb and flow of their status as both a social taboo and a taboo science, or what some scholars in science and technology studies have called "undone science" (Hess 2007; Frickel et al. 2010), referring to scientific research that is deliberately left out of the agenda. Undone or taboo science is when a particular topic of research, the production of a particular kind of knowledge, is marginalized or sequestered. In the era of private funding, the confluence of elite institutions and corporate interests, rather than the public interest, determines what sort of research is considered fundable or doable, and these often defend established markets, despite risk factors or better alternatives. For instance, over the past century the chemical industry has repeatedly sidelined research into nonchlorine alternatives to chlorine use and production, despite documented (by some researchers) damages to the environment and to the health of people exposed to chlorine (Frickel et al. 2010). Social movements and public organizations often try to direct attention to or away from these abandoned or taboo areas, as is the case with research on animals, which has been limited by the animal-rights movement, and stem-cell research, which was banned and/or limited thanks to a coalition of movements in the United States.

In the case of cryonics, the Society for Cryobiology, riding the wave of social taboos, explicitly produced a taboo research zone. On at least one occasion during my fieldwork, cryonics organizations approached a university to collaborate on a research agenda funded by cryonicists themselves. The program would examine some aspect of research relevant to cryonics and would be a part of existing departments and not a separate program on cryonics. Though the university began the conversation, it also ended it relatively quickly. Uncharacteristically, they turned down the offer of money. After poking around, a legal advocate involved in the negotiations made this claim to me: "I found out that they turned it down because a couple of professors that they talked to did not want to be associated with cryonics because they don't think that it's real science, and I think that that's a shame. I always heard that university campuses were a place where real research can be done regardless how crazy the idea is but I'm wrong. I know that."

Mishandled cases in the early years along with the gory public reputation of cryonics have had much to do with its sequestration. But organ transplantation was in a similar position in its early days, with many naysayers and appalled populations. In its later days it was beset by numerous cases of fraud, lack of consent, and abuse within established medical settings—from the gigantic case involving thousands of organ-stripped and discarded corpses at UCLA (Nelkin 1998), to organ black markets in the United States and abroad (Scheper-Hughes 2000), to the trade between organ procurers and coroners for cornea removal and tissue extraction (Timmermans 2006, 232–44). Although trading organs or tissue is a crime, the National Organ Transplantation Act allows for a "handling and processing fee." By 2006 the handling and processing burgeoned into a $500-million-a-year tissue trade (Timmermans 2006, 243).

Organ transplantation involves many of the same scientific, medical, and cultural processes that are involved in cryonics, from the most general one of "redefining death"—successfully achieved, in the case of organ transplant—to the specifics of organ cryopreservation and transportation. While some rising cryobiologists who have made contributions to the field and are also avid cryonicists have maintained a *public* distance from their cryonics activities for fear of jeopardizing the credibility of their lab-based cryobiology research,[23] cryobiologists with long involvement in cryonics have developed what appears to be one of the more advanced and least toxic cryoprotectant solutions, M22, at a research lab called 21st Century Medicine (21CM), funded in part by cryonics advocates. M22 is currently being used in both mainstream organ preservation *and* for cryonic preservation at Alcor (Fahy et al. 2004). Dr. Greg Fahy, who has been interested in cryonics since his high school days, has been a bridge between cryonics and cryobiology. Additionally, over the years 21CM has developed a number of other products (e.g., ice-blocking agents) some of the research for which has been published in the journal *Cryobiology* (Wowk et al. 2000). Back in 1981, Fahy was also one of the first cryobiologists to offer "vitrification"[24] as a solution to the problem of ice formation (M. Darwin 1981). Given the interests and backgrounds of the people involved, these products were in a very direct sense the results of *cryonics* research—which therefore can be said to have been making contributions to cryobiology at large and to general medical science by extension.

THE ETERNAL RETURNS OF THE "I" WORD

My goal is not to advocate for the legitimacy or desirability of cryonics but to show how cultural notions of immortality, especially its persistent relation to religion, have historically influenced the scientific research agenda specifically and orientations toward continuity more generally. Materialism buried the soul and its immortality, thereby creating a marker for the divide between religion and the properly scientific. That boundary is largely managed within what I have been outlining as a secular tension:

immortality is science, immortality is faith. In that divide, other categories have been implicated, such as mind and person, human and nonhuman, as well as cells and species and society.

Immortality flows and eddies across domains because it runs through forms of continuity and projected futures that are unstable and sometimes ephemeral and for which materialism must constantly invent new procedures and measures, as well as new (soulless) entities as the key, often abstract, often transcendent, units of continuity. Immortality has to be warded off, but it keeps coming back, as much in science (the realm of the nonmetaphorical!) as in social and cultural domains. It keeps being reinvented through the practices of science and the concepts that drive them, through materialism itself, through the unsettled tensions between it and rationalism, which does not assume the material reducibility of mind, and finally through questions about continuity: What is it that has, in fact, continued in time (life itself, but not individual species, according to Pearl and Weismann, e.g.)? What is it that must be continued? The self? Life? Society? Nation? Human species? Human consciousness? DNA? Genetic information? Progeny? Great works? Pyramids? Ghosts? You? What is your unit of continuity?

2

Immortalism

THE HISTORY OF A FUTURISTIC MOVEMENT

The Ramada Inn in Plainville, Connecticut, is your basic small-town motel in the Northeast, right off a highway exit, just before Main Street, a long four-story building at the far end of a parking lot the size of two football fields. It's the kind of hotel that turns its generally dim, unoccupied restaurant/bar, called Crooked Street Station, into a Saturday-night disco for local guys with souped-up engines and dolled-up girlfriends who cup their colorful cocktails like gourds of magic potion warding off who knows what cast of fear. The carpets are faded, the towels are frayed, and the old ice machine down the hall whirrs a little too loudly. It was not the kind of place I had imagined. I was expecting a Sheraton convention center, at least some sort of conference facility, something anesthetized and affluent enough to fit what I imagined should be the home of the first semiofficial meeting of the Immortality Institute.

In the lobby I came across people I knew from the web. Seven or eight of us had arrived at just about the same time, 3 p.m. on a mild Saturday in early December 2006, and most of us didn't know each other except by usernames, by pre-Facebook-era identifiers like "Lazarus" and "Osiris" and "Reason" and "Mind" and "Caliban" and "Bill O'Rights." That was one of my first lessons in immortalism: metaphors matter.

Some of those present I recognized from their 72 dpi pictures on the discussion board IDs. People introduced themselves by their real names, and that created an awkward lag while we put faces on usernames and then real names back on those same faces. The double adjustment took a while. In fact, it took all night, as members continued to call each other by their usernames. When you write passionate emails every day about life and death to Lazarus, Osiris, and Reason, it's hard to suddenly call them Ted or Jamie or Chris.

Entering the hotel, Lazarus enthused in his big voice: "I saw Bill O'Rights' car outside. The big red one. I recognize it from one of his posts. He must be here. Is that his car?"

Bill O'Rights was actually right there next to him. Caliban, the German bioethicist, assumed his organizing role immediately, tactfully gesturing to Bill the human being.

"This is Bill," he said. "He is right here."

"Oh, hi Bill," said Lazarus.

Bill O'Rights had a big diamond-studded cross on a hefty chain swinging down from a white hooded sweat top. His white, shaven head gleamed, as polished as his car. Some people said he had largely paid for the cost of the gathering at the motel. The rumor was whispered because it was not clear how, or even why, O'Rights would do this. O'Rights, who was as knowledgeable about supplements and the U.S. Constitution as anyone, would go on to become increasingly intriguing over the years: after disappearing for a short stint, he showed up in prison, taken in on drug-related charges, then, after his release, succumbed to a fatal illness. He had no money to pay for a cryonics policy. The immortalist community banded together online to collect the necessary funds for his cryopreservation. That was another lesson: despite the individualism, communalism mattered here.

None of these gestures or identities was what I had expected at a gathering of the members of the Immortality Institute. But then again, what might one expect from an organization whose motto is "Conquering the blight of involuntary death"?

What people did think or say about them—that they're nuts, weirdos, scam artists, godless fools, devils, immoral, wealthy, and selfish extremists—was a major preoccupation. I had been introduced to the personal depth of this preoccupation earlier that morning when I went to meet my ride in Hoboken, New Jersey. A biology student, S.R., and I were being driven up to Connecticut in Randy's truck. Randy is an old-time gay activist who was into cloning as a method of survival, his preferred form of nonsexual reproduction and continuity.[1] He called it immortality through cloning. He was there for Stonewall, and now he was proud to be the world's first cloning activist, heckling bioethicists at conferences and getting interviewed on TV. As we walked to the truck, S.R.'s phone rang. It was apparently a friend trying to get together for the weekend. S.R. used to be a real estate assessor. He couldn't stand his job and felt smarter than the job required. He'd finish two hours early and spend the rest of the time reading science books and magazines under the table. That's how he came across such topics as nanotechnology and immortality, by reading a science magazine, like a great majority of immortalists and cryonicists. He saw the light, decided to quit his job, lose the salary, and go to school again to study molecular biology. He joined the Immortality Institute and became one of the forum coordinators on the web. His girlfriend, who did not join us, was an English major. Once, they promised each other to exchange books in their respective fields, since he doesn't read anything but science and she doesn't read anything but fiction. He gave her Richard Dawkins; she gave him Harry Potter. Neither managed to get through to the end of the other's gift. While we waited for him in the car, S.R. tried to get off the phone gracefully, telling his friend that he had to go. His friend on the other side obviously asked him why he was going to Connecticut.

S.R. answered, "My website is having a meeting." Not the Immortality Institute, not my organization, but *my website.*

His friend must have been inquisitive, asking for a little more detail, because S.R. then said, "Oh, science and biology, you know, it's a social gathering."

The fear of mentioning one's connection to immortalism in general, or to cryonics in particular, was common. The repercussions were not just verbal and emotional. A number of people have given me accounts of being professionally ostracized when they discussed their involvements (see introduction). As the ideas behind the movements have gained backers, some credibility, and more visibility, the impulse to hide has waned. In Connecticut, the self-consciousness was still very much present. What do people think when they hear the word "immortality"? What about the name "Immortality Institute"? What, for example, did the people at the front desk of the Ramada Inn think when they booked rooms for an "Immortality conference" organized by the "Immortality Institute"?

Bruce Klein, then director of the institute, pointed two fingers over his head like the devil and made as though addressing the receptionists. "See, we don't have horns," he joked, and everyone laughed.

The receptionists, it turned out when I asked them, didn't think much either way—as long as someone's booking rooms every once in a while, they're fine, they don't inquire beyond the spelling. "Immortality," with an I and double M. For all they cared, it may as well have been the Rotary Club. But Bruce's joke worked like an inoculation against the outside world. It's OK to be here because everyone's together now. They scotch-taped signs near the elevator, announcing "Immortality Institute Meeting, 4th floor."

The fourth-floor suite had a round, chipped table, stained sofa, and a coffee machine. An undersized cover thrown over the table concealed its blemishes, and this became the conference room. The predominant feeling around the airless room was of something clandestine, radical, marginal, socially unacceptable, and underfunded. This was not the image I had had either. This was not the community of wealthy, hubristic pharaoh-men seeking another way to immortalize themselves. Nor was it the community of well-heeled, institutionally supported hi-tech scientists.

The group was almost all white, middle class, and male. The exceptions were the Latina half of one of the couples present and the Portuguese girlfriend of one of the scientists. There were two scientists in the room, one a researcher at Harvard, the other at a private biotech company in search of a cure for the flu. An Anglo-German bioethicist was the secretary of the group, and mocking his own Germanness, he kept the discussion organized and confined to the allotted time. The others were designers, students, teachers, small-business owners, and the gay activist from Hoboken, Randy. The ages ranged from late twenties to late sixties, aging to aged.

The discussions that ran passionately through the night were mainly procedural—how to phrase the mission of the institute, how to improve the forum, and expand it, how to get more people, and more prominent people, to think about the topic or take it seriously. How and where, for example, should a conference be organized, and whom should be invited to speak? Names from the immortalist pantheon, like Aubrey de Grey, Eric Drexler, Ralph Merkle, Ray Kurzweil, flashed across the room. Topics from calorie restriction to biogerontology to cryonics funding to AI and mind uploading were proposed.

But these practical matters also implied social and political issues. How can one talk about immortality without evoking a long history of yearning and myth and Christian tradition? How do you remain secular in this realm? What can one do to counter the publicized attacks of people like the bioethicist Leon Kass (then the chair of the President's Council on Bioethics) or the journalist Bill McKibben (who had recently published a book against radical technological alterations of human biological nature)? Immortalists present felt besieged, surrounded by hostile forces, hampered by government regulation and religious atavism as well as by narrow-minded scientists who simply labeled the venture impossible, by the interests of "big pharma" (which, it was said, did not want cures for anything), by academic sluggishness and traditional thinking in general— in effect, they felt ostracized by the culture at large, which they called "deathist." It would take concerted, programmatic effort to overcome the assumptions and habits of such a culture. That would be the task of the Immortality Institute.

A year and a half later, the institute's first official conference, in Atlanta, Georgia, attracted 150 people. The attendees included researchers from prestigious institutions, engineers, doctors, journalists, and academics. Among the speakers were Michael Rose, a professor in the Department of Ecology and Evolutionary Biology at the School of Biological Sciences, University of California, Irvine, and author of the book *Evolutionary Biology of Aging* (1991); Ralph Merkle, professor of computer science at Georgia Tech College of Computing; Aubrey de Grey, an engineer and gerontologist at Cambridge University who was developing a series of strategies to defeat aging through molecular biology and had been recently vilified in a cover article in the *MIT Technological Journal*; James Hughes, a sociologist at Trinity College who is a Buddhist, an ecosocialist, and executive director of the World Transhumanist Association; Martine Rothblatt, a trans inventor, writer, lawyer, and activist who launched the first satellite-to-car

radio company (Sirius) as well as the first nationwide vehicle geoloca-
tion system (Geostar), led the International Bar Association's project to
develop a draft Human Genome Treaty for the United Nations, founded
and runs the biotech company United Therapeutics, and still managed to
find time to write books with titles like *The Apartheid of Sex* and *Unzipped
Genes* and to start a nonprofit organization called Terasem, dedicated to
joyful immortality and geoethical nanotechnology; Eliezer Yudkowski,
cofounder of the Singularity Institute for Artificial Intelligence (SIAI);
and Ben Best, onetime director of the Cryonics Institute. The conference
was chaired by Bruce Klein, founder of the Immortality Institute, who
would soon leave the institute and briefly join forces on the West Coast, in
Silicon Valley to be precise, with one of the speakers, Ben Goertzel, an AI
researcher, to lead an attempt to build artificial general intelligence (AGI),
in part as a bid to produce the immortality of the mind. Eventually, Klein
would drop off that venture as well, lose money, become insolvent, and
attempt various scamming schemes for which he would be apprehended
by law enforcement.

The list of speakers demonstrated the range of fields or groups involved
in this emerging network of people pouring in to the catchment site of
technoscientific immortality, bringing together people and disciplines that
don't regularly come together: computer engineering, AI, transhumanism,
molecular biology, nanotechnology, gerontology, psychology, cryonics,
cloning, mathematics, finance, space colonization. Many of the speakers
already knew each other, as the overlap of interests, ideas, and practices
between, say, nanotechnology, cryonics, AI, and transhumanism had
long been mapped. The Immortality Institute had provided a common
roof for a multidisciplinary gathering where institutional researchers met
and exchanged information with DIY researchers. It wasn't at the glitzy
Sheraton conference center that housed the Atlanta-headquartered CNN
conferences, but compared to the Ramada Inn in Plainville, Connecticut,
this could be ranked a few rungs up.

Immortalism, or radical life extension (RLE), would slowly gain more
mainstream visibility, gain more members, and find its way into more
mainstream media as well as established scientific settings. Yudkowski's
Singularity Institute got linked up to celebrity futurist Ray Kurzweil and

a number of Silicon Valley players, including one of the founding investors of Facebook, and it hosts annual summits attended by luminaries in
biotechnology, philosophy, and neuroscience. De Grey would quickly
rise to become the most visible star in the immortalist firmament, appearing on countless magazine covers, in a Barbara Walters special, and,
perhaps most importantly for the movement's sense of credibility and
vindication, at a number of speaking engagements at established scientific
institutions, including the prestigious New York Academy of Sciences.
Serious articles on topics such as cryopreservation, calorie restriction,
and cyberimmortality would appear in magazines and papers from the
New York Times and the *New Yorker* to *Rolling Stone. National Geographic,
Oprah,* and *60 Minutes* would air television programs on life-extension
strategies, including conversations with immortalists and the exploration
of cryonics as well as calorie restriction. Cyberimmortality, Kurzweil, and
the Singularity would appear as cover stories in *Rolling Stone* and *Time*
as well as many others.

And most people would end up in Silicon Valley, the cradle of technocivilization. Kurzweil, who already had connections to DARPA, would
help set up the Singularity University at NASA and get hired by Google.
Google would start its own company to do research into extending lives—
the California Life Company (CALICO)—and a host of tech billionaires,
most prominently Peter Thiel, would fund similar ventures into defeating
disease and death or for advancing brain-mapping and mind-uploading
options. The science of RLE gained legitimacy through this unlikely network of contacts as well as through scientific work in stem-cell research,
life extension in animals, and a host of species that suddenly were being
reported to be able to "reverse the life cycle" or rejuvenate themselves.
Instead of stories mocking the "corpsicles" of cryonics, it was now these
stories that were making headlines. As a 2012 *New York Times Magazine*
headline asked, "Can a Jellyfish Unlock the Secret of Immortality?"[2]

Until recently, cryonics appeared in the media and in science publications as the butt of jokes or an occasion to delight in scandals, gore,
zombies, and decapitation. Today it no longer surprises me to see prominent mainstream science publications put out serious pieces on cryonics
as a credible scientific project. In its July 2, 2016, issue the *New Scientist*

carried a cover story called "The Resurrection Project," with three full features on various aspects of cryonics. In the fall, the *MIT Technology Review* had published a piece called "The Science Surrounding Cryonics," written in response to a piece published a month earlier called "The False Science of Cryonics"—which in turn was a response to a very popular front-page *New York Times* story documenting the last days of a young woman who, having been diagnosed with terminal cancer, had opted to be cryopreserved upon death.

When I point to the mainstreaming of a once-taboo science, I am not suggesting that everyone on the West Coast or in Silicon Valley today is walking around with cryonics contracts, waiting to be taken down to metabolic stasis in liquid nitrogen temperatures. Far from it. Still, all the alliances and new formations show why and how the credibility of research into the indefinite extension of life, including cryonics and its affiliated sciences, is not as easily dismissed with a smirk and a Rip Van Winkle quip.

One last stop for now, taking us to the summer of 2016 at the Town and Country Resort and Convention Center in San Diego, where, in a gray suit, Dr. José Cordeiro kept bouncing around a large well-lit stage, pumping his fist, shouting "revolution, revolution, revolution." Dr. Cordeiro is an upbeat techno-optimist and transhumanist, founding faculty and energy adviser at Singularity University in NASA Ames Research Park, Silicon Valley, and an adviser to a long list of organizations, including the Brain Preservation Foundation and the Center for Responsible Nanotechnology. On this night he was moderating a panel of scientists speaking about "Current Age Reversal Innovations." A little earlier a band called West of Berlin had performed a decidedly amateurish but upbeat set called "Prelude to Revolution." The next day, somewhere between talks on the fusion of cells and circuitry, there would be an experimental theater piece about gene therapy and a dance performance on immortality. Clearly, this was not going to be your regular life-extension conference, of the sort I had been accustomed to over the last decade or so. The conference name conveyed a little of the taste and the agenda, for this was not billed as a convention or a conference or symposium, but as a festival titled the Revolution against Aging and Death, or RAADFest for short.

Organized by an unprecedented partnership of groups gathered under the umbrella of the Coalition for Radical Life Extension, RAADFest indicated the increasing acceptance and mainstreaming of the once taboo, marginalized, and radical ideas of technoscientific immortalists. Ray Kurzweil rolled onstage in the form of a digital screen mounted on a self-propelled robot transmitting his live image from, well, from wherever he was situated, most likely around Silicon Valley. The next day, Suzanne Somers, onetime actress (Chrissy Snow on *Three's Company*) and current author of the books *Ageless* and *Sexy Forever,* appeared as a featured celebrity speaker. Exhibitors and sponsors included companies using sophisticated technologies to measure and even elongate telomere length to peddlers of seed oil and the powerful vibrator Sybian, the "leader in self-gratification products for women," promising that "Ageless Sexuality is about staying juicy, vibrant and vital . . . having all our body parts working as much as possible, for eternity." It was the coming together of those who, at one end of the spectrum, want their selves to be uploaded to a nonbiological computer-based platform and those on the other who want a few supplements to go along with some nip and tuck around the eyelids and the tummy. With almost a thousand participants, RAADFest now bills itself as the largest gathering of its sort.

With this much pop-culture traction, some of the old life-extension crowd who had been used to the technical and male-dominated gatherings were excited by the upbeat and popular vibe. Others less so. On the one hand, their message seemed to have finally reached a large, mainstream audience. On the other, it all seemed too superficial: it's a bunch of "motivational speakers," one researcher told me, while another called them "Hollywood types who want to look good." When one person added, "We'll see how long they'll last," it came off as an unintended pun on longevity of another sort, the patient commitment to the slow work of research. For the tastes of a hardcore technoscientific lot, it was a little too like a gathering of folksy New Age types promoting the power of positive thinking while invoking a little too much "spirituality."

Still, RAADFest was indicative of the increasing legitimacy, in both scientific and popular arenas, of ideas generated by the RLE communities. It is now common to see people like Aubrey de Grey appear on the same stage as Brian Kennedy from the Buck Institute and other regular anti-aging researchers, who had always warned against immortality re-search. Whereas a few years ago the RLE advocates would berate the anti-aging researchers for their narrow vision regarding the possibilities of breaking through the human life span, and in turn the anti-aging re-searchers would roll their eyes at claims often made by others regarding defeating death or at least producing a good healthy thousand-year-old, now everyone is speaking of "health span nudging up life span."

Although in the introduction I made a distinction between immortal-ism and other ventures in health and long life, it is also important to not overstate the separation between immortalists and the rest of American society or the global medical–scientific contexts, since immortalism, far from being a marginal and isolated curiosity, is very much part of those contexts. Immortalism is a story about groups with a very American, DIY, science- and science-fiction-based orientation, with roots in the white futurism of the 1960s and 1970s, that were later hitched to a tech and biotech boom in hyperdrive, with immense economic growth and technical achievements, reigniting a techno-utopian optimism that has constituted a major current through the course of American history, espe-cially from the nineteenth century onward. Immortalism must be viewed as part of America's history of biomedicalization, of the convergence of the informatic sciences, of modern institutionalized attempts to manage death medically, psychologically, and legally, and part of secular structures of future reckoning, that is, of facing future prospects in terms of such notions as risk, survival, and purpose.

Immortalism's movement from forbidden toward permitted science, from nonscience to, well, possibly some kind of science, has primarily to do with how its projects are embedded in other related, and sometimes eccentric, technoscience ideas and projects, some of which also have moved out of their shadowy beginnings into the limelight of legitimacy. Additionally, developments in biotech have tilted if not flipped the onto-logical ground around life and death, thereby making RLE or immortality projects a plausible part of the general technoscientific imaginary. All this has also meant that immortalism is now squarely part of Silicon Valley's cartography of power; although in their origins, cryonics and life extension were hardly the adventures of the 1 percent, now all Silicon Valley's rich and powerful players have started some version of their own life-extension ventures. What they will yield is not something I can predict, and Silicon Valley itself pretends to an ethos of disruption that holds within it the promise of overhauling the old social faultlines. Yet they seem unable to account for the cumulative and structural inequalities that undergird their own social and economic sphere (also see chapter 6): Silicon Valley itself, just like these immortalist projects, has coalesced along recognizable axes of power, along traditional racial and gender lines,[3] and through the neoliberal concentration of wealth in the hands of a few powerful actors, who despite their libertarian penchants work in tandem with large governmental agencies working on policing and military matters.[4]

SUBZERO ORIGINS

I will trace a particular history in the following pages, one that is not comprehensive by any means. Among the important strands it elides is a set of connections to the former Soviet Union and Russia today, where immortality has been part of ongoing scientific research and state projects from the late nineteenth century onward (see Bernstein 2015, 2019; Gray 2012). There has been some crossover in the histories of superpower immortality ventures, linked to imperial ambitions and scientific prowess. One of my first tours at a cryonics facility was conducted by a Russian cryobiologist, Dr. Yuri Pichugin, who joined the Cryonics Institute in 2001 (he has since returned to the Ukraine). Dmitry Itskov's projects and presence between the United States and Russia, described in the introduction, are another example, as is the adoption of Russian cosmism by American transhumanists and immortalists (see, e.g., Goertzel's *Cosmist Manifesto* in chapter 5). Anya Bernstein's recent research (2019) illuminates some of the differences in both the histories and the formations of the immortalist ventures. But when I started my research there were no cryonically preserved bodies anywhere in the world except in the United States, and KrioRus, the Russian cryopreservation company, had only just established itself. It has since grown to have over fifty bodies in cryopreservation. In any case, the United States continues to have the largest cross-section of technoscientific immortality projects—from biological to computational — as well as the largest number of groups and individuals pursuing these projects.

American immortalism in its contemporary form was largely launched through the practice of cryonics. In late 1962, the same year Crick and Watson won the Nobel Prize for their discovery of the structure of the DNA molecule, two Americans independently articulated the idea of cryonics in self-published books: Evan Cooper, under the pseudonym Nathan Duhring, penned *Immortality: Physically, Scientifically, Now,* and Robert Ettinger wrote *The Prospect of Immortality.* Gathering together a range of results from scientific experiments, a handful of principles, and some new insights, each proposed essentially the same thing. Ettinger put it clearly on the very first page of his book:

> *The fact:* At very low temperatures it is possible, *right now,* to preserve dead people with essentially no deterioration, indefinitely. . . .
>
> *The assumption:* If civilization endures, medical science should *eventually* be able to repair almost any damage to the human body, including freezing damage and senile debility or other cause of death. . . .
>
> Hence we need only arrange to have our bodies, *after we die,* stored in suitable freezers against the time when science may be able to help us. No matter what kills us, whether old age or disease, and even if freezing techniques are still crude when we die, *sooner or later* our friends of the future should be equal to the task of reviving and curing us. This is the essence of the main argument. (1965, 1)

Though the procedures and technologies, as well as visions of the task awaiting our "friends of the future," have changed, the "essence of the main argument" has remained the same.

For Ettinger and Cooper, who had read about recent experiments in cryobiology in which animals had been temporarily taken down to low temperatures and then revived, two things were clear and essential: that freezing stops the decay process, and that the future would almost certainly acquire the technological power to cure and reanimate. Whatever the biology of death, that American view of the future, freed from assumptions about fate and tied to technologies that assert human control over matter and human destinies, has been crucial to immortalism from the beginning.

Evan Cooper, known as Ev, was something of a romantic. An obsessive sailor with a penchant for the great works of literature, he numbered Odysseus and Don Quixote among his literary companions but seemed to admire most the "great books" of the twentieth century (Kent 1983). He led a discussion group on this subject and published his own book through a small imprint called 20th Century Books. More a man of quick action and grand speculation than of careful analysis, he had a flash of insight on the possibility of achieving immortality through science as he synthesized a number of influences. These ranged from literature to the social sciences to psychology, but it was cybernetics that provided the true enabling spark, specifically his reading of Norbert Wiener's *The Human Use of Human Beings* (1954).

Wiener's influence on cybernetics and hence on today's world of computation and informatics can't be overstated, but Wiener himself began in a guarded manner, claiming his project as merely the investigation of metaphors. He fielded two paradigms: that a biological organism could be conceptualized as a message, "a pattern," and that the living individual had a parallel in "some of the newer computational machines." His so-called metaphors harden into real propositions as the book progresses, but throughout Wiener continues to use one striking trope to advance his case: death. It is a trope he starts out with: "The metaphor to which I devote this chapter is one in which the organism is seen as message. Organism is opposed to chaos, to disintegration, to death, as message is to noise." He was right to protect himself by talking about this as metaphor, because his use of the term "death" was nothing if not rhetorical. It served no function in the theory; neither has it served a function in cybernetic practice. He equated chaos and noise with death, because he saw noise as disintegration and death as the end state of disintegration. But these are incorrect analogies—for example, noise and chaos can be, indeed are, continuous and infinite and quite fully part of life, even generative of it; they certainly do not lead necessarily to an end state.

Still, it was Wiener's dramatic use of death juxtaposed with the cybernetic idea of biology as message that focused Cooper on a different sort of vision: "Being a fervent addict of Joyce and Freud, I was well-conditioned to reading into any author, especially Wiener, strange and unusual meanings. I claimed that Wiener's book contained either intentionally or unintentionally . . . a message about immortality" (1962, 5)—a message that did not disintegrate on Cooper, whose own book is an impassioned, broad plea, mixing Frazer, Camus, Wiener, Freud, and scientific experiments in low-temperature preservation into a proposition on personal immortality via the preservation of the "message," that is, the content of an individual. He considered immortality the apotheosis of "the individual's freedom to choose his destiny" (41). Being a sailor, he may have been attracted by that freedom more than by the scientific venture; being a white American, he may have had the privilege to assume it possible.

Few copies of Cooper's book were printed,[5] and its patchy science and literary forays did not aid its chances of survival. The emerging life-extension community preferred Asimov to Camus, for instance, and would not be attracted to the idea that the best place for freezing bodies would be in a "permafrost region" in "natural storage," with the United Nations handling their freezing and resuscitation documents.

But there was a practical side to Cooper, too, along with an impulse for action. In 1963, just after he published his book, Cooper founded the first organization dedicated to promoting the idea of human cryopreservation and reanimation, the Life Extension Society (LES), operating out of the Washington, D.C., home he shared with his wife, Mildred. Wasting no time, he set about his work. The LES issued the first newsletters on the subject, eventually named "Freeze—Wait—Reanimate," and Cooper began to recruit people, asking them to participate in seminars and set up local LES chapters. He somehow managed to get seminars going at places like the Pratt Institute in Brooklyn, New York, with the participation of medical doctors and researchers initially attracted to the ideas of cryonics (Kent 1983). At the time, he was attracting a motley crew of adherents: a cryobiologist briefly entered the scene, along with some students, a medical doctor, a lawyer, an industrial designer, a TV repairman, and a wigmaker in Phoenix, future home of the largest cryonics operation, who bore the appropriate name Ed Hope. Hope thought cryonics could make him some money. He formed a company called Cryo-Care Equipment and built the first capsule for the purpose of holding frozen bodies (R. Nelson 1968).

Despite the burst of enthusiasm from these early participants, though, no one came forth to be frozen. The general lack of response always surprised early cryonicists, as it continues to mystify today's members. How could this age-old dream not have adherents?! In the LES newsletter of December 1964, Cooper cried out in disbelief: "Are we shouting in the abyss? How could 110 million go to their deaths without one, at least, trying for a life in the future via freezing? Where is the individualism, scientific curiosity, and even eccentricity we hear so much about?" (quoted in Perry 1991, 11).

Cooper persisted, galvanizing people's attentions and organizing LES meetings and conferences that brought together some of the main actors in the future of cryonics. One of these was Saul Kent, a student who would become a central figure in the development of cryonics and its institutions. He had been introduced to the idea through Robert Ettinger's better-known book but met up with Cooper because Cooper had the organization and that is what Kent wanted. Kent took the bus down to Washington, D.C., with two friends to meet Cooper and "discuss immortality." Cooper had gathered over a dozen members by then, some of whom were as eccentric and romantic as himself, and all of whom "clearly looked up to Ev as their leader" (Kent 1983, 12). Among those whom Kent and his friends were introduced to were the "Vice President John Prince, a tall (6'7") black man who dressed in 3-piece suits; Bill Albaugh, who was about to run for Congress in Maryland on a 'Freeze-Wait-Reanimate' program (the title of the LES newsletter); and Phil Carlson, who was intrigued with the concept of personal identity and how it might change in the future" (Kent 1983).

Kent and his friends returned to set up the New York chapter of LES, but they would splinter off within the year to form their own group, "the first organization to compete with LES," as Kent put it (Kent 1983, 11). At an early meeting of the New York offshoot, Karl Werner, a Pratt student who had attended Cooper's early seminar at the institute, coined the term "cryonics" to describe the preservation of humans through cryogenic procedures. Thus was born the Cryonics Society of New York in July

1965. Four years later, Werner would drop out of the cryonics movement to join the Church of Scientology.

Like so many of the early cryonics organizations, LES proved to be short-lived, and Cooper quickly became disillusioned with the whole enterprise. Even though by 1969 a few people had been cryopreserved in California, Cooper dropped out of cryonics altogether, dedicating himself to his other passion, sailing. He was in his forties by then, without a desire for a fixed job or a stable home. He spent the next decade and a half without a fixed address, sailing his *Pelican* up and down the eastern seaboard, finding odd jobs, mainly as a carpenter. Sometime at the beginning of the 1980s he collected all his papers, letters, and documents and destroyed them. It is not clear why. On October 9, 1982, while moored at Nantucket, Massachusetts, the *Pelican* was damaged in a severe storm by the wind-dragged anchor of another, larger vessel, whose insurance company, he learned, would probably not pay damages. On October 17, having made some makeshift repairs, Cooper set sail from Nantucket, plying his way around Martha's Vineyard Island, bound for a warmer winter at Beaufort, North Carolina, hundreds of miles to the south. On October 21 he was spotted and hailed by a coast guard patrol boat off Wood's Hole, heading south for Block Island, Rhode Island, via Vineyard Sound. Everything was fine, he said before moving on.

Cooper was never heard from again. The coast guard conducted a search, but no trace of the *Pelican* or its pilot could be found. One of the founders of contemporary immortalism could only be presumed dead, no trace left, not even his papers. Perhaps too much Camus had gotten to him.

Cooper wrote the first book and set up the first organization, but Ettinger is generally credited with fathering cryonics. In part that is because Ettinger stayed on at the core of cryonics, founding one of its two most important and lasting organizations, the Cryonics Institute, and writing other books since his first. Also, having earned master's degrees in physics and mathematics, he was more attuned to the necessities of scientific, rather than literary, rigor. His book was much more thorough from a scientific point of view; it did not sound speculative like Cooper's and was much more grounded in the biology of the matter rather than ideas about the self.

Born in 1918, Ettinger almost lost his life in battle in World War II.

"On the day I was wounded," he wrote, "I was a second lieutenant, directing mortar fire. I had to stand up to see. The rifleman next to me was prone—as of course he should have been—but the same shell that got me through the legs got him through the head and killed him. It was just dumb luck" (1994, 28). That incident is cited often by Ettinger and seems to have been formative in his moral and scientific thinking. If life, death, and survival were the function of nothing but "dumb luck," well, "dumb luck" had better be eliminated from the mix. Honored with a Purple Heart, Ettinger spent four years in hospital recovering, reading, thinking—and writing his own science fiction story about freezing and immortality, published in a small magazine in 1948. But he was not satisfied with stories and was determined to "do something about this freezing business," since nobody else would. That determination led to research, which led to his groundbreaking book.

According to Ettinger (1994), he first outlined his idea in a brief pamphlet of three or four pages that he sent out to a random list of people taken from *Who's Who in America*. There was no response from any of them, so Ettinger decided he needed to explain the matter more fully. That's how he got to writing a book instead of a pamphlet. In 1962 he published his own book and began to correspond with Cooper. In the meantime he sent his manuscript out to several publishers, and an editor at Doubleday finally thought the book could sell some copies but wanted it to "be stamped kosher on the scientific side" (Ettinger 1994). To have its *science* evaluated, the editor sent it out to a *science fiction* writer, Isaac Asimov, then at the zenith of his career.

Asimov, like most other science fiction writers, liked cryonics as an idea to play with but never became involved in the movement. In fact, according to Ettinger (1994), he was against it, mainly on the ethical grounds that immortality would entrench power even further in existing individuals and institutions. Still, the science seemed sound to Asimov and he gave Ettinger the thumbs-up. Doubleday, a major publisher, issued the book in 1964. For a short while after its publication the book enjoyed wide popularity, eventually getting translated into eleven languages. Ettinger got some media and press attention, too, his first major TV appearance being on the *Tonight Show*. What better introduction could the idea hope for? As a result of the exposure, a range of people seemed to have expressed interest, among them Arthur C. Clarke (author of *2001: A Space Odyssey*) and Stanley Kubrick (director of the film adapted from Clarke's book), who passed Ettinger's book around to his friends. But in the end Ettinger's frequent TV appearances and all the early articles on the subject did not yield any significant gains in terms of the number of adherents for the budding movement, which could not shake off labels like "science fiction fantasy."

IMAGES OF POSSIBLE FUTURES: A SCIENCE NONFICTION

Immortalism and cryonics are commonly dismissed as science fiction. It is true that science fiction was a formative influence on cryonics and is a continuing part of the culture. Ettinger grew up an early fan of science fiction, as did many other early cryonicists. Many current members refer to science fiction as their portal into the practice of cryonics and life extension, and cite it as an important influence on their current thinking. I found myself assisting in emergency preparations to the daily roar of the theme music to *Battlestar Galactica* alongside a very powerful woman who once used to make props for *Star Trek* TV shows. Versions of "cryonic" freezing and reawakening have appeared in American fantasy and fiction works since the late 1900s, often mixing a number of themes about the frontier, exploration, and futurism. The earliest may be the 1887 story "The Frozen Pirate" by the explorer William Clark Russell, in which sailors in a pirate ship are discovered frozen in blocks of ice. In 1931 a writer by the name of Neil Ronald Jones sent a frozen protagonist, an eccentric professor, into space, where his body after death floated, preserved by the cold vacuum until an alien race found him, revived him, and stuffed him into a robotic exoskeleton that allowed him to live forever in an aging-free environment. The story, "The Jameson Satellite," was published in the July 1931 issue of a pulp science fiction publication, *Amazing Stories*. That issue made it into the hands of a boy by the name of Robert Ettinger. It was his first exposure to the idea of low-temperature preservation and immortality.

In remarking on the connection between science fiction and cryonics, the point is not to invalidate cryonics or other immortalist strategies as science fiction fantasy adventures. Rather, I want to emphasize the porous boundaries between science fiction, science, and the cultures of immortalism. Science's necessary cordoning off of its domain as a privileged realm of pure facts and evidence that ought not be tainted by such things as fantasy does not occlude the reality that fantasy has always been part of scientific inquiry and that science fiction was (and continues to be) an effective strategy and medium for projecting possible futures that do not yet have corollaries in the material world.

Through the twentieth century, a good number of scientists and engineers have not only been reared on science fiction but have had parallel careers as SF writers.[6] Of those relevant to the themes of this book, one might cite J. D. Bernal, physicist, mathematician, and Marxist philosopher, best known in science as the originator of X-ray crystallography, a field dedicated to the study of the molecular composition of crystals through the use of X-rays. But he was also obsessed with immortal life and the improvement of human biology through engineering (Parry 2004). In a 1929 treatise mixing science, prediction, and social prognosis, Bernal presents the first rendition of what we now call a "brain in a vat"—the extraction of the brain and its preservation in an "artificial" environment in which it might continue to function and indeed be improved upon. The idea that the brain is the true seat of the self and, more, that the brain can function independently of the body if properly preserved is entrenched in immortalist ventures today, both in the form of whole-brain emulation and neural nets and more literally in cryonics' neuro-option, in which only the head or brain is cryopreserved—a literal brain in a vat.

Asimov and Heinlein popularized science fiction in the 1940s as a quintessentially American genre, making both space and robotics central features that admittedly had their roots in the older *War of the Worlds*, with its Martian landing and hi-tech equipment. Orson Welles's prank reading of H. G. Wells's text[7] was a prelude to contemporary UFO and flying-saucer phenomena, whose origins—including their terminology and iconography—can be traced back to the same magazine that triggered Robert Ettinger's imagination, *Amazing Stories*, and its editor, Raymond Palmer, who placed the first image of what would later be called a "flying saucer" on its cover in September 1946. Palmer had accidentally tapped into the needs and fantasies of thousands of Americans who were writing letters to the magazine giving testimony to their own version of encounters with space aliens. In 1947 the merger of science fiction and science non-fiction found consummation in the form of an alien spacecraft whose crashed remains were said to have been found in the desert of Roswell, New Mexico. Science fiction, once a genre of storytelling, became the generator of research (Ufology), expeditions, conferences, and paranoias that persist to this day at that site and many others. It became science nonfiction.

Space travel remains the most important and widespread crosscurrent of science and science fiction. By the 1960s, when Ettinger and Cooper began writing, Yuri Gagarin had become the first man to enter orbit, followed by John Glenn, the first American to do so, and the Soviets had already landed an unmanned craft on the moon. For the first time, going into space, reaching the moon—such things as had only been the subject matter of fantasy and mythology—were all of a sudden within the reach of technology, migrated from the realm of the imagination, of science fiction, to the realm of the real and the possible. The blurring of the boundary between the *merely imaginable* and the possible has been paradigmatic for the American imagination since the early days of colonial settlement and the development of the frontier. Space was "the final frontier," *Star Trek* would claim in 1966, and humans could "boldly go where no man had gone before." Today, space travel and space colonization are an integral part of immortalist strategies as they imagine scenarios for survival in the distant future.

In part, of course, the problem is with the future, which, distant or not, does not exist yet. Potential futures have always been part of invention and technology, but in the second half of the twentieth century the boundaries of possible futures were being expanded by rapid growth in scientific experiments and techniques. A range of unprecedented man-made biological objects and hybrids was changing the understanding of biology, life, matter, and the universe in radical ways. They were changing the way the world could be thought of and the way it had been posited. As Cooper wrote, "More and more we rely on 'images of possible futures' to achieve our goals" (1962, 60). Appropriating a phrase used by one of Debbora Battaglia's subjects, one could say that twentieth-century technoscience produced a series of "epistemological shocks" (2005, 10). Images of a landing on the moon would qualify, as would the reality of one person's heart beating in another person's body.

Secular and scientific strategies to see into the future (vs. religious ones such as prophecy or augury) developed to facilitate, manage, market, and promote technoscientific work and biopolitical governance. These techniques and concepts have been mainly analyzed in terms of risk assessment (Beck 1992) and disease prediction and prognosis (N. Rose 2007; Jain 2013), but "images of possible futures" come in a variety of interdependent forms, combining scientific achievement, science fiction imagination, probability and risk assessment, and technological prediction (Farman 2010). The latter takes us beyond what has been called "anticipatory regimes" (V. Adams et al. 2009), in which contemporary cultures are always poised probabilistically waiting for a calculated future to arrive. Today, technoscience has a monopolistic claim on the future such that it is hard to find future imaginaries, even dystopian ones, in which certain kinds of technology—computational, biological—do not play the main role. Technoscience's claims on the future not only mobilize risk and worry as the grounds of present experience and intervention but encourage a futurology in which prediction, as secular, technological modality, and promise, as an affective modality, are wrapped together. Technoscientific immortality is both promise and prediction.

Again, the technoscientific future (prediction plus promise) that has captured the greatest share of immortalist and indeed current scientific imaginaries has been the synthesis of AI and biology, the blurring of the boundary between machines and cells, mind and matter, that was launched by cybernetics at that time. In the 1940s the mathematician and early computer scientist John von Neumann proposed a theory of cellular automata, merging biology and systems theory, while Norbert Wiener was writing at the same time about parallels in communication systems between animals and machines, and mathematician Claude Shannon had imagined human genetic composition in terms of information. The Macy Conferences, organized explicitly to explore foundations for a scientific understanding of the human mind, brought psychologists, sociologists, and anthropologists (Gregory Bateson and Margaret Mead) together with mathematicians and systems theorists who were already discussing how computers could "learn." The early cybernetics pioneer Grey Walter, who had made goal-oriented electromechanical turtles seeking out power outlets to "revive" themselves, published *The Living Brain* in 1953, and Marvin Minsky at MIT was writing about AI in the 1950s. (He later became a transhumanist and was an avid cryonicist until his death.) The merging of man and machine was being proposed in more popular publications such as *Scientific American,* which published several articles on the topic in the 1950s. So long the seat of human exceptionalism, the marker of human consciousness and will, the ontological boundary separating what humans were and what nature was, the human brain was being presented as a performing machine, turning the conscious subjective human into nothing more than a set of response mechanisms, "performative stuff in a performative world" (Pickering 2009). And if a merger between mind and machine, organic and nonorganic, were possible in the future, if on some level human identity could be translated into other processes, then our concepts of personal identity could be, had to be, reimagined. While most secular humanists recoiled from the prospect of the computational reductionism of mind, Cooper and Ettinger tapped into the potential offered by this line of thinking. Each suggested in his own way a sort of technological shape-shifting, a continuation of personal identity beyond biological death through some version of nonorganic or artificial intelligence (Ettinger 1965, 129–33; Cooper 1962, 17–19) instantiating mind/self on nonbiological platforms. In other words, from early on, cryonics and immortalism moved beyond simple biological survival to imagine and claim a posthuman future.

BEYOND BIOLOGY

The earliest manifestation of immortalism, cryonics was a futuristic idea without a theory for that future. What would it take to revive a frozen "dead" person? How would you even start thinking about restoring biological and mental health? Regarding the body as a series of parts, being familiar with information theory, physics, as well as biology, Ettinger had intimated the idea of body-repair technologies but had been unable to propose what that might actually entail in the future.

The future gained its strongest theoretical foothold through the work of an MIT graduate by the name of Eric Drexler. Today, Drexler is considered the father of nanotechnology, although his own vision of nanotechnology has been sidelined. Mainstream nanotechnology is understood to encompass any technique that deals with matter on a very small scale, on the scale of a nanometer (one billionth of a meter). But Drexler's vision was more all-encompassing and not just a matter of scale. Introduced to a 1959 lecture by Richard Feynman titled "There Is Plenty of Room at the Bottom," in which the famed physicist argued that much more can be done at the small scale of atoms and molecules, Drexler spent the late 1970s imagining the possibility of designer molecules, especially biological ones such as protein molecules. If white blood cells can travel around the body independently but with a clear goal of countering foreign materials and repairing the potential damage done by infectious diseases, why shouldn't designed biomolecules potentially move around and repair whatever needs repair? That was the essence of Drexler's vision: designed molecular robots, or assemblers, carrying out specific goals by moving around the body, or around any matter actually, and assembling other atoms or molecules in optimal arrangements. With that single idea it suddenly seemed like anything could be done, that anything could be manipulated and manufactured more or less out of anything else. It was just a question of shifting the atoms around into the desired layout, an intervention into the heart of matter itself, whereby alchemy would yield to engineering the promise of transmutation, turning any material structure into any other, over wide limits—from good blood cells to good energy to good societies. Utopia went nano. That at least was the vision flashing before Drexler's eyes.

In 1981 Drexler published his first paper, outlining the possibility of creating "molecular machinery" that would "open a path to the fabrication of devices" according to "atomic specifications." Five years later he published *Engines of Creation* and, with his then wife Christine Peterson, founded the Foresight Institute, which today carries on the work of promoting and educating the public on nanotechnological issues. The Institute's forums and meetings are also, as they were then, a place for like-minded futurists to gather and feel at home in the exchange of ideas that would seem outlandish in other settings.

Nanotechnology is everywhere: there are summits at the Department of Energy, congressional funding, National Science Foundation conferences, corporate research, and cutting-edge products. But most of the nanotechnology we witness would be and often is disavowed by Drexler himself for the simple reason that nanotechnologists today are concerned with marketing stronger, smaller materials and have almost completely renounced the idea of molecular assemblers.[8] Immortalists, for their part, are generally avid (though not entirely uncritical) fans of Drexler's version of nanotechnology. And with good reason. Drexler's nanotechnological vision scooped cryonics out of a theoretical void and gave it a molecular engine to hook future revival on. Recall that there were two main problems with cryonics revival: How do you repair freezing damage to the cells, and how do you ensure that brain cells would not only come alive but also maintain memory and other functions crucial to personal identity? Molecular repair, it was thought, might be able to solve both of these problems, the former with greater ease than the latter. Drexler included in *Engines of Creation* a chapter on cryonics in which he outlined the possibilities of repair.

The implications were taken up and advanced by the computer scientist Ralph Merkle, whose ideas also changed cryonics and introduced a new criterion, the "information-theoretic definition of death," by which it was held that if sufficient information about the atomic arrangement of the brain is preserved—on any substrate, carbon or silicon, in the brain itself or a computer chip—then the person cannot be called *irreversibly* dead (see introduction for more). The hope of future reanimation in cryonics, as well as utopian projections of enhanced capacities and abundance, are still fueled by the Drexlerian vision of transmutation.

Nanotechnology formally and theoretically wedded cryonics to a network of engineers and computer scientists like Merkle, who, recognizing common interests, began to align themselves into groups and networks that would become crucial to the development of immortalism, futurism,

AI, and eventually the NBIC convergence. Indeed, Drexler had been introduced to cryonics through the early sproutings of such a network. Before conjuring up nanotechnology, Drexler had entered MIT interested in space colonization. On the heels of the space program, the idea of space colonization had become imaginable and a group of West Coast enthusiasts, led by the Arizona-based couple Cynthia and Keith Henson, became its vocal organizers. In 1975 the Hensons formed a group called L-5 to explore and advocate for space settlements. This was the 1970s. The oil crisis and the dire warnings of the Club of Rome had initiated a worry over the finitude of Earth's resources and survival on Earth—both of concern mainly for the highly developed and mediatized industrial world, Westerners who felt there were important things to protect.[9] At the same time, in 1975 the U.S. Apollo and Soviet Soyuz spacecraft connected in space, heralding not just the prospect of superpower cooperation outside Earth's atmosphere but also performing the possibility of human sociality in space, a performance prefigured and perhaps predetermined by science fiction, and especially *Star Trek*. Three years later, *Star Wars* was released and *Battlestar Galactica* aired on TV.

L-5 gained its power from this mutually reinforcing context: space was crucial to future survival, so the future would have to be geared toward space. The enthusiasts who gathered around L-5 saw space both as a heady adventure where the greatest promises of science would unfold and as a solution for the troubles of earthly finitude. With its thematic mixture of science fiction, science, engineering, and exploration, as well its social mixture of amateur enthusiasm (an aspect of American pragmatic sociality that would be later termed DIY) and institutional linkups, the group soon became a locus of activity and exchange for all futuristically minded people. Early members included Timothy Leary, Isaac Asimov, Marvin Minsky, and Freeman Dyson, as well as the computer scientist and writer Hans Moravec, the cryonicist Saul Kent, Eric Drexler, and the future sociologist, cryonicist, and NSF director William Sims Bainbridge. At its peak L-5 was said to have had sixteen thousand members (Regis 1990; Bainbridge 2007b, 37).

As the space program lagged, L-5 lost its momentum, but for futurists and immortalists it set the tone of things to come. It was where space, immortality, and molecular technologies came together to produce a particular orientation toward the future that still informs the move-ments today: survival, long life, technological utopia. It also served as an incubator for ideas and social relations. Into Keith Henson's vision of space colonization, Moravec and Minsky threw mind uploading, Drexler sprinkled some nanotechnology, while Saul Kent brought in cryonics. Many became active cryonicists while computation would lead to the idea of cybernetic immortality. As chapter 6 will show, it is precisely this fusion of ideas and people that provided the impetus behind the NSF's "NBIC convergence" projects.

The new futurist blend rode on a past of American frontierism and a promise of unprecedented potential for transcending the physical limits of human existence. It was this idea of transcendence in the physical rather than the spiritual sense that in this same period reinvigorated the term "transhumanism" and the movement associated with it. Its main purveyor was the Iranian-born author Fereidoun Esfandiary, better known by his adopted name of FM-2030. A diplomat's son and a onetime officer at the United Nations Conciliation Commission for Palestine, FM-2030 quickly grew disillusioned with Cold War politics, nationalism, and even the contemporary framework of human rights. Through his novels as well as opinion pieces for the *Nation,* the *New York Times,* and the *Village Voice,* he advocated for a world in which national identity would play a lesser role, older models of kinship would yield to mobile and flexible determinations of kin, and local humans would transform into global citizens. But it was his later writings and teachings, focused on the role of technology as the medium taking us to a utopian posthuman future, that sealed his place in the futurist firmament. Starting with *Up-Wingers: A Futurist Manifesto,* published in 1973, he moved from a global earthly politics to a cosmic politics: neither left-wing nor right-wing, but up-wing. Through the late 1970s and early 1980s he taught futurist-related courses first at the New School and then at UCLA, where he became another nucleus around which the futurist movement would cluster. His futuristic predictions and plans gained enough fame to get him invited for consultancies with large corporations as well as for a multitude of TV appearances. It was in the 1980s, on the West Coast, through work with his coterie of up-wingers that the term "transhumanism" began to reemerge. In 1985 he officially changed his name to FM-2030, and in 1989 he published what would become one of the first official texts of the transhumanist movement, *Are You a Transhuman?*, a manifesto at once challenging the status quo and galloping off with optimism toward a utopian world of limitless energy, limitless food, limitless raw materials. It is worth noting that in the same year, the father of cryonics, Robert Ettinger, self-published a reprint of another seminal transhumanist text, *Man into Superman* (originally published in 1972). The techno-optimism infusing these visions was an American antidote to the Club of Rome's dire predictions, a technologically enabled version of Thomas Paine's exhortation to Americans that "We have it in our powers to begin the world anew." FM-2030's life was taken by cancer at the turn of the millennium, on July 8, 2000. He is currently awaiting the year of his name under cryopreservation at Alcor.

As a metaphor for jettisoning the past (judging from interviews, "jettison" was FM-2030's favorite verb), changing one's name became a common strategy, adopted as a performative act by a number of futurists, including FM-2030's early collaborator in transhumanist ventures, Natasha Vita-More, and a British philosopher who came to be known as Max More. Max T. O'Connor had taken up cryonics advocacy while still in Britain, but his activism flourished when he moved to the University of Southern California in 1988 to complete a PhD in philosophy. A year later, he changed his name to Max More, signifying both abundance and a dissatisfaction with what is given. Along with Tom Morrow, another man with a signifying name, More started a journal of ideas called *Extropy,* so named to provide a counterconcept to entropy, that is, to signify greater organization, life, energy, and vitality. This latter—vitality, or life—formed the central core of Natasha Vita-More's assemblage of a name, while the last part of her name is a social artifact of her marriage to Max. Both have continued to be avid cryonicists; indeed, More became president and CEO at Alcor in 2011, following a few stressful years of leadership crisis at the company.

Founded in 1992, More and Morrow's Extropy Institute became the new L-5 of futurism, with the advantage that it was not focused narrowly on space colonization but widely concerned with enhancement technologies that, in the late 1980s and early 1990s, were beginning to hold up a new set of promises—control over biology, control over the brain, control over all matter in the universe—a set of promises that would become the pillars of the NBIC convergence. Some of the older names from L-5 also appeared on the roster of speakers at the Extropy Institute (such as the ubiquitous Marvin Minsky). A host of new names signaled not only a wider social berth for the futurist movement but also the arrival of the new sciences, specifically the biosciences and genetics, exemplified by figures like molecular biologist Cynthia Kenyon. Many current futurists and immortalists trace their roots and early sense of transhumanist excitement back to the extropian meetings. Thus, the Extropy Institute, defunct since 2007, both created and rode the crest of a wave of techno-utopian optimism that, in 1998, would see the formation of the World Transhumanist Association (WTA), cofounded by another philosopher, Nick Bostrom. The WTA splintered off from Extropy Institute and eventually replaced it as the main transhumanist organization, leaving many extropians still bitter over the separation.

That futurism would sprout on the West Coast seems to make sense, for more reasons than one, but especially given the context of the scientific and technological explosions of the late 1980s and 1990s. The 1990s were a defining decade for contemporary science and technology. The internet became ubiquitous, and the tech boom (and bust) pushed hi-tech into private lives, creating a new financial sector as well as a significant class of science and tech workers; the same occurred for biotech as its projects and successes created new financial, cultural, and scientific sectors, exemplified by the human genome project, begun in 1990, and more or less completed with great fanfare as the U.S. president announced a "working draft" of the human genome in 2000.

In general terms, the spread of the internet and the biotech revolution normalized the optimistic view of the "boom" as a major shift in human history. The International Monetary Fund's World Economic Outlook 2001 called the growth of information technology a revolution, comparing its growth to that spurred by the industrial revolution, a comparison that became ubiquitous. On the social level, the tech boom expanded the field of people working for and interested in the technology and biotechnology sectors. Just as the tech bubble was bursting, the American Electronics Association report for 2002[10] summed up the comparative growth of the tech sector in the United States: "Since 1995, manufacturing jobs have increased by 46,000, while software and computer services jobs have increased by 1.2 million." The total number of people employed in the high-tech industry had reached 5.6 million, with 440,000 new tech jobs added in 2000 alone. Biotechnology—to be distinguished from high-tech medicine by its focus on genetic, cellular, and molecular manipulation— was a much more recent and much smaller sector, but it was already rapidly growing by the end of the 1990s, its growth marked by its ability to create products and attract capital more than its ability to create large labor pools. According to a Brookings Institution report (Cortright and Mayer 2002), while most biotech firms were small, employing less than a hundred people, between 1995 and 2001 they attracted over eleven hundred venture-capital investments in the biotech sector for a total of over $10 billion—a sum that now feels like pocket change in Silicon Valley. Altogether, in the 1990s the pool of people who would be more open to tech-based ideas and solutions grew enormously, especially on the West Coast thanks to the concentration of companies in Silicon Valley. The futurist groups were largely populated through this pool of self-defined "geeks."

There was a conceptual and psychological side to the growth of the futuristic movement as well. The "biotechnology revolution," with all its uncanny triumphs—its mapping of the human genome in 2000, successes in the manipulation and transfer of DNA, finding molecular solutions to medical problems, and the production of previously unimaginable entities (transgenic animals such as cows that produce insulin in their milk; the DuPont-copyrighted OncoMouse; Dolly the cloned sheep; the production of a cure to hemophilia, the protein factor VII, from hamster cells grown in the lab)—increased the credibility of ideas and hypotheses in the domain of biology that previously had been deemed fantastical or science fiction.

In public debates, scientists and ethicists went from ridiculing the claims outright to saying that they were *feasible* if not desirable. So despite the ethical gaps in people's views of biotechnology and its products, the credibility gap has closed significantly in the second decade of the twenty-first century. In fact, quite likely, the opening of the ethical gap is proportional to the closing of the credibility gap, as unfamiliar entities and possibilities became more and more imaginable. If the intervention of the biosciences into life's matter created public debates over cloning, tissue harvesting and growing, genome editing and reproductive technologies, and expanded the purview of bioethicists in the political arena, it also suddenly thrust the marginalized technofuturist organizations into the limelight, since they had been thinking and talking about such matters for years.

BIOGERONTOLOGY, OR STRATEGIES FOR ENGINEERED NEGLIGIBLE SENESCENCE

The rise of Immortalism, especially its bioengineering version, can be partly contextualized in the wider envelope of longevity and anti-aging research that surged in the 1990s in part in response to the concerns of the graying baby boomer generation. Though the actors in the broader field of anti-aging have not liked the association, the various strains of mainstream anti-aging research and the more radical approaches of the budding immortalist strains were part of the same current that began to see aging as a disease and biotechnology as a cure. The difference is that anti-aging physicians do not aim for immortality and immortalists don't much care for human growth hormone. But both, and indeed the whole wider field of anti-aging, came to share the view that aging is not a condition but a disease.

Work on forestalling aging dates back to the turn of the century and then early research on calorie restriction in the 1930s (Park 2016). But it was the new genetic sciences that gave anti-aging research a theoretical and strong practical basis, opening it up to possibilities of treating aging as a disease and eventually of defeating death. This sort of approach had largely flagged after Leonard Hayflick put to rest the myth of Alexis Carrel's immortal cells, but it restarted in 1986 when Michael Rose first proposed an evolutionary theory for the occurrence of aging, thus taking death out of the realm of the inevitable once again. Another biologist, Michael West, swept aside a psychological and practical barrier put in place by Hayflick's limits of cell duplication and the telomere ("tail" or endcap of a chromosome); rather than taking the Hayflick limit as a natural and inevitable law for the aging process, West used research on telomeres to attack aging. In 1990 he formed Geron, one of the earliest companies funded by big venture capital, to do anti-aging research with telomeres and stem cells. Similarly, Cynthia Kenyon, who had doubled the life span of nematode worms in 1993 by altering a single gene, turned to venture capital to start her own company, Elixir Pharmaceuticals (Hall 2003). West is now CEO of a company called BioTime, which started out as a cryonics company; and Cynthia Kenyon is part of Google's longevity venture, Calico. Although the money continues to flow and all these scientists are considered important award winners, no major human anti-aging triumphs have emerged.

Still, a great number of promises and new approaches, sometimes categorized as rejuvenation therapies, have appeared on the horizon. New research on mice shows them living much longer on calorie-restricted diets, getting younger and stronger muscles when given boosters of NAD (nicotinamide adenine dinucleotide), a molecule in human cells; similarly, research on telomere length as controller of senescence in cells also led to the identification of telomerase as a possible counter to telomere shortening during cell division. Between NADs and telomerase and their technoscientific cachet as genomic science (as opposed to, say, fuddy duddy gerontology), anti-aging has revved up into high science and profit. Scientists like MIT's Lenny Guarente not only do research but start their own anti-aging supplements companies, selling "metabolic repair" products. BioViva, a company founded by Elizabeth Parrish, carried out the first human experiment offshore (to avoid legal issues), with Parrish volunteering herself as Patient Zero. Telomerase was delivered to tissue and cells in her body through a viral vector in order to repair her shortening telomeres and thus avoid or even supposedly reverse the senescence of cells. Parrish presents at cryonics and RLE conferences.[11]

In the midst of this hype of bioinformatic technologies, many scientists and physicians expressed problems with life-extension efforts in principle. In 2003, Sherwin Nuland, a highly visible physician and well-respected professor of clinical surgery at Yale, with a number of *New York Times* best sellers under his scrubs, including *How We Die* (1995), published an essay in *Acumen Journal* called "How to Grow Old" (later published as chapters 7 and 8 in Nuland 2007). Aging and death, he wrote, are "the condition[s] upon which we have been given life. The aging and eventual death of each of us is as important to the ecosystem of our planet as the changing of the seasons. When William Haseltine, PhD, the brilliant biotechnology entrepreneur who is the CEO of Human Genome Sciences, says, 'I believe our generation is the first to be able to map a possible route to individual immortality,' we should cringe with distaste and even fear, not only at the hubris of such a statement but also at the danger it poses to the very concept of what it means to be human" (2007, 157).

Less than two years later, the highly regarded *MIT Technology Review* asked Nuland to write an essay assessing the claims of a certain Dr. Aubrey de Grey. At the time, de Grey was relatively unknown outside a limited circle of researchers and ardent immortalists, who themselves were only then beginning to hear about his ideas. He was based in Cambridge, UK, married, and not getting out much. He was just starting to build his program and setting up his foundation. His claim was that with an engineering approach to biomedicine we could defeat aging and death; his program has become more detailed since, but he already had outlined the theoretical approach: aging is an effect of cumulative damage due to a number of different processes; damage can be repaired if properly identified and disaggregated; therefore, one must identify the key sites of damage in the process called aging and conduct research to find solutions for them. He suggested that we need not have a theory of aging or a full understanding of an overall metabolic approach; we just need to find the damage and deal with it. A mechanic does not need to be able to build or even know the physics of combustion engines to repair a car. Extending life another ten years by attacking specific breakdowns would give technology another ten years to develop the next required pushback, and so on. At some point this would achieve what he has called "escape velocity": the technology would push back the effects of aging so far that in effect many of the younger cohort would end up with extremely long life spans, on the order of hundreds or thousands of years. This might not be of great help to those older members of society whose engines have already gone to rust in large part, but for those who were only in their teens it held great promise. Indeed, de Grey regularly claimed that the first thousand-year-old is alive today somewhere, perhaps sitting right next to you.

In his *Technology Review* article, Nuland did not engage the substantive or theoretical ideas put forward by de Grey. Instead, he ridiculed de Grey for being a rogue scientist and likened his long red beard to Rip Van Winkle's. He ended his essay by suggesting that this sort of blind science would spell the end for humanity, concluding with the following quip on de Grey's proposal to defeat death through biomedicine and engineering: "With the passion of a single-minded zealot crusading against time, he has issued the ultimate challenge, I believe, to our entire concept of the meaning of humanness" (Nuland 2005).

An accompanying editorial by Jason Potin, an editor at the journal, was even more scathing in his characterization of de Grey, calling the whole idea "science fiction" (what else?). Potin had recently joined the *MIT Technology Review* from *Acumen Journal,* which under his watch had published Nuland's first essay on aging and death.

The Immortality Institute's forums immediately linked to the articles, asking members to write to the *Technology Review*; news also spread to other listservs. Enraged letters flooded in from immortalists as well as other readers who had noticed that no substantive point had been made. The article became the magazine's most popular piece ever (Bartlett 2005), but apparently, Potin was surprised by the reaction and felt somewhat remiss. After discussions with de Grey he decided to ask scientists to review the substance of de Grey's claims. Several scientists accepted, then declined, so Potin took it upon himself to offer $10,000 to any scientist who could *prove* to an independent panel of scientists that de Grey's claims had *no scientific merit,* that is, that they were scientifically implausible. De Grey's foundation sweetened the pot with another $10,000. No one came forward with any accepted, definitive counterclaims. From that point on, de Grey's reputation began to change, and this has become a sort of myth of origins for de Grey and the SENS strategy. Building on this coup, he managed to gather high-profile immortalists such as Ray Kurzweil around his project, explicitly telling me that while he is not concerned with AI as such, he does want to appeal to the shared interest in technologically driven longevity in the AI and Singularity communities. At the same time, he has tried to make inroads into more mainstream academic scientific settings by setting up research programs within university labs and appearing at conferences and academic centers. His invitation to speak at the illustrious New York Academy of Sciences in 2009 demonstrated how far he had come. He has rolled back his claims regarding attainable ages to numbers that appear more realistic in the public's image of future worlds. But what has been at stake in these disputes, as Nuland's article clearly shows, is some notion of humanness and humanity tied to biology and death, to finitude. I argued in the introduction that these notions came out of a history of secularization and were tied to secular identities that held on to the finality of death as a marker of their secularity against religious notions of the afterlife. What has been consistently true, as in the case of Nuland and Greenblatt and many others, is that immortalist challenges to finitude tend to get secular humanists very upset.

ARTIFICIAL INTELLIGENCE, AUTHENTIC IMMORTALITY

The story of cybernetics and AI has been told many times (for two versions relevant to this project see Dupuy 2009 and Gleick 2011), but the link between cybernetics and immortalism has been less frequently explored (exceptions include Geraci 2010). I will not reinvent that history here, but I will emphasize two points: first, that cybernetics, as Hans Jonas ([1966] 1991) understood, was highly preoccupied with the problem of purpose (a subject I will explore in chapter 5); and second, that the idea of mind as a material and replicable, but substrate-independent, system was already imagined in the earliest days of cybernetics, though initially its mechanical and replicable aspects were not considered in structural terms but rather in communicative terms, in terms of information or messages being transmitted. Later, computer engineers such as Hans Moravec combined the structural and the informatic to develop the concept of "mind uploading," which clearly implied the survival of the mind on nonbiological platforms and the potential continuation of a person.

But here, too, it was the infotech and biotech booms that reinvigorated the mind-uploading scenarios and generated a social movement centered on its promises. This part of the story must be understood in the context of information technologies, dry lab work, and computational models not just of mind but also of biology. Through the 1990s and early 2000s genetics was becoming an increasingly computational matter, and scientists were prescribing molecular cures and doing biology by code rather than by pipette. At the same time, the development of cognitive science and neuroscience into the great sciences of the late twentieth century led to various attempts at brain mapping. The Decade of the Brain was declared by U.S. president George H. W. Bush just as the IBM computer Deep Blue got switched on in 1996 to play (and defeat the following year) the chess master Gary Kasparov, giving a big boost to flagging AI enthusiasm. By the first years of the new millennium these sciences were converging in what was given the acronym NBIC: nano, bio, info, cogno. The NBIC convergence really stood on the ground of information sciences and again brought a number of futurists together in the fold of legitimate science. It was that strange thing called information, neither material nor immaterial, that was enabling it all. Everything is information, after all, so it seemed like computation and digitization could represent and manipulate the whole world, from the cell to the mind. That was and continues to be the main promise driving the future and the futurists, some of whom are now associated with what is known as Singularitarianism.

The Singularity surged with another fusion of science and science fiction in the figure of Vernor Vinge, a mathematician and computer scientist at San Diego State University who is also a Hugo Award–winning science fiction writer. But it was in a science nonfiction lecture at a 1993 symposium sponsored by NASA that Vinge made his lasting contribution to the futurist and immortalist project. He laid out a predictive scenario in a half-troubled, half-ecstatic tone, declaring that "within thirty years, we will have the technological means to create superhuman intelligence," after which events will become completely unpredictable and "the human era will be ended" (Vinge 1993). Vinge called his threshold the Singularity, a term from mathematics and physics that describes an event horizon past which things will change so radically that the consequences are incalculable to current forecasters. The Singularity is now not only a popular concept but also a technosocial movement with links to the NSF and NASA.

The movement gained popularity through its association with Ray Kurzweil, its unofficial leader and spokesman. He assumed this mantle with the publication of *The Singularity Is Near* (2005), a book that analyzes the curve of technological development from humble flintknapping to the zippy microchip. The curve he draws rises exponentially, not linearly, and we are sitting right on the elbow, which means very suddenly and very quickly this trend toward faster, smaller, more powerful, and smarter technologies will yield greater-than-human machine intelligence. That sort of superintelligence will proliferate not only by self-replication but by building agents with even greater intelligence than itself, which will in turn build even more intelligent ones, resulting in an "intelligence explosion" so vast and so fast that the laws and certainties with which we are familiar will no longer apply.

However, since our brains are wet, messy, relatively inefficient products of evolution and, as one Singularitarian put it to me, were "not designed to be end-user modifiable," we biological humans may or may not become part of this intelligence expansion. But if we don't survive at all, well, at least thanks to us the universe itself will have been saturated with something of independent value, all its "dumb matter" transformed into "exquisitely *sublime* forms of intelligence" (Kurzweil 2005).

Dropped like a futurist gauntlet, the Singularity meme was picked up by Eliezer Yudkowski, who, along with programmer Tyler Emerson, set up the Singularity Institute for Artificial Intelligence in 2000 (changed to Machine Intelligence Research Institute in 2012). A bearded, convivial prodigy who speaks in highly formal sentences and is proud to not have a PhD, "Eli," as he's known, was excited by the prospect of superintelligent and superpowerful artificial agents, but the humanist in him worried that such an agent might end up, willfully or accidentally, destroying all the things we care about—like human lives.

Earlier, the Singularity Institute was essentially an online mail list called SL4 (for Shock Level 4), with a small subscription base of futurists from groups such as transhumanists and extropians, as well as a few researchers tracking the holy grail of AGI. The early discussions sound exploratory now, but they already contained the quasi-schizophrenic *sublime* element that characterizes Singularitarian conversations today:

on the one hand, worry; on the other, excited anticipation for the "most critical event in human history."

The subject of occasional discussion on the SL4 list, Kurzweil was well known back then, already consulting with the government and appearing on TV shows. His previous book *The Age of Spiritual Machines* (1999) had laid out the first exponential curves of an accelerating technology trend, presenting a utopian future of unlimited energy and great sex enabled by conscious machines. But it did not mention the Singularity.

It was only after the turn of the millennium—when else?—that the Singularity increasingly moved to the center of Kurzweil's platform. He started out by debating Vinge and then spoke about the Singularity in symposia at two futurist institutes, Foresight and Extropy, where Yudkowski politely lambasted him from the audience. Yudkowski asked Kurzweil whether the Singularity was good or bad, and what people could do about it. Kurzweil had no answer. On his SL4 list, Yudkowski would later write, "What Kurzweil is selling, under the brand name of the 'Singularity,' is the idea that technological progress will continue to go on exactly as it has done over the last century."[12] Kurzweil's "pseudo-Singularity" depended on the inevitability of his predictions and was leading to nothing more than "luxuries" such as superintelligent servants. Yudkowski's true Singularity, by contrast, was potentially dangerous and demanded intervention. It could lead to a land of posthuman bliss, but only if handled properly. It would need a vanguard of rationalists and scientists to attend to it. In short, Yudkowski was calling for a movement.

His beef with Kurzweil was about engaged activism versus passive predictivism, but it was also a bid for attention. He upped it by issuing a paper about safe AI timed to coincide with the release of the film adaptation of Isaac Asimov's *I, Robot*. Kurzweil, who was about to release his book on the Singularity, joined the SIAI as adviser and then board member. It benefited both. Kurzweil couldn't have assumed a mantle of authority without a strong link to the young challengers at the institute, while the institute needed his clout and his connections. The nonprofit institute soon found itself on firm financial ground, thanks in part to Kurzweil's friend and libertarian financial guru Peter Thiel, who made his money cofounding PayPal and investing early in Facebook. Thiel also coauthored

a nativist book called *The Diversity Myth,* reportedly donated $1 million
to the anti-immigrant group NumbersUSA, and funded James O'Keefe,
the not-so-independent guerilla videographer responsible for undermin-
ing Planned Parenthood and ACORN while lambasting minorities,[13] and
was a backer of Donald Trump's presidency.

Kurzweil's book has sold hundreds of thousands of copies. There
are Singularity activists, Singularity blogs, Singularity T-shirts, and even
Singularity bashers, all of which confirm the status of the Singularity as a
social movement, though Singularitarians themselves sometimes call it a
revolution, an inevitable one in which you or I don't have much of a say.
"The problem with choosing or not choosing to be a part of our 'revolution,'" Michael Anissimov, cofounder of the Singularity Summit, used to
repeat over and over in stately claims on the future, "is that, for better or
for worse, there probably is no choice. When superintelligence is created,
it will impact everyone on Earth, whether we like it or not." Anissimov,
who went on for a period to work for the Machine Intelligence Research
Institute, ended up coming out as one of Silicon Valley's alt-right figures
penning a white nationalist manifesto, fueling the suspicions that what
has developed in Silicon Valley is, in the words of one blogger, a white-
supremacist futurism.[14]

WHO IS AN IMMORTALIST

What I have described is the interlinked development of several groups
and networks that came together at a specific point in time around a common set of concerns and goals, among them radically extended human
lives but also libertarian politics. The groups are linked together in the
catchment site of immortality not only because they have elective affinities in their theoretical armature or ontological views on reality or their
goals for the future but because they share genealogies: they developed
through a contingent confluence of actors, discourses, tools, and ideas.
The membership also shares social backgrounds, continuing to be mostly
white, male, libertarian, and, in their own words, rationalist geeks—a fact
that has an impact on their imperial claims on the future and their neglect

of structural matters (such as race and gender) that shape the technologies and societies in which they work.

The following figures are meant to provide a sense of the membership profile of the various groups that in total or in part I have gathered under the umbrella of immortalism. That is all they are meant to give, a sense, for the movements are new and rapidly changing, and the figures for one group may not properly represent the membership of another.

In 2018 the two main cryonics organizations, Alcor and the Cryonics Institute, had over a thousand members each and more than a hundred "patients" in cryopreservation (Alcor 165; Cryonics Institute 170). The growth has not been exponential, as the figures had not changed dramatically in the previous eight years. Alcor receives more than 150 requests a month for information packets but adds an average of five members a month. Based on data from 2008,[15] Alcor's membership was mostly male, with females constituting only about 22 percent of the total. However, in an indication of a slight demographic shift, almost half of those women had signed up in the previous five years, versus 40 percent of men. Average age for both men and women is fifty, but average age at the time of signup (i.e., when they become members) is forty-one for women and forty-three for men. However, women who had signed up in the previous five years tend to have signed up at an earlier age, with most being in their late thirties.

There are few data on other aspects of social identity among the scattered membership. Official income and race data were not available. The WTA periodically gathers data through online surveys, the most recent one dating from 2007, with 760 respondents.[16] The numbers are indicative, but it must be kept in mind that, unlike cryonics, transhumanism has much greater reach beyond the United States and that not all transhumanists are immortalists. Indeed, based on answers to a survey question that asked "Do you think it would be a good thing if people could live (in good health) for hundreds of years or longer?" it would be safe to say that 13 percent are *not* immortalists, as they responded in the negative. The gender gap was much greater in the transhumanist population, with 89 percent male and only 10 percent female; and transhumanists were by and large younger than cryonicists.

Politically, transhumanists identify mostly as "technoprogressives" (16 percent) and libertarians (10 percent) or libertarian socialists (7 percent), with 11 percent claiming to be nonpolitical. The rest represent a wide range of political positions, from anarcho-capitalist (2 percent) to European liberal (5 percent) to green (4 percent) to "upwinger or advocate of future political system" (7 percent). With regard to religion, the 2007 WTA survey yielded the following data: 64 percent identify as atheist, secular humanist, agnostic, or nontheist; 2 percent as pagan, 5 percent as spiritual, 1 percent as Raelian; the rest as a smattering (1–4 percent) of faiths from Buddhist to Mormon[17] to Protestant to pantheist. An internal poll (421 respondents) conducted in 2003 by the Immortality Institute,[18] mainly acting as an organized online forum for proponents of RLE, showed similar results, with 58.5 percent identifying as agnostic, secular, or atheist, 3.24 percent as pagan, 19.20 percent as "other," and the rest a smattering of faiths. Depending on how one interprets categories like pantheism, paganism, spiritualism, or "other," the nontheistic membership can be anywhere from 72 to 85 percent.

Clearly, immortalism cannot be viewed as an isolated project, divorced from the rest of society. It is a cultural, social, and economic catchment site for a number of different technoscience projects through which specific versions of continuity are being formulated and instantiated. The convergence of old alliances and new research formations in the cradle of technocivilization have legitimized research into the indefinite extension of life. The strength of technomogul social and financial capital matters too, as major players in Silicon Valley, from Peter Thiel to Ray Kurzweil, place their heft behind these ventures. The links to institutions of U.S. security and empire are notable: Kurzweil, for example, is adviser to DARPA and the U.S. military, while Singularity summits attract an audience from Lockheed Martin (as well as philosophers such as David Chalmers), and Thiel cofounded Palantir, a secretive data-mining and surveillance company backed by the CIA with big State Department and Defense contracts.

Nevertheless, it is too easy to simply dismiss immortalism as the playground of very rich, powerful people with a good life who want to continue being rich and powerful and inhabiting a good life. Transhumanism, like immortalism, is not a uniform project, and there have been a series of splits and tensions within the membership along political lines. While recognizing that James Hughes (2012) makes much of the distance between the Peter Thiel billionaires and the socially progressive transhumanists, I heard more agreement on libertarian grounds and the ultimate desirability of all advanced technologies. And although most liberals and secular humanists find it hard to admit, the conditions that have shaped immortalism are the same ones that are shaping larger currents in secular and liberal societies that produce contemporary notions about the value of life and the form of a good life. The economy, moral and financial, generated by the mainstreaming of immortalism is going to mean that these projects will materialize in ways that, even if unpredictable, cannot be ignored. Most of the debate in recent years has been either technical or bioethical. Usually, people take their transhumanism with great slops of either enthused or accusatory dogma. They either think the technologies are going to ruin humans or make them better. I am not sure which one should occasion greater grief. My own proclivity is to think that better humans are worse humans, and yet ruining humanity altogether is difficult to defend as well. In thinking about the conceptual, political, and material coevolution of immortalist projects in the secular crucible that shapes our—not just their—dispositions vis-à-vis basic matters like life and death, I am also trying to cut a different path from the humanist past into a posthuman future, and eventually, down the road, recuperate other outcomes, open up other imaginaries, some of them made possible by immortalist visions that break down, despite themselves, some founding secular liberal positions about self, personhood, autonomy, reason, and matter.

3

Suspension

STRETCHING TIME BETWEEN
THE FINITE AND THE INFINITE

It is difficult to imagine a human sense of time, or temporality, without the end point of death. Human perceptions of time and temporalizing practices owe a lot to the materiality of our bodies as they develop and move into the ill-understood processes of decay and finally come to rest in a state that in most places is marked, though marked differently, as death. In "On the Relation of Time to Death," Helmut Plessner (1983) wrote that the awareness of time passing and the awareness of death rise from the same horizon. More recently, Rane Willerslev, Dorthe Reflund Christensen, and Lotte Meinert (2013, 1) have made the same point, that "bodily death and material decay are a central point of reference that offers key insights into human perceptions of time."

The complex ways in which physical processes, cultural concepts, and social practices interact make me skeptical that the human awareness of mortality can be abstracted so cleanly as an existential or transcendental essence. There exist other temporalities and temporalizing practices that may not have so much to do with human death itself—those, for example, that have to do with nonhuman unfoldings: the cycles of the moon, the sun, the seasons, but also the age of the universe, of the planet, of fossils, measures of which change our feeling of a horizon. Changes in processes of death and dying, whether material or cultural, affect the awareness of death as well as the sense of time so intimately linked to it. What's more, timekeeping technologies, concepts of time, and the experience of temporality undergo changes, sometimes radically so as to create new conditions of temporality. All of these affect the awareness of mortality. Rather than use the proximity of death and temporality as a fixed ground for theorizing about particular practices or concepts, I will consider the dynamic, often contradictory relation between the particular material and conceptual conditions of death and the multiple material and conceptual practices of "temporalization" (Fabian [1983] 2014, 74). Specifically, I will explore how time and the body are mutually mapped and how a particular secular history, set of cultural practices, scientific experiments and ideas, and medical dispositifs have generated temporal experiences.

As a set of techniques, procedures, and data, the biological sciences have found new ways to string together time and biological matter, only in part by measuring the length of life in living organisms or projecting the probable advent of the end (i.e., mortality rates, life expectancy). Today, research into aging, life extension, and rejuvenation medicine—by any name the "I" word goes—produces new kinds of time and continuity out of biological matter: measuring telomere length, cellular senescence, DNA methylation, metabolic exchange, stem-cell regeneration (Belsky et al. 2015; Horvath et al. 2012; Sanders and Newman 2013). One of the most important ways in which biological temporality has been changed has been through control over temperature, specifically by achieving very cold temperatures, a process that is being theorized recently under the rubric of cryopolitics (Radin and Kowal 2017). Freezer technologies have deep political and economic consequences, having made possible the cold chain infrastructure that is responsible for global foodways, as well as tissue and blood storage and transportation with wide medical and humanitarian applications. Inducing cold is a way of controlling important aspects of biological matter, by manipulating the rate of change in matter, specifically, slowing down its metabolism or energy exchange. In this sense one might also say there is an exchange between the state of matter and temporality, such that the manipulation of matter through temperature affects the temporal relationship internal to that matter (slowing down its rate of decay) as well as in relation to its environment. Since human temporality is in important ways also social and ideological, the manipulation of matter through temperature is also entangled with structures, meanings, and experiences of time, and especially of endings—what later I will discuss variously as eschatological matter and after-lifetimes.

Originally proposed by Michael Bravo and W. Gareth Rees (2006) to discuss the politics of melting ice caps and the role of low temperatures and thermodynamic exchange in the planetary sphere, cryopolitics has been suggested more recently by Radin and Kowal (2017) as an analytic extension of Foucauldian biopolitics that applies to certain areas of control over life and death. Whereas biopolitics is organized around the imperative to make live but let die, cryopolitics begins with the imperative to *not let die.* By using temperature to manage the time of life, cryopolitics is conceived of as a temporal prosthesis to preserve or perpetuate life indefinitely such that "death appears perpetually deferred" (2017, 7). As parsed by others in the same volume, cryopolitics denotes the way future *potential* becomes arranged as a form of salvation—but one that promotes, as Radin and Kowal note, "the abdication of responsibility for action in the present" (9). Additionally, the technique of freezing or cold suspension fits right into a racialized and colonial imaginary of preservation where a rush started for taking biological samples from peoples categorized as endangered. Kim Tallbear (2017) is characteristically trenchant in her critique of the preservation impulse in the racialized politics of cryotechnologies deployed in domains such as genomics and the biological sampling of populations under the auspices of programs such as the International Biological Program (also analyzed by Radin [2017]). Tallbear's critique is simple but cutting: "Cryopreservation Aims to Preserve Indigenous DNA, but Is Predicated on Indigenous Death" (2017, 182). The way "perpetually impending death" is attached to modern and scientific imaginaries of and projects related to "indigenous peoples" supports the power of technoscientific states and institutions to rearticulate indigenous life in their own ways, raising a question about who owns the power to preserve what, and who has a claim on survival in the future. Those whose death is already implied in these imaginaries may be ignored in projections of the future—a point that will become more salient in the final chapter.

The politics of suspension derives specific forms of value out of the exchange between life and time—that is, such a politics is predicated on prior valuations of life, time, survival, the possibilities of continuity, and imaginaries of the future. Potential is not potential for everyone, or at least it's not the same kind of potential. For example, Radin (2017) adds another angle to Tallbear's critique by showing that indigenous communities perceive the frozen and suspended blood samples of their now deceased relatives and/or ancestors not as samples taken for future improvements to their lives but as "incomplete deaths." Clearly, different notions of continuity are at work, different immortality regimes regulating the time of life and the time after. If cryopolitics blurs the distinction between life and death, continuity and discontinuity, what were the conditions of their separation to begin with? What are some of the notions and histories that underlie the valuations in these technoscientific exchanges between the biological matter of life and notions and experiences of temporality, or between biology and claims on the future? If the suspension of life produces potential time as a kind of salvation for some, what cultural conditions closed time off in such a way that it could be redeemed through this cryobiological salvage project? What made the present perpetually insufficient? And what specific sorts of futures are at stake?

I will explore these issues through a focus on cryonics, but the argument expands out to immortalism in general and other modalities of suspension. As a techno-utopian project, cryonics often has been dismissed as yet another instance of modernity's unhealthy relationship to death, a simplistic and technological form of death denial. I have, rather, taken the cryopreserved figure seriously not to legitimize cryonics or simply to situate it; rather, the cryopreserved figure and the schisms it produces point to contradictions and tensions *within* modern secular formations and imaginaries that structure lives and lifetimes for many inhabitants of this secular age. The secularization and naturalization of time, along with biological understandings of cellular life and the ability to manipulate it, changed the relationship between death, the body, and time in ways that shed light on the practice of cryonics while cryonics illuminates ways in which secular temporalities and relations to death may be contradictory, a tension I parse in terms of lifetimes and after-lifetimes. I argue that suspension is a space of translation between biological and chronological time, one that *emerges as meaningful* through the broader historical and cultural relation to finitude and infinitude, to secular eschatologies that on the one hand confined time to the body and on the other expanded the time of the world in which the body finds itself thrown.

A THIRD STATE OF BEING

There is nothing fancy about the conference room at the Alcor Life Extension Foundation in Scottsdale, Arizona: the usual brown laminate table at the center, surrounded by some chairs and, as at many regular organizations and companies, framed portraits of members lining the wall. But there is something odd about the framed photographs, as they represent neither the consecrated dead nor the living guides of the organization. For the Alcor Life Extension Foundation, named after a dim star in the handle of the Big Dipper, these members are neither alive nor dead. They are only legally dead. As far as Alcor is concerned, these members are cryopreserved; they are temporarily deanimated. So underneath each member's photograph, there is a date of birth, followed by another date, not the date of death, but the date of "deanimation."

Cryonicists push reductive materialism to its logical end: since in a materialist view the self, unlike the soul, does not exceed the material conditions that give rise to it, and a person's consciousness is assumed to be reducible essentially to his or her brain states, then, cryonicists argue, it follows that the self too may be preserved and revived if the body and, most important, the brain structure are properly safeguarded to the smallest scale. It is not that all cryonicists believe in the certainty of reanimation; most use cryonics as a backup plan, and many place the likelihood of reanimation at no more than a few percentage points, some even lower. The common argument is that when it comes to life and death, any chance above zero is infinitely better than zero. Zero is the chance that all those people who opt for burial or cremation give themselves. An old quip attributed to the cryonicist and nanotechnologist Ralph Merkle has become a stock phrase: "Cryonics is an experiment. The control group is not doing so well." So while cryonics is not exactly the community of faith it is often described to be, members place much store in the power of science and technology to achieve dizzying heights in an advanced future in which they may be reanimated.

In the virtual space between the actuality of a frozen body and the projected future of technoscience, they have created an actual human person who is deemed neither dead nor alive but resides in what they themselves call a "third state." That is what I refer to as the "suspended figure"—a figure through which the translations of time and life and the future are projected and understood.

During this period of indefinite and "uncertain waiting," as cryonicist Mike Darwin (2005) has called it, the important thing cryonicists focus on is keeping the brain, the seat of the self, in as good a condition as possible, with its structures in place, and without any risk of further decay. It is in order to prevent further damage and cell death that cryonicists are cooled down upon legal death and stored, or suspended, in gleaming dewars called cryostats. And in their period of uncertain waiting, all those cryonicists in suspended animation cannot be treated as dead people. Secured under lock and surveillance cam, those cryopreserved bodies are not called corpses; they are called *patients*. They have patient files with patient numbers and are maintained in the "patient bay." The conference room has a viewing window onto the patient bay through which you can see the vats of liquid nitrogen holding the bodies of the members whose pictures line the walls. It makes for a recursive loop of stillness: time frozen in the photographs, time frozen in the tanks.

Here in the conference room, I, the anthropologist, get my first training in standby and stabilization procedures. All the "cryonics natives" (as some called themselves when they heard I was an anthropologist) are pretty much white and male. "Anita" is an exception only in her gender. I mention Anita because I think of her often, the way she stood looking at those pictures and then at the cryostats and back. How she gestured at the people in the photographs and said, "Oh, I envy them. I want to be up there already. I've been thinking about which picture I would put up, and I think I have one."

Like other cryonicists, Anita, so impatient to not die, knew there is no other place in the secular realm to come to die, if you hope to live again. In her urgency—to be up there already—there are revealing temporal conflations and contradictions: the finitude and infinitude of time, as well as history and destiny, where in modernity both are temporalities of human making, and both are haunted by the specter of endings. Anita's is an affect that betrays life's ongoing dissatisfaction with the present (decay), the idealization of the future (possibility), and the poetics of suspension (how to manage the distance between the two), relying as much on large institutional shifts as well as subtle adjustments of perspective.

The equipment laid out on the conference room table consists of a chest thumper, to circulate blood, and a CPR mannequin, with a blue torso and a mouth half open. The training was led by a former EMS nurse, "Jenny," who was then in the employ of Alcor but was not a cryonicist herself, and by the firemen from the local station, who, thanks to Jenny's experience, had slowly warmed to cryonics, meaning mainly that they had reduced their giggles to an occasional snicker or joke. Jenny was a thorough instructor, taking us through our paces in the art of resuscitation: chest compressions, pinch the nose, lift the head, mouth to mouth. For cryonicists this is crucial because it keeps the body, and more specifically the brain, oxygenated so that even if the heart has stopped, the brain cells don't kick into apoptosis. "Apoptosis" is the scientific word for "cell suicide," a somewhat mysterious process by which cells understand that the end is near and decide to throw in the towel. To be administered to a cryonics patient after their death is legally declared, our chest compressions and mouth to mouth would help prevent the cascade of apoptosis, helping to save the legally dead person from more cellular dying while the rest of the crew takes the body down to stasis.

Suddenly Jenny stopped the whole thing.

"Not too hard," she warned. "What you wanna do is, you just want to circulate the oxygen and the blood, but you don't want to go *too* hard. It'd be bad if you got the heart going and ended up with a live person again. It would defeat the purpose."

We, the novice trainees, looked up a little confused at first. It was one of those ethnographic moments when the space–time fabric rips open and you find yourself in a parallel universe, where everything looks the same but is not.

"That's why we call it cardiopulmonary *support,* not cardiopulmonary *resuscitation,* CPS instead of CPR," Jenny explained.

At the end of the training we also received a certification in basic CPR, but what stayed with me was the single difference between the world of resuscitation and the world of suspension. The distance between R and S, that, as in the alphabet, is no more than a breath apart but opens up an infinite bracket between two states, allowing us to think of—or think with—a third, liminal state that may be considered neither alive nor dead, an undetermined state of uncertainty that exists and operates in the cracks between the rules and regulations, the normalities and certitudes of the entire social apparatus that determines the process of dying, that sorts being from not being. To view and understand existence as cryonicists do, you must be able to occupy this third space cracked open in time between R and S and inhabited by the suspended subject.

What is the temporality of cryonic suspension? What sort of time is a time in which death may be reversible but the arrow of time itself is not? Does the suspended figure have biological time? In part the imagined answers are a function of liquid nitrogen temperatures and technologies and the trouble with freezing techniques. The goal of cryopreservation or cryostasis is simply to prevent further decay, degeneration, and decomposition in the body and brain structure. It's a familiar process in many ways. We are always "preserving" biological matter for later use. Strawberries in the fridge, hamburgers in the freezer, and semen in a fertility clinic are being preserved for later use. Hamburgers are kept at mildly subzero temperatures where biochemical activity is minimized but not entirely halted, which means that eventually the meat *will* go bad, whereas semen or embryos in deep freeze at liquid nitrogen temperatures (–196°C)[1] have been kept frozen for years, because at that temperature there is almost no biochemical activity: biological time virtually stops.

The problem with freezing is ice. When you freeze a fresh strawberry, for example, and then defrost it, you will notice that the strawberry is mushy and discolored and not quite in the same fresh shape as the strawberry you froze. This is because the water in the strawberry cells turns into ice, and ice molecules form into hard, knife-sharp crystals like ninja stars. These molecular ninja stars form in the space between cells and cut into the strawberry cell membranes, damaging them for good.

As described earlier, cryobiologists have used glycerol and other agents with similar properties to achieve ice-free freezing. These agents minimize ice formation and only harden or vitrify into a glasslike state rather than forming sharp crystals. Cryoprotective agents have been successfully used to protect semen, eggs, and sometimes animal organs during the freezing process. Upon thawing, the cryoprotectant is washed out and, ideally, the sperm continues on its path to new life as though nothing had happened in between, during that state of suspension. In actuality it is hard to achieve consistent results. Experiments show that frozen sperm is not as active, fertile, or lively as fresh sperm.[2] Nevertheless, the process for embryos and semen is functional enough to produce acceptable results for human fertility clinics and animal breeders.

In the meantime, success in long-term cryopreservation of whole organisms has been elusive, though some advances have been made with organs. Protecting a *whole* human body, and the person occupying it, from freezing damage is extremely complicated. As serious cryonics researchers admit, there are problems with cryoprotective agents and preservation processes in cryonics, mainly having to do with toxicity, permeation, and chilling injury. Cryoprotective agents are known to be toxic, causing some damage to blood cells, cell membranes, DNA, and proteins. The extent of the damage is unknown, but researchers at the Cryonics Institute and 21st Century Medicine, which is linked to Alcor, have patented their own cryoprotective agents, claiming lower toxicity and improved "vitrification" (solutions that vitrify do not form crystals, but transition quickly into a glasslike solid state). It is assumed and hoped by now that whatever damage is done by the cryoprotective solution is much less than the damage done by ice crystals or older and more toxic cryoprotectants.

Liquid nitrogen, at least, has the advantage of being very, very cold. As mentioned, biochemical processes are almost completely suspended at -196°C, a temperature at which virtually nothing happens. But since this *is* a matter of life and death, it's best to be precise about *how much* nothing *actually* happens. According to an article by Ben Best (2008), a cryonics researcher and former director of the Cryonics Institute, biochemical reactions that at a normal 37°C occur over a time span of six minutes would, at a temperature of -196°C, take 100 sextillion years to unfold.

In this example, the choice of a six-minute timescale is not arbitrary.

Conventionally, six to eight minutes is considered the time it takes after cardiac arrest for ischemic damage to result in final brain death. It is the conventional window of irreversible dying.

In many cases, however, to reach irreversibility the cascade of biological events leading to death takes longer than the six to eight minutes assumed by hospitals, laws, and some experts. Lower temperatures help: the longest case of submersion in an ice-cold river was a two-and-a-half-year-old girl, who had drowned for at least sixty-six minutes (Bolte et al. 1988). Indeed, "induced hypothermia" is increasingly used in what is now called "resuscitation science," as well as in complex heart surgery and other health-related projects (Kumral et al. 2001).[3] What's more, despite the difficulties, small organisms and some organs from larger animals, such as rabbit kidneys, have been frozen and revived (Fahy et al. 2004), while a host of cells and some smaller species are now understood to be beyond the apparently natural processes of senescence and expiration.

Cryonicists feel vindicated by these developments, since they have long claimed that biological time is elastic time: you can stretch it, and if you get in there fast enough, right after the proper secular authority pronounces death, and you start stabilizing the body, cooling the head, and circulating blood and oxygen so that organs and especially brain cells remain alive, you can eventually, perhaps, turn the window of dying into a window of not dying, expand those six minutes into, well, 100 sextillion years.

A cryonics researcher I'll call Richard described the stasis of cryopreserved patients in these terms: "They're settled in their storage spots and they can just hang out there." Indeed, among those he was referring to, the patients just "hanging out there," awaiting reanimation in the distant future, is Richard's own father, a reluctant cryonicist at best, whom Richard convinced to undergo cryopreservation. When he speaks of the cryopreserved, then, it is not in the abstract. At the same time, there is no way of determining the state his dad is in—how good the preservation was, how much damage his brain might have suffered, whether he has any memories left. And Richard says he would rather not know anyway. He just wants to wait and see. For now, "it's like somebody hit the pause button," he says, "except the pause button can last decades or hundreds of years and that's how I see it. Everything's on pause."

Here is how another researcher justified this temporality of the pause: "To me the distinction between a person and a corpse is whether or not there is a reasonable hope of them regaining consciousness. With a cryonics patient I think there is a reasonable chance. *The time gap is irrelevant.*"

In this light, one might say of the patients in the Alcor patient bay that while they may not have much of a life, they have, at least, a whole lot of time ahead. I mean this literally. For the figure of suspension presents a new conjunction of a life-form with a life-time, a patient with an indefinite, almost infinite, bioviability (100 sextillion years far exceeding the universe's estimated existence). Suspension is characterized by the translation of life into time, and its possibility produces new senses of time in life, beyond death. To repeat, my argument is that suspension is made possible by *and emerges as meaningful through* the broader historical and cultural relation to finitude and infinitude that on the one hand limited time to the finite body and on the other infinitely expanded the time of the world in which the body finds itself thrown. Put another way, the cryopreserved body's material continuity in indefinite time is made possible through the technical suspension of its biological life, but the cultural meaningfulness of that suspension—the possibilities, desires, and horrors it evokes—emerges in the ongoing time of the world which is slated to keep moving on after lives expire, leaving everyone, every life-time, behind. These relations to endings, to death and not-dying, affect how time, as a kind of pressure at the end of life, is felt and lived sometimes as an opening, at other times as a foreshortening or an incompleteness.

THE TIME GAP AS SECULAR ESCHATOLOGY

The body not dead, the body stuck between life and death, suspended between the animate and inanimate: these are familiar figures for anthropology. From E. B. Tylor and George Frazer on to Robert Hertz's seminal manuscript on secondary burial, anthropologists have long documented and theorized the various ways in which the dead don't simply die but continue to be materially and psychologically present in social life, how the dead continue to move back and forth between death and life, as Michael Lambek writes, creating in that movement a realm populated by "beings who do live beyond or outside death—and yet not in some entirely different transcendent world" (2016, 641). From the preserved bodies and animate statues of rulers to relics, mummies, photographs, and bodies possessed and animated by the dead, these figures and their after-lifetimes have made endless appearances in the literature of cultural anthropology. Ever since Margaret Lock's (2002) groundbreaking book on organ transplant and brain death, the anthropology of science and medicine has in turn been preoccupied with the liminal bodies of technoscience. Tethered to technologies and modern medical institutions that are theorized as being unable to accommodate death and so tend to push life at its end into the realm of the monstrous, these zombie bodies—the brain-dead, the comatose, the cryopreserved (Lock 2002; Kaufman 2000, 2005; Farman 2013; Bernstein 2015)—hover in a "zone of ambiguity" (Rabinow 1999; Kaufman and Morgan 2005) blurring the boundary between death and life.

Often displayed as an exemplary outcome of the confusions caused by modern medicine's manipulations of life, the zone of ambiguity has in fact been a source of power for biotech projects, asserting technological sovereignty over the literal matter of life and death. Hannah Landecker (2005, 2007) was among the first to point out that the power of biotech—I would add, its symbolic, magical, and instrumental power—can partly be seen as its ability to stop and start life, to freeze eggs and sperm and then return them back to life later, to isolate tissue culture so that it can continue "living" indefinitely. Work done since then has examined the implications of a politics of latency (Romain 2010; Radin 2013; Roosth 2014) and potentiality (Taussig, Hoeyer, and Helmreich 2013), remaking life as a discontinuous process of interruptions.

However, my concern with suspension here is not in relation to the definition or status of life and death, their social determinations and biological realities; and my example of cryopreservation is not another instance of the "extreme, problematic, or limit cases of what counts as life" (Roosth 2014) or the quandaries of pinpointing a "coherent" notion or form of human life (Roosth and Helmreich 2010) or death (Lock 2002;

Sharp 2007). Rather, I want to follow Lambek's suggestion to consider the "connected processes of animation, de-animation and reanimation" (2016, 643), which I will consider in terms of chronobiopolitics (the coincidence of temporal experience, biology, and politics).

Adding "time" to the workings of Foucauldian biopolitics seems crucial because so much of the ordering of life depends on the ordering of time, mapping the temporalities of the body and the temporalities of the social onto each other. The coincidence of biology, time, and social institutions in the body, the way the body in time and time in the body are experienced subjectively, must be considered as part of the temporality of biopolitics. Bringing these together is the term "chronobiopolitics," used by Dana Luciano (2007) as an analytic with which to examine the affective and sexual arrangement of the time of life on the level of populations. Luciano does this through the "affective chronometry" of new forms of grief in nineteenth-century America, whereby a slow, gendered, and emotional time was permitted in times of mourning, standing in contrast with the fast, linear, production-driven timeline of modernity. But one might also think of the curve of Medicare eligibility, debates over retirement age, and pension payouts, as well as the age of sexual consent, the appropriate time for marriage and childbearing, the life cycle, and so on as these meet and shape the requirements of social existence. Tiffany Romain (2010) has observed that cryogenic techniques used in embryo preservation and cryonics disembed the rhythms of work and social life from biological cycles, allowing users to "buy time." Thus women who want to have children but also want to build their careers now have the option to freeze time for a while by freezing their fertilized eggs until they are ready to dedicate time to pregnancy and parenting—thereby bypassing what has become the culturally gendered term for time in the body, a woman's "biological clock." Through chronobiopolitics, one can see the way bodies become time-bound in specific ways, as the temporal rhythms of medical or state institutions govern various economic, social, and political horizons toward which the male and female bodies are variously angled. Given the secular finality of death, the uber horizon of chronobiopolitics is the horizon of death. The body, in this secular version of chronobiopolitics, becomes a kind of time-reckoning device rather than a timekeeping device. That is to say, it is not merely marking time; rather, it feels it must make sense of time in relation to assessments of a horizon, an ending. The cryopreserved figure, in suspension, shifts that horizon.

In thinking with the cryopreserved figure, I also want to move away from the rhetoric regarding modern death denial that since the middle of the last century has characterized work on death and dying and more recently the work on cryopolitics. What interests me is the contemporaneity of that cryopreserved figure as it continues to exist, unconscious but coeval, alongside other living beings in time, a life-form whose functionality moves from the biological to the temporal and for which the disconnect between lifetime (time in life) and after-lifetime (any time thought to be left after a life has expired) is at stake. Combining Landecker and Luciano, conjoining the molecular to the molar, the cellular to the corporeal, the cryopolitical to the chronopolitical, we might consider the implications of a chronobiopolitics in which life and time are united in and limited to one body; in this secular conjoining of life and time, bodily crises get translated into temporal crises, and this, in turn, has important consequences in terms of how pressures of life and time are felt within a lifetime and how various end-of-life issues are dealt with (including clinical trials, drug development, experience of prognosis, precarity, and so on [Jain 2013]).

Not surprisingly, cryonicists often talk about the distant rather than the near future (Guyer 2007), worrying not just about biological viability but also about maintenance of material structures (such as buildings) and social ones (such as bank accounts, contracts, or kin relations). Sometimes they talk about cryonics as a kind of time travel, imagining the future they might wake up in, returning after what was characterized above as a "time gap." A new, very large center being designed to store all sorts of cryogenically preserved matter—from tissue to stem cells to DNA to full humans—makes this explicit in its name: called Timeship, its somewhat megalomaniacal architecture refers to ancient Egypt as much as to futuristic space voyage, to the age-old human dream of immortality. With the current practice of suspension and the future conquest of death, say the project's designers, "our notions of time will have to change completely." The Timeship is meant to embody at once "both time and timelessness" (S. Valentine 2009, 45). Yet cryonicists, like other technofuturists, speak

of an "accelerated future" in which exponential technological development will quickly bring about changes that will radically transform human
existence in the future. Ironic it is that this desired technological acceleration moves in contrast to the steady decay of the biological body, so
that as machinic time speeds up with faster and faster processing rates,
human bodies and brains operate at the same old biochemical speeds.
To catch up with exponentially accelerating machine time is to become
machine. Until that happens, there is cryonics, taking one in the opposite
temporal direction.

That contrast between the suspended, cryopreserved figure metaboli-
cally slowed down to a timeless halt, on the one hand, and the modern
world of acceleration and space–time compression, on the other, reminds
me of another image of the body in time, a negative image of sorts,
projected at the cusp of secular modernity by one of its early theorists.
Probing the entailments of a godless and disenchanted world, Arthur
Schopenhauer described the "vanity of existence" in these temporal
terms: "A man finds himself, to his great astonishment, suddenly existing,
after thousands and thousands of years of non-existence: he lives for a
little while; and then, again, comes an equally long period when he must
exist no more" (Schopenhauer 2005). During that short period, we can
but chase after time: "We are like a man running downhill, who cannot
keep on his legs unless he runs on, and will inevitably fall if he stops."
Yet we are organisms, and despite "the cunning and complex working of
its machinery," the human organism, running or stopping, "must fall to
dust and yield up itself and all its strivings to extinction—this is the naïve
way in which Nature, who is always so true and sincere in what she says,
proclaims the whole struggle of this will as in its very essence barren and
unprofitable. Were it of any value in itself, anything unconditioned and
absolute, it could not thus end in mere nothing."[4]

Note the bifurcation between the time of the organism making us run
in futility and the eternal and ongoing nothingness—this is the eschatological gap produced by the secular as it tries to order its relation and
produce its norms regarding the finality of death and what follows. If I
call this eschatology secular, it is because, as explained in the introduction, two important shifts affected the relationship between body, time,

and death. The materialist elimination of the soul and of the afterlife, understood as religious and "primitive" illusions regarding the body and the person, led to a particular modern experience of the finitude of life, limiting the time of existence to the time of the body, to the conjunction casually called a *lifetime*. At the same time, the naturalization of time (Fabian [1983] 2014), coming via extensions in geological and evolutionary time, led to the expansion of time frames on either side of the present, an infinite and open timeline moving past any individual lifetime. The secular scientific worldview delegitimized temporal orders that included resurrection, cyclical time, apocalyptic time (Robbins 2007), and other "metahistorical" time frames that were said to characterize religious or "primitive" temporality (e.g., see Kravel-Tovi and Bilu 2008). It is worth emphasizing again that the naturalization of time and the finitude of existence were simultaneously secularized and marked racially from the get-go as they got established as the proper modern attitudes to assume in relation to time and being.

In the face of a finite existence that ends in nothing but nothing, Schopenhauer's "ever-fleeting time" generates a different experience of lived time than, say, for a being subject to death but whose life is embedded not in infinite linear time but in cyclical time, or a being whose death represents a break but nevertheless accepts some sort of continuation after death. In this new secular temporality, time becomes "that by which at every moment all things become as nothing in our hands," and so "life presents itself chiefly as a task—the task, I mean, of subsisting at all, gagner sa vie. If this is accomplished, life is a burden, and then there comes the second task of doing something with that which has been won. . . . The first task is to win something; the second, to banish the feeling that it has been won; otherwise it is a burden. . . . Human life must be some kind of mistake." It is through a new view of humans as part of a unified biological and rational species that these passages on nature and extinction make sense, for it is not simply a specific life that is absurd but the existence and strivings of the self-aware species that falsely theorizes away its biological inevitabilities. The absurdity of the rationalizations becomes even more pronounced in contrast to the cold, immense unfolding of time. All the "bustle" of daily life in "the little span of human life" appears absurd as all "this terrible activity produces a comic effect." A life lived toward a final death in an infinite universal timeline presented a new set of problems and experiences: the fear that time was constantly running out emphasized the modern anxiety that existence itself was pointless.

It was Tolstoy's grappling with these matters, via Schopenhauer, that formed the foundation of Weberian disenchantment—a theory that over the last century has heavily influenced understandings of the modern secular world. Weber's analysis of science, disenchantment, and progress has received a lot of attention; what has been less remarked upon is the affective force behind his view and the outline of what appears as a new secular disposition, for he enters the discussion through an existential weighing of modern life and death in a scientific world. "Scientific work is chained to the course of progress" (Weber 1958, 137), yet progress has an ambivalent valence, because even if science leads, *in principle,* to a mastery of all things of this world and ever more efficient technical achievements, it cannot guide people as to the ends (what should be mastered and why). The kind of progress Weber outlines is literally open-ended—that is, there can be no known end to aim at, and each achievement is superseded by the next one.[5] "In science, each of us knows that what he has accomplished will be antiquated. . . . That is the fate to which science is subjected; it is the very *meaning* of scientific work. . . . Every scientific 'fulfilment' raises new 'questions'; it *asks* to be 'surpassed' and outdated. . . . In principle, this progress goes on *ad infinitum*" (1946b, 138).

Since "scientific work" is potentially unending and each achievement will only be surpassed, the dilemma of the conduct of daily life as a problem of modern temporality is first raised for the scientist and gets posed clearly as a problem of purpose: "Why does one engage in doing something that in reality never comes, and never can come, to an end? . . . What does he who allows himself to be integrated into this specialized organization, running on *ad infinitum,* hope to accomplish that is significant in these productions that are always destined to be outdated?" (1946b, 138). The condition of the secular scientist is to live with a sense that is the opposite of fulfillment; it is always open, unfinished, liable to be outstripped sooner or later.

Such an indefinite sense of progress is markedly different from earlier ideologies of progress, the Enlightenment's controlled, planned advancement toward social perfection, in which (as per Comte and his academic progeny down to Weber's time and after) the social sciences, the natural sciences, and the state could work together to plan on and produce the good life as the obvious telos of European and later North American civilization. That was the original sense of "progress." It was directed, teleological, historicist, moving away from a "primitive" past, in Marxist theories as much as in liberal theories, toward Kantian perpetual peace or utilitarian states maximizing Mills's grab bag of general happiness. Reinhardt Koselleck (2002) identifies structures of modern, secular time with progress: "Since it was progress that conceptualized the difference between the past so far and the coming future . . . progress is the first genuinely historical definition of time that has not derived its meaning from other areas of experience such as theology or mythical foreknowledge" (2002, 120). Contrary to this claim, I am suggesting that progress and modern time more generally came to be constituted and understood through their opposition to theology and premodern temporalities associated with non-European cultures, from which modernity was ostensibly breaking.

Delivered on the heels of the devastations and disappointments of World War I, Weber's "Science as a Vocation" was also a comment on what he called "civilized life." Immediately after the famous passage defining disenchantment, Weber, taking off from Tolstoy, situates all "civilized men" in the infinite line of progress; but that also means the path of obsolescence and meaninglessness:

> For civilized man death has no meaning. It has none because the individual life of civilized man, placed into an infinite "progress," according to its own imminent meaning should never come to an end; for there is always a further step ahead of one who stands in the march of progress. And no man who comes to die stands upon the peak which lies in infinity. Abraham, or some peasant of the past, died "old and satiated with life" because he stood in the organic cycle of life; because his life, in terms of its meaning and on the eve of his days, had given to him what life had to offer; because for him there remained no puzzles he might wish to solve; and therefore he could have had "enough" of life. Whereas civilized man,

placed in the midst of the continuous enrichment of culture by ideas, knowledge, and problems, may become "tired of life" but not "satiated with life." . . . What he seizes is always something provisional and not definitive, and therefore death for him is a meaningless occurrence. And because death is meaningless, civilized life as such is meaningless; by its very "progressiveness" it gives death the imprint of meaninglessness. (1946b, 140)

In premodern life, as in art, each life is its own fulfillment. In "civilized" life, each life is surpassed. Why? Because despite mastery by calculation, an individual life does not appear as "definitive" but rather as something "provisional," hence always incomplete and unbounded. The isolated, finite individual with no recourse to an afterlife or a cosmology stands in an infinite stream of progress which, he is aware, will surpass him, making his particular life appear insignificant or meaningless. Whereas Schopenhauer applied his view on the absurdity of existence to all of human history—that is what humanists and literary authors came to call the "human condition"—Weber sees this as a problem that has arisen at a particular time and place in human history, one indeed that implicated the very notion of history as a secular temporality in which the action of intentional bodies—as opposed to, say, gods and spirits—characterized the only proper form of agency and site of meaning in this immanent world (Chakrabarty 1997, 36).

Weber's observations on disenchantment recognize an important secular shift in temporality, a shift that affects expectations and experiences of modern living and dying in significant ways. In this secular age, the human is a finite body dropped into an infinite timeline, running against it. In a materialist world of scientific progress, in which personal death without an afterlife is final and in which everything we "achieve" in this life will be surpassed, daily life and work will feel provisional and incomplete; life becomes a project, but, as a project with no greater purpose, it becomes subject to failure. From the wider perspective taken up by Weber, *all life* in a disenchanted world is ultimately subject to this failure, regardless of its current—merely current—worldly success. All life will be surpassed. Despite this absurdity, despite the certainty of cosmic failure, we must get on with the workaday, without knowing why, without even asking why.

The new relation to death as a final point of life also influenced secular knowledge production; as Foucault demonstrates in his own excavation of finitude in *The Order of Things* ([1970] 1994), time becomes a suddenly important factor in economic calculations as well as biological and demographic theories, including Malthus's. His paradigmatic case here is economics, with Smith's analysis of labor marking the turning point wherein value is no longer related to need as such, either in use or in exchange, but to time, that is, to the amount of toil in time, time being a limited good. Rather than need, Smith "unearths labour, that is, toil and time, the working-day that at once patterns and uses up man's life. The equivalence of the objects of desire is no longer established by the intermediary of other objects and other desires. . . . [I]f there is an order regulating the forms of wealth, if this can buy that . . . it is not because men have comparable desires; it is not because they experience the same hunger in their bodies, or because their hearts are all swayed by the same passions; it is because they are all subject to time, to toil, to weariness, and, in the last resort, to death itself" ([1970] 1994, 225).

A similar historicity, also brought on by dying, permeates biology as soon as function takes over taxonomy as the guiding principle in the investigation of nature. If taxonomy is the classification of visible signs (organs, morphology), function is the labor of the interior, concerned with efficiency, maintenance, and ultimately survival—that is, with life. This view of function also accounts for death as an integral part of life, inasmuch as the death of the individual is responsible for nothing more, and nothing less, than perpetuating life itself. Life, then, is conceived of as a force encompassing the individual bodies endowed with it, the living beings that are "no more than transitory figures" ([1970] 1994, 278) in the "infinity of life." This dynamic view of nature, balanced between transition and continuity, necessitating death in the act of constant transformation of life, provides the grounds not for a natural history, contingent and mythical as that was, but for a history of nature, that is, for a theory of evolution.

Although he is concerned with its anthropological implications, Foucault does not specifically explore the kinds of social experiences finitude eliminates or other kinds of experience it may give rise to. If both life and labor had become infused with an analytics of time, then certain experiences of time, certain social structures of time, must have changed.[6] Where Foucault claims that the metaphysics of infinity has become useless in an epistemological sense, as a way of grounding knowledge, he diminishes the fact that a sort of infinity nevertheless and by his own account engulfed the whole condition in a particular, phenomenological way—not the infinity of God or transcendent forms of knowledge but of the apprehension of time as an ongoing, endless sequence of events into which we are thrown and of which we constitute only a minor, finite fragment.

Similar tensions ring out around Darwin. Some commentators have argued that notions of immortality were secularized through Darwinian theory and transferred to the idea of survival and continuity through the perpetuation of the species. If that had been the case in the popular or eugenicist imaginations, it was not exactly what Darwin himself had proposed. What is embedded in evolution, as in progress itself, is closer to obsolescence. What "succession by generation" entails, in this ongoing progress within an "inappreciable length" of time, is the transformation of the human—as of other species—beyond recognition. In Darwin's words: "Judging from the past, we may safely infer that not one living species will transmit its unaltered likeness to a distant futurity" (1859, 489). Thus, in defining the development of species, Darwin was also laying the groundwork for its natural future extermination. Humans, in time, would no longer be human. Or at best they would be obsolete.

I am not suggesting that changes in the sense of time and finitude are only caused by the secular transformation of life and death. I am arguing that different eschatological frames and conceptions of the afterlife (or lack thereof) have an effect on lived time and that the afterlife is experienced and used also as a temporal strategy. Take, for example, changes in Christian afterlife doctrines in relation to the medieval invention of purgatory. While conceptions and doctrines of the afterlife were based on ideas of justice and reward or punishment, of good and evil, they also applied varying pressures on the *end* of life (Aries 1975; Binski 1996; Le Goff 1984). Aries narrates a change in Europe from collective to individual death. He suggests that there was in early medieval Europe a perhaps inherited sense that death was a "collective destiny." Everyone ended up there, and the only difference was that if you had entrusted your body to the church then at the Second Coming (i.e., at the end of time), you'd be awakened for eternal life in the heavenly Jerusalem. As he writes: "There was no place for individual responsibility, for the counting of good and bad deeds." A number of ideas and material practices (iconography, deathbed rituals) transform the time of death into the time of judgment, and in particular of individual judgment: the weighing of the souls, inspired by the book of Matthew, and then the *liber vitae,* which Aries calls "the formidable census of the universe" wherein good and bad deeds are chalked up, and finally the joining of the traditional deathbed ceremony to this moment of judgment (1975, 32–33). For Aries, this was crucial in making of death a mirror in which the self was revealed, the *speculum mortis.* But in passing he also pays brief attention to the transformation of destiny, the notion he started out with. He remarks that in this movement, the eschatological time between death and the end of the world was eliminated. Your destiny would be made right there and then (1975, 46).

It is in the harshness of this model that, according to Le Goff, purgatory was invented. Before purgatory, the Last Judgment took place at the time of death, bringing with it a sense of terror, which in turn was used as a means toward self-improvement—the moral force of a retributive model of the afterlife. The reaction against this "eschatological fanaticism" (Le Goff 1984, 231) led to the invention of purgatory as a middle time, a time of waiting, wherein the dead became the responsibility of the living, who could pray and do good deeds for them. While this intercessionary strategy might not have been vital when it came to the few saints and the few more damned, for regular people in a state of venial or forgivable sin it came as a great relief. They could now transition to heaven from purgatory. As Le Goff argues, the invention of purgatory resulted from a need for hope and justice. Such a hope was created by altering the temporality of lived life through a manipulation of eschatological temporality. The temporal (and hence moral) pressure of death on life was altered by expanding eschatological time. The Christian afterlife, once established, creates temporal pressures, and so at various times various new models were created as pressure valves.

Though it makes no reference to finitude in this sense nor to the afterlife, I read the famous work by Robert Bellah and his colleagues, *Habits of the Heart* (1985), as an example of the effect of temporal pressures at the nexus of individualism, the American dream of success, and the midlife crisis, "especially for middle-class American men"—a group that constitutes the great bulk of the immortalist movement. The frustration of a sense of progress in individual careers at around the time of "midlife"—itself a secular temporal designation—"marks the 'end of the dream' of a utilitarian self" (Bellah et al. 1985, 69). "As these dreams die, the possibility fades of a self that can use work and its rewards to provide the matrix of its own transcendent identity." As a result, the American notion of "making it" in life "loses its meaning" (69). Echoing the idea of missing a life's goal, the psychologist Robert Kastenbaum, who has analyzed the relationship between death and meaning in personal lives, writes: "How we project (or fail to project) purpose and fulfillment into the future will be a major influence on how we construct both life and death" (1997, 378). Similarly, the social philosopher Norbert Elias ([1985] 2001, 62) suggests that the valuation of the goodness or badness of death "depends on how far the dying person feels that life has been fulfilled and meaningful—or unfulfilled and meaningless. . . . [W]e can perhaps assume that dying becomes easier for people who feel they have missed their life's goal, and especially hard for those who, however fulfilled their life may have been, feel that the manner of their dying is itself meaningless."

The temporality of modern living has been stuck between contradictory forces, between two futures that have opposing effects on human life: an endless, open future into which life in general moves, and an ultimately closed future that pushes back against life. Again, these pressures vary for different people depending on the historical precarities and survival conditions they face. For those who have been politically subjected as a group to conditions of premature death (Ruth Gilmore's [2006] definition of racism)—that is, for those on whom the future is foreclosed—the utopian imaginary of open progress may not be as convincing to begin with (also see Sharma 2015).

The point here is that especially for those who have claimed the future as their utopian possibility, the secular future may feel not only like an opening on the timeline ahead (progress) but also as a force (finitude, death) coming at them, acting against them. Because in the context of radical finitude there is nothing after death, nothing of value at any rate: the future *qua* death appears as a kind of pressure on life, and the *liber vitae* will chart not the good and bad deeds but achievements and failures. In the context of considerably extended average life expectancy in industrialized countries, Zygmunt Bauman (1992) sums up this tension in the temporality of finitude as a paradox: we live longer but feel short of time. In short, different eschatological frames, based on ideas about death, endings, and continuity, including secular ones, affect how time, as a kind of pressure at the end of life, is felt and lived. It is not that humans once felt infinite and immortal but suddenly came to feel engulfed by mortality; rather, the awareness of mortality and its consequences was transformed by the change in the conditions of knowledge and experience.

BIOTECHNO-PURGATORIO AND THE CHRONICALLY INCOMPLETE LIFETIME

Work on modern, medical temporalities of dying and biopolitics (Kaufman 2005; Lock 2002; Lavi 2009) generally assumes the frame of finitude, finding little analytic or ethnographic room for other (equally secular) temporalities. More recent work on biotechnology has focused on technologies such as resurrection via nuclear transfer, cloning, and freezing that challenge the ineluctable birth-to-death trajectory of finitude in cellular life, opening up biology to suspensions of life and deferrals of death (Battaglia 2005; Heatherington 2012; Landecker 2007; Romain 2010; Roosth 2014). But what kind of continuity is suspension? What are its effects? What happens when one kind of time is stopped (biology) but another continues?

I want to return briefly to Schopenhauer's striking precinematic image of the man falling downhill, where the sense of speed and the feeling of meaninglessness arise specifically in contrast to the very long brackets of "nonexistence" that envelop him: the man stumbling downhill to his fall is a man whose lifetime abides no after-lifetime. Similarly, the temporal dissonance between the variable but finite internal biological time, on the one hand, and the even rate and vast depth of external time, on the other, is reflected in some of the ideas and the discourses of biology. In chapter 1, I briefly pointed to the ways in which work on life duration and biology was shaping ideas about the relationship between the biological processes of the body and the duration of a life, how finitude as cultural background and immortality as biological project were constituting each other. This interplay was strikingly captured in 1931 by the eugenicist and Nobel Prize winner Alexis Carrel in an article titled "Physiological Time."

In experiments with chicken heart tissue, Carrel had produced cultures in vitro that remained alive and dividing far past the life span of a regular chicken, which seemed like time enough to consider talking about the category "permanent life"—a concept he used in his first publication of the experiments in 1912. The use of the symbolically laden organ, the heart, as source of tissue and the fact the tissue was said to be pulsating gave the experiments the sort of cultural weight Carrel seemed always to pursue.

Despite the fame surrounding the "immortal cells," his research on immortality alongside the famed eugenicist aviator Charles Lindbergh, his gravitation to the esoteric, and his pronouncements on the possibility of suspended life and people preserved in storage, Carrel himself seems to have held a sober view of biological aging and mortality. The "process of aging remains irreversible," he would write. Ironically, for someone who is best known for his ideas on immortality, Carrel was more inclined to think the end is already contained in the beginning. "The process of aging," he wrote, "starts simultaneously with embryonic life. It is expressed by irreversible changes progressing during the entire span of our existence. The decrease in the rate of growth during infancy and youth, the occurrence of puberty and menopause, the lowering of basal metabolism and the modifications of the skin and hair, etc., appear as the stamp of time on the organism" (1931, 619). The Darwinian phrase imports evolutionary time into the body itself, into the developmental process. A few years later, Carrel would come to think, following another eugenicist, Raymond Pearl, that the size and agility of the body determined much about the length of life. In *Man the Unknown,* Carrel concluded that although "man" would always thirst after immortality, "he is bound by certain laws of his organic constitution. . . . Never will he vanquish death" ([1935] 1961, 142).

Carrel's sense of duration was dependent on a distinction between what he called "physiological time"—the time of the body or internal time, the time of decay and death—and "physical time": "Physiological time is part of the body while physical time is foreign to it" (1931, 620). Unlike the even uniformity and abstraction of physical time, physiological time is uneven and personal: "Our duration does not flow at an even rate." Like Pearl, Carrel saw the variability of physiological duration as a function of metabolism—and claimed equally that the rate of aging depends on size and metabolic activity. But he posited a telling relationship between the two sorts of time:

> Let us suppose two trains starting with the same speed and running on parallel tracks. The first train represents physical time and moves at a constant speed. The second train, on which we travel, represents physiological time and moves at a decreasing speed. At the beginning, the first train remains immobile, because we run as fast as it does. Later, as we advance less rapidly, its speed increases. Finally, when in maturity and old age we slow down, the train symbolizing physical time acquires great velocity and flies away. (1931, 621)

The bifurcation of temporality in this image of physical time speeding away from us and physiological time coming to a final, insignificant Schopenhauerian ending is exemplary of modern sensibilities in its juxtaposition of the mechanical and the biological and in its breaking up of the unity of time. This was the time of Einstein's relativity (see Canales 2015) and—less noted!—new thought experiments regarding "asymmetric aging," an analytic fiction critiqued by Bergson ([1922] 1965) and launched by a French physicist who in a philosophy convention in Bologna in 1911 argued for the first time that a space traveler would return to Earth younger than his homebound twin brother. But in an age in which time was no longer unified, these oppositions of duration versus time were reproducing an already-existing secular temporality of finitude—one marked, as I have been arguing, by the discontinuity between bodily or human time (physiological time) and the world's time (physical or chronological time), which was foreign to it. Thus temporality in the secular is conditioned not just by the finality of death and our Heideggerian leaning into it but also by the fear that we will end while the universe goes on and on. Without us.

The speeding up of time, then, is not only an effect of timekeeping devices, infrastructures of movement, and modes of prediction, as has most frequently been argued (D. Harvey 1989; for an exception see May and Thrift 2001); rather, as an effect, the "foreshortening of the future" should be seen within the cultural frame of a secular eschatology in which individual endings and the continuity of the world are in constant tension, in which the time of the body and the time of the world are constantly being bifurcated and yoked together in ways that modulate the temporality of existence.[7]

Alongside the linear, infinite, and open timeline of modernity, in which bodies move always away from an irreversible past toward a future, the secular also produced a temporality of finitude, of hard endings in which the time of bodies became confined in all respects to themselves alone. One way to characterize it within a secular context is to think of it as a *life without an afterlife*—and I think as Foucault was well aware, though he never engaged the secular as an analytic, that it is this unit, this body, this *afterlife-less life,* that is the object of the operations of biopolitical power. With the elimination of the afterlife and of transcendent, eternal, or recurring time frames, secular temporality ended up with the ever-presence of hard endings, in which life ends with the body and without the soul, and, what's more, even civilizations and the Earth itself will end too. These eschatologies, themselves immanent, are interruptions in the homogeneous unfolding of an infinite and linear timeline, pockmarking progress with interminable and inevitable endings, personal and collective. The anxieties of meaningless civilizations and useless progress are part of the anxieties of finitude where lifetimes and after-lifetimes, biological and physical time, are discontinuous.

While I argue that this eschatological gap, the play of finitude and infinitude embodied by the cryopreserved figure as a finite body in indefinite time, is in fact constitutive of secular temporality, I am not reifying secular time or time as such. Temporalities are widely acknowledged to be multiple and heterogeneous (May and Thrift 2001; also see Adam 1995; K. Davies 1994). Despite these complexities, I maintain that there are specifically secular premises regarding hard endings that make time and death in the secular scientific setting be experienced and made sense of in particular ways, as pressures, disjunctures, discontinuities, and so on. Suspension here functions as a kind of purgatorial strategy, although I do not equate it with purgatory as a Christian doctrine and strategy, for even if there is a sense of waiting and future redemption implicated in suspension, it is not in relation to God, sin, or heaven, nor is it a matter of intercession.

With that caveat, let me return to the biotechno-purgatorial space of Alcor and its conference room, lined with pictures of cryopreserved members. Underneath each member's photograph there is, as I mentioned, a date of birth and a date of deanimation. But there is also space left at the bottom, the space of optimism, for one to inscribe a third date in the future, the date of *re*animation.[8] The movement past death, or deanimation, into that open, empty time span, linking the finiteness of a lifetime to the continuity of secular cosmological time, makes cryonics a chronobiopolitical practice (Freeman 2007) that materially bridges what I called the eschatological gap. That is, the cryopreserved figure, as well as architectural projects like Timeship, become the material index of a chronologically indefinite bios, an open future relinked to the body whose time was supposed to have been up. "Timeship is a vehicle to carry its passengers to the future," write the project's organizers, and "will eventually make the distant future accessible to its patients" (S. Valentine 2009, 44–45).

The imaginaries of that distant future are an amalgam of daily life and hyped-up science fiction. Like many in this community, Richard had been a science and science fiction fan from an early age. He dreamed of becoming an astronaut and going to outer space. "I was born when the Apollo went up," Richard told me. "The Apollo went up and I came out. I joke that that was my birth cry." His dreams of navigating the cosmos were dashed by a torn ACL. It was, he says, largely in response to this that he became interested in cryonics. It could in some sense give him a second opportunity in the distant future, turning the dashed American dream into a dream placed, to use his own words, on pause. But cryonics became both a scientific end in itself—humans have been trying to achieve immortality from the very beginning—and an opportunity for much more than his lifetime could contain. It is the latter that interests me here, as I heard versions of it from many immortalists who asserted that their motive behind seeking radically long life is not death avoidance in a simplistic manner. It is neither redemption of a dream nor a neutral form of bodily survival that is the dominant disposition here, but the desire to participate in progress, to participate in the secular after-lifetime; this is what turns the prospect of death into a Weberian sense of chronic incompletion in the present, the impossibility of being both finite and fulfilled (see C. Taylor [2007] on flourishing and disenchantment).

That difference between personal end and universal continuity kept reemerging in discussions with cryonicists and immortalists in general. Hear how Richard, again, explains his own motivations: "It's not that *death* is the driving factor as such, but there are things I want to do. My interest is in exploring, in understanding, learning. I want to see the universe and I can't do that in seventy years here."

It's also a *very* common refrain to hear immortalists say there is just not enough time to do all they want to do in this fascinating universe of ours, that seventy years isn't enough to walk on the moon, visit Andromeda, learn twenty languages, and master quantum physics.

Opening up the lens even wider, Richard lamented: My "ceas[ing] to exist . . . is a totally neutral state for the universe. Our galaxy is on a collision course with the Andromeda galaxy, and in roughly 300 to 500 million years they will collide. So there are two fates for the Earth, either it will be flung off or it will be recycled. . . . I will die, the Earth will die, the universe is going to go on and on."

That difference, then, the eschatological gap between the lifetime and the after-lifetime, appears as the very tension at the heart of modern death, one between the body in time (where cosmic time goes on independently) and time in the body (in which finitude is measured through the time of the body, its lifetime). If, as I argued in chapter 1, science and biology shifted finitude from a metaphysical proposition to a physiological condition and underlined the separation of the time of physics from the time of the body, cryonics and the cryopreserved figure produce a reunification of the two times. The cryopreserved figure presents a chronobiopolitical counterstrategy to the limitations of secular lifetimes by creating a kind of after-lifetime, allowing for that infinite cosmic future to be embodied and graspable, affecting the "meaning horizons" (Munn 1992) of immortalist presents.

But the eschatological gap also relates to a particular way in which beginnings and endings have been collapsed in national narratives in the United States. The imagined futures forming around the possibility of turning death into life, of turning endings into beginnings, may be new, technologically enabled reactions to features of finitude, but they are wedded to old, white American ideals and rhetorics of pioneering, frontierism, and ahistoricism (or atemporality)—and outer space, as the final frontier, is the embodiment of that fantasy of limitless expansion. Whereas the space venture is often justified in utilitarian terms—resources!—it is obviously more than that. David Valentine (2016) writes eloquently about many of its nonutilitarian aspects. What I want to briefly comment on is the ways in which space travel and the fantasy of space itself require temporal warps and woofs. Outer space is as much a spatial as a temporal—and as we'll see later a psychological—fantasy or concept, one with thick roots in post-European U.S. history. "In the beginning," John Locke wrote, "all the world was America." This ethos of renewal—via Native genocide and African slavery, that is, with the negation of racialized futures—which conflates beginnings and endings keeps getting played out in retellings of U.S. history as much as in cyclical narratives of white American lives, with their second and third and fourth chances. In part this is why Native claims on American history feel like a crisis to non–Native Americans, as they upend the historyless origin story and thus the possibility of constant and costless renewal. While *Star Trek* calls space the final frontier, in its colonial rhetoric and deliberate imaginary of a land without history or even time, the American frontier was the first "outer" space for the white pioneers.

The American frontier as conceived (while being eulogized) by Frederick Jackson Turner ([1893] 1920) was as much a concept in geographic space as one in temporal space. "American social development has been continually beginning over again on the frontier," wrote Turner, giving expression to a sense of "new" American life outside history, not subject to the passage of time yet wedded to perpetuity, to a planarian penchant for constant renewal. "This perennial rebirth, this fluidity of American life, this expansion westward with its new opportunities, its continuous touch with the simplicity of primitive society, furnish the forces dominating American character." His intellectual heir, James Truslow Adams, forty years later popularized the term "American dream" by building a theory of the dream based on Turner's theory of the frontier, but focusing now fully on temporality. In his seminal book *The Epic of America* (1931), Adams described the American link between time and space: "There is never any past on the frontier, only a future, and one of the most radical changes which frontier mentality undergoes is precisely this complete shift of orientation in time. To dream solely of the future instead of the past" ([1931] 2012, 138). By focusing on "the future," on time instead of space, Adams turned the American frontier into the American dream, inscribing Turner's "perennial rebirth" onto the "American character." The American dream, then, sits on this same foundation of colonial erasure, collapsing endings and beginnings to produce dehistoricized futures.

No surprise, then, that the frontier continues to be an active trope, not just for groups that project space colonization as the best option for the future but more generally for immortalist and futurist discourses, in all their whiteness. The pioneering ideal becomes explicit in new experiments such as the Peter Thiel–backed Seasteading Institute, founded in 2008 by a transhumanist and immortalist "to establish permanent, autonomous ocean communities to enable experimentation and innovation with diverse social, political, and legal systems."[9] At a public conference I attended, organized mainly by the institute, one of the founders spoke explicitly about establishing a new version of the frontier beyond territorial waters: "We need a frontier and we don't have one. Space *is* a frontier, but it is not realistic yet. . . . Whenever we've had a frontier, humanity has flourished." The lesson and the commitment coincide with a long narrative thread in the story told by white America about itself: the possibility of redemption may be realized when pasts are erased and new lives wrought of new space and new time.

Richard's engagement with cryonics as a way of redeeming pasts through futures, of inventing new lives in this perceived eschatological gap, is echoed by a large number of other immortalist life stories. Adam, who has had "half a dozen different professional directions" from aerospace engineering to advertising and environmental consulting and independent dot-com entrepreneur, described how his attitudes toward employment, life trajectories, and death opened up after he turned to cryonics:

> There was actually a time in the late eighties when my mantra in life was "everything now," and now I feel like "everything whenever I get around to it." So I feel like—I'm almost, I'll be fifty this year—and I wanna have a lot of . . . I wanna have a very good nice life, and if it takes me five, ten, twenty, thirty years to get it, that's fine because I'll still have a long future ahead of me. I wanna do things like walk on the moon! I wanna do all these things and I feel like now I have the time to do it. . . . I have a sense, since I signed up with cryonics, like that's a real possibility.

Or take Randy, the gay immortalist who has been active on many transhumanist fronts, reading, speaking to the press, organizing meetings. "For twenty-nine years," he said, "I sold calculators and then lightbulbs to conservative couples coming in to decorate their homes. I'm only beginning to live now. I'm vlogging. I'm active. I want to organize more and protest more. I want to raise some hell." Randy was seventy-one years old at the time.

For him, as for many other cryonicists, life extension has already become time extension, shifting the pressure of the eschatological horizon, recalibrating secularity's life without an afterlife, *through* the manipulation of matter itself—by freezing or suspending it, turning the cryopreserved patient into a material semiotic agent (Haraway 1992) and cryonics into a temporalizing practice.

Cryonicists can thus relate to the suspended figure and the time it embodies on a symbolic and affective level, imagining through it a future self in future time. Steve Bridge ([2006] 2014), a longtime activist and former president of Alcor, echoes the words of many cryonicists I spoke to when he wrote on the Alcor blog: "Even though I had read science fiction for many years, this was the first time that I actually envisioned myself as part of the future."

In the production of the new social subject that is the secular immortal, certain features stand out. In other chapters I will explore "techniques of the self" in relation to algorithmic rationality and to the cosmos; here I will focus on the temporal aspects of becoming a "future self." How does one prepare ethically and physically for the values of life extension and the promises of the future? One answer comes through a series of typical recommended actions from an article titled, somewhat tongue-in-cheek, "Things to Do with Your Body While You Wait for Immortality":

> For those who would live longer, it's wise to face challenges every day, and put oneself in uncomfortable situations that require growth—physically, mentally, and emotionally. Part of the fun of being a transhumanist can involve measuring your progress across an increasing number of variables and seeing how improvement in one area can create unexpected and even synergistic progress in other areas. My advice would be to establish baselines and sophisticated objectives for strength, speed, endurance, stamina, balance, coordination, agility, skill (perhaps in a growing number of activities), days free of illness or injury, as well as body weight and, of course, the number of years spent being alive.[10]

Emotional, mental, and physical strength are to be charted informatically as measurable goals, optimization of the material body measured as an outline of "progress" in the "growth" of the self. The author then suggests a number of actual measures having to do with a variety of tests (e.g., body fat analysis), nutritional intakes (the Supershake), and daily innovations in exercise (the treadmill desk, so that you can exercise while working). For each there are baseline measurements and therefore improvement goals to work toward. One is left to assume that committing oneself to the goal of immortality, working on the body with that in mind, will in itself effect the making of an immortalist self. That is, your life can be envisioned as a waiting for immortality rather than a struggle against premature death.

The possibility of producing alternative ways of being a self in the future is, as Kevin, another transhumanist life-extension researcher, told me, "a big motivation." Future selves would include people whose current capacities would be enhanced or optimized, as well as digital selves, operating simultaneously on a number of different substrates or networks, not confined to a particular material substrate and certainly not to naturally given limits of our current biological bodies. In a few cases the "technology" is there already, for example, the common use of enhancement supplements, such as modafinil to enhance memory and cognitive function. For the most part, the technology is not there to achieve the most interesting of the imagined future selves. But even when the technology is not there, the excitement is, because based on certain precepts about information and mind, as well as about space travel, their potential is *felt* to be quite real. Organizations such as the Singularity Institute and Terasem were set up precisely to deal with and explore the implications, explicitly stating their goals to be the preparation of guidelines for changes that are coming in the future, including preparing for transformations of our present selves and our value systems ("Joy, Growth and Choice," according to one book [Goertzel 2010]) and finding ways to provide "concrete guidance" to the issues we will face when our *selves* will have changed. The World Transhumanist Association explicitly idealizes "future beings," those "whose basic capacities so radically exceed those of present humans as to be no longer unambiguously human."[11]

In effect, a "future being" in the immortalist sense redefines the future just as it redefines the being. A future being is not just *any* being some distance away on the linear timeline. It is a being with certain characteristics, first and foremost the characteristic of no longer being unambiguously human. In the more detailed transhumanist and immortalist imagination, the future being has the ability to cross species at will, building fins and gills or wings, say, into its molecular structure; it has the ability to "mind meld," that is, unite its subjectivity with others'; it has the ability to travel at the speed of light and colonize space; it has superintelligent capabilities, which by definition are unimaginable by our sub-super abilities. Similarly, then, "the future" is not just any point ahead of us on the timeline, but a very specific point. In the abstract it is that point at which things change enough to transform the human world and the limited way the world is known by humans; in the concrete, it is marked by the appearance of specific technologies, mentioned above. Thus, the "future being" is not a being in the future but a specific sort of being in the present, perched toward *that* future, anticipating *that* future, building *that* future in which biologically inherited human limits will have been overcome specifically through the combination of digital and biotechnologies.

Moving from anticipation to desire, the WTA adds that "many transhumanists wish to follow life paths which would require growing into posthuman persons." It goes on to say, "At present, there is no manner by which any human can become a posthuman. This is the primary reason for the strong interest in life extension and cryonics among transhumanists. Those of us who live long enough to witness currently foreseeable technologies come to fruition may get the chance to become posthuman." And from desire to prescription: "Although there are no guarantees of success, there are some things that can be done." It then lists five main actions a transhumanist might take:

1. Live healthily and avoid unnecessary risks;

2. Sign up for cryonics;

3. Keep abreast of current research and save some money so that you can afford future life-extension treatments when they become available;

4. Support the development of transhuman technologies through donations, advocacy, investment, or choosing a career in the field; work to make access more universal and to make the world safer from existential risks; . . .

5. Join others to help promote transhumanism.

The constant assessment and management of risk and the optimization of somatic health have in themselves become the immortalist "form of life." The self-discipline and "self-techniques" (N. Rose 2003) in the immortalist community extend to their extreme ends the more general and mainstream strategies of the biopolitical citizen. Immortalists engage heavily in personal research on life extension, exchanging notes online; they consult longevity specialists, or "life-extension doctors," and read books on how to extend health; they have developed intense supplement[12] and dietary regimens and frequently take "enhancement" drugs. Those who take their supplements seriously often take upward of twenty different supplements a day, amounting to forty, fifty, or a hundred pills. Not, of course, all taken at the same time. Supplements need to be "timed correctly," as one woman told me during breakfast at an AI conference. "Some need to be taken in the morning on an empty stomach, some should be taken at night, et cetera. I buy an eight-week supply at a time, six times a year." There are also online discussions regarding their benefits as well as notes on the optimal ways of ingesting the supplements. These are all based on observations made by users themselves who report on the effects of the quantity, combination, and timing of the supplement ingestions.

So just as with the cryopreserved body given to future science, here too immortalists themselves are the subjects of their own experiments—experiments aimed at transcending not just the self but humanity, its limitations, and its histories. And yet, because of secular finitude, the white American ideology of perennial rebirth, what Bellah called the entitlement to a second chance constitutive of the American dream, draws tension with the "midlife crises" and the "fading of the dream": rebirth will eventually face failure, the dream will face history. In the mode of suspension, transhumanism and immortalism are channeling claims on the future by vaulting over any reckoning with the past.

REMAKING TIME AFTER LIFE

The problem of modern death has been glossed as one of simple finitude, of a final death without the possibility of extensions, returns, or personal afterlives. This is all there is, and after your death it all ends: there is no more. The adult modern nonreligious reaction must simply be to accept this life-without-an-afterlife, even though it makes the future feel like it is "looming simply as a blank eternity of nonexistence" (Scheffler 2013, 29). I have argued that the problem of death (or its terror, as Aries [1975] and Unamuno [1972] have put it) in a secular setting was not simply one of nonexistence, that you died and there was nothing more; to the contrary, the problem was precisely that there was so much more, that everything else went on and on getting better, but it did so without you; indeed, progress-without-you was rending cosmic destiny from individual destiny altogether, turning life into a project that, as Weber (1946b) recognized early on, is bound to feel chronically incomplete.

Compared to temporal orders that align personal, social, and cosmic processes (Adam 2004; Luciano 2007), often through cycles of life and death, secular temporality is marked by this Weberian *breach* between personal horizons and cosmic ones. It is also this very bifurcation of time into the temporality of biology and the temporality of the universe that allows for the increasingly common way of speaking about biological and chronological time as distinct processes, moving toward distinct ends. This conceptual separation of time is favored by today's researchers working on radical life extension; it has also become part of the image of the self. For example, Ray Kurzweil has consistently claimed a discrepancy between his biological age and his chronological one. His coauthor, the physician Terry Grossman, also spoke to me of the possibility of affecting biological age through therapies and disembedding it from chronological time. Elizabeth Parrish, who volunteered as the experimental subject of her own company's product (see chapter 2), claimed to me that she felt her body younger and has continued to say publicly that biomedical assessments show her telomeres have not shortened. Today, gerontologists and anti-aging researchers frequently invoke the unyoking of biological and cosmological time. In this, Aubrey de Grey's notion of an escape velocity, described in an interview, is not an anomaly: "If you're sixty and you get a therapy that makes you biologically thirty, then, yes, you will be biologically sixty again by the time you're chronologically ninety. Sure enough, the therapies won't really work anymore, because the damage that has made you biologically sixty again is, by definition, the more difficult damage, the damage that the therapies don't repair. But this is thirty years on, and that's an insanely long time in any technology, including medical technology."[13]

My grappling with the cryopreserved figure is meant to illuminate some of the broader technoscientific, affective, and historical textures of suspension in which the biological body itself has become the ultimate timepiece, ticking with the sound of its own end. It is just this sort of lifetime, trapped by metrics of the body, that is a precondition of the pause-and-play powers of many current cryotechnological and de-extinction projects. Among them one might cite the Lazarus Project, which has become well known for "reviving" the genome of an extinct species of frog (Kirksey 2017); the cloning of the Sardinian Mouflon sheep (Heatherington 2012); the Frozen Ark, otherwise known as the doomsday vault, where DNA samples of everything are being stored for future reanimation; the "resurrection biology" of Silicon Valley's Long Now Foundation (for a critique see the Extinction Studies project in Australia, especially D. Rose and van Dooren 2011); and even "the salvage genetics of the Human Genome Diversity Project," in which scientists in the International Biological Program collected and preserved samples of unpolluted nature, including blood from "primitive humans," populations that were deemed to be both unchanged and endangered (Radin 2013; Kowal, Radin, and Reardon 2013). The politics of suspension, then, is not linked to the technology or politics of freezing alone; it is part of the temporal underpinnings of biopolitics, whose probabilistic measurements have depended on the finitude of a lifetime, the statistical certainty of the final horizon called death, through which measures such as life expectancy, death rates, and survival rates become calculable and which, as a result, create particular lifetime conditions such as living in diagnosis, being a presymptomatic patient, or being terminal (Jain 2013; Farman 2017b)—conditions that must in turn be overcome by modern medicine, for otherwise there would be no viable notion of progress in which the individual might participate.

Yet the finality of death as the very condition of finitude is also predicated on the impossibility of other temporal forms, such that the secular death, for example, may not be overcome through returns, survivals, or displacements to eternity and infinity and other transcendental time frames. If suspension—or the cryopolitical bypassing of death in the translation of life into time—is a key figure here, it is because finitude, as Foucault pointed out, must "also promise . . . that very infinity it refuses," such that "man's finitude is outlined in the paradoxical form of the endless" ([1970] 1994, 314).

Like Le Goff's account of purgatory as a kind of moral and temporal pressure valve, secularity's *afterlife-less life* can be said to have changed the pressure valve of life and death, by creating its own "eschatological fanaticism"—a final end contrasted with an ongoing future, a lifetime cut off from an after-lifetime. Immortalist narratives are not just revolts against death as such, nor do they simply entail, in the words of E. P. Thompson (1967), the "flouting of respectable time-values" like previous revolts against modern time such as the beatnik or bohemian or anarchist versions. The technologically driven movements to transcend biology and the chronic imperfections of the present are a reaction against secular finitude widely considered, where temporal crises get framed as biological limitations, where biological decay and social experiences of incompleteness fold into each other, and where the infinite progress of knowledge leaves the finite individual behind.

Suspension is the secular space in which the promise of the indefinite future can assume material embodiment in the figure of the cryopreserved patient and under a scientific canopy negate the absurdities of death. The most common visual symbol of cryonics and immortality is the gleaming dewars that, like monuments to the future, contain the bodies of the suspended—and thus embody their future personhood. The politics of suspension involves the constant and ongoing translation of life into time, or rather of chronically incomplete lifetimes into promise-bearing forms of future life. These include the brain-dead and the comatose but also seeds frozen and preserved in the doomsday vault in order to prevent, well, doomsday; wooly mammoths and passenger pigeons on the brink of return from extinction; and the cryopreserved figure stored in cryostats in places like Detroit or Phoenix. That the times are out of joint is an old affair; that time should be running so fast and so out of control that we feel we cannot catch up with it is a modern affect; the contemporary sense, embodied by suspension, is that since we can no longer catch up with it, we can at least wait it out. As Déborah Danowski and Eduardo Viveiros de Castro note, today we live in a world of "too little too late" (2017, 8)—time has bypassed us, because unavoidable catastrophe is already on its way. Suspension characterizes the very idea of human destiny as it appears today, hanging between a future of technologically induced extinction and a future of technological liberation from the confines of human finitude. We are suspended in this space of waiting. Wondering what will unfold. What ending will come at us from the future.

4

Deanimation

MATTER, MATERIALISM, AND PERSONHOOD BEYOND DEATH

"The state of 'death' is nothing more than a physician admitting he is unable to restore a person to life, that this decision as to when a person dies will vary from doctor to doctor, place to place and most important from time to time," wrote veteran cryonicist Mike Darwin in 1983. "A person suffering a cardiac arrest in a hospital cafeteria will have a radically different prognosis than the same person with the same condition on the crowded streets of Bombay. . . . We understand that as progress in our understanding of physiology and medicine is made and translated into improved medical technology we are able to recover more and more people from so-called 'death.' In short, we believe that today's physicians are likely mistaken when they apply the label 'dead' to cryonics patients. We believe that a more accurate statement of fact is that cryonics patients are neither dead nor alive, but in a third state: that of uncertain waiting" (1983b).

Things can get really dizzying when you understand death differently, when you think of it not as a final, irreversible moment but as a series of potentially reversible events. Can the cryopreserved person keep any money in the bank in case of future reanimation? What happens to life insurance? Is the cryopreserved body subject to the regulations of the state funeral board? What does the medical examiner enter on the death certificate? If you die in California and must be transported to Arizona for cryopreservation, what rules apply? What if you have Alzheimer's and want to be cryopreserved before your brain is entirely laid to waste?

If the person is indeed reanimated, will he or she be the *same* person? What criteria might determine that? What kind of implications does *not dying* have? What attendant anxieties?

These are not just speculative questions. From an ethnographic standpoint they are questions that cryonicists, like many others in the radical life-extension community, live with and work with everyday, questions that affect their selves, their practices, their ideas and conduct. They are also ethical questions, requiring judgments and determinations that have larger implications regarding the changing relationships between the categories of death, personhood, and life.

When I met him two years before his death, the founder of cryonics, Robert Ettinger, was devising ways to orchestrate his own death and suspension in a well-timed but legal manner. He was old, and both his wives had already been cryopreserved; his health was deteriorating, and he wanted to take his own life and be suspended before the onset of harmful mental diseases that would make his survival that much less probable. For cryonicists, an organized, well-timed suspension would increase chances of a proper preservation. Ideally, such a suspension would have to take place before "legal" or "medical" death is declared, which might take a long time to come, in the process of which the brain might accrue excessive damage. However, the cryonics organization cannot suspend a person before his or her legal death, for that would be considered a wrongful death, a homicide, even though for the cryonicist it would be a life-saving technique. Ettinger was struggling to find a doctor in Detroit who would cooperate with his plan and appear in a timely fashion to officially declare his death as quickly as possible so that he could be rushed next door for immediate care by the cryonics team. While Michigan had no law that would charge anyone with having failed to prevent suicide, anyone—doctors, cryonicists, funeral directors—participating in the plan nevertheless would be at least exposed to charges of homicide, though they would be helping, in Ettinger's view, to save his life for the future. That is the cryonics bind.

Cryonics produces its suspended subject in relation to legal, ethical, and medical procedures and categories in a liminal zone that heightens the inherent instability and indeterminacy of the life–death transition.

Tracking this process, I hope also to demonstrate the various ways in which the conditions for understanding, representing, and altering the categories of life, death, and continuity (including things such as the will, organ donation regulations, and the status of the cryopreserved figure) appear to us through the nodes of the immortality regime, in places such as the office of vital statistics. Since notions of personhood and their ends are not explicit and precise ("the ungroundedness of what is meant by 'persons'" [Wolfe 2010, 61]), but embedded in and produced through immortality regimes, immortalist disruptions of the existential normalcies entail both the production of new forms of personhood and the production of an existential crisis within dominant regimes, expressed, as we'll see, through institutional language and legal questions.

A SECULAR IMMORTALITY REGIME

Until 1969, no state in the United States had a statute defining death (Zaner 1988, 2). When the introduction of the ventilator and the possibility of anatomical gifts pushed the issue to the fore, states dealt with the matter piecemeal. So there was a time when one could be dead in Kansas but not in New Hampshire.

The ventilator was initially developed to help with surgery and infectious diseases that prohibited breathing, such as polio, but as the equipment got introduced into more hospitals, moral obligation followed technical routinization (Koenig 1988), and its "heroic" use for general "life support" soon became automatic. It also enabled the development of successful techniques for organ transplantation, a procedure that fit the medical and cultural imperative to "save lives" (Timmermans 1999). But saving lives in this way proved to be controversial. Ventilators and organ transplants launched a set of ethical dilemmas around the problems of when, why, and how to ascertain the finality of death.

One consequence (of saving lives, one might say) was to force a new definition of death—what came to be known as brain death. The federal government sought to create some uniformity and settle the controversies by passing the Uniform Determination of Death Act (UDDA) in 1981. Scientific studies and presidential committees lent the effort a great deal of social authority, and the criteria for brain death determination were codified, alongside the traditional cessation of heartbeat and breathing criteria, in the UDDA, eventually adopted by most states (President's Commission 1981; Lizza 2006). The UDDA enacts the following criteria for the determination of death:

(1) irreversible cessation of circulatory and respiratory functions, or
(2) irreversible cessation of all functions of the entire brain, including the brain stem.[1]

The key word, *irreversible,* is not further defined by the UDDA, and this too has led to ethical dilemmas. At what point does the threshold to irreversibility get crossed?

One problem is that though they were presented as complementary, inclusive, and noncontradictory, the assumptions holding up the two UDDA criteria oppose each other. The somatic or bodily criterion of death (heartbeat, breathing) takes *bare animal life* as the significant measure, an absolute view of being as opposed to not being, regardless of capability. By contrast, the acceptance of brain death is based on the assumption that *consciousness* is the important measure of value in human beings, marking human life as the life of the mind. But consciousness is a notoriously difficult issue to settle, because there is no satisfying account of how it can arise from (or be reducible to) material or biological causes. Yet, the measure of consciousness (its presence or absence) at the end of life is taken through material indicators, such as electric signals in parts of the brain. There is nothing absolute or stable in the scientific criteria, and legal concepts guiding the determination of death have changed considerably (and controversially) over the years (Zaner 1988; Lizza 2006; Merkle 1992). Attempts to clarify the issue by using concepts like "somatic integration," "whole-brain death," and "higher-brain death" have been inconclusive and inconsistent. As techniques for detecting brain activity have become more precise, the controversy has only increased, initiating other debates— for example regarding activity in patients in a persistent vegetative state (PVS) or regarding distinctions between whole or neocortical brain death (Shewmon 2004; Cranford 1988).[2]

To ease worries in organ transplant cases, some jurisdictions have specified a temporal measure to deal with the problem of irreversibility, *just to make sure*: an eight-to-ten-minute period of circulatory arrest is required before death is declared and organs removed (Whetstine et al. 2005). Of course, this only underlines the arbitrariness, ambiguities, and insecurities that accompany the determination of death and its aftermath, particularly in instances that see the introduction of technology into the life–death continuum. As many have already noted, technological interventions have destabilized the boundary between life and death (Kaufman and Morgan 2005) as well as traditional ideas of personhood (Lizza 2006) and of the person as "conscious, reflexive, embodied and ontologically distinct from machines or technological procedures" (Kaufman 2000, 70).

Before these technologies, death appeared more unified, with a number of phenomena, including the cessation of breathing and heartbeat, following each other within a relatively short span (Ewin 2002). Today, biological death has been fragmented into a series of discrete events (Strathern 1992), each one producing its own figure (PVS, comatose, cryopreserved, embryonic stem cell, etc.) and its own controversies over the relation of life, death, personhood, and continuity.

Secular law often turns to religion as a source of value (Feldman 2005)—as in related cases on abortion, stem-cell research, and brain death—but when pushed to the limit, secular law, which characterizes its legal subject nonmaterially as person, nevertheless grounds its secular authority regarding these matters in scientific materialism. The trouble is that these are precisely the sorts of matters that materialism cannot ground, in part because they in turn depend "on values articulated in the broader social milieu" (Lock 2002, 37). Where on the continuum of physiological viability you select your point for death reflects not a natural fact about death but prior ethical, social, and political notions of personhood (Youngner, Arnold, and Shapiro 1999; Crippen and Whetstine 2007; Veatch 2005). What's more, the definition of "person" is not delineated in U.S. law, even though the law assumes the person as its subject and recognizes different kinds of persons as being subject to the law—natural persons, for instance, and unnatural persons, such as corporations (Glenn 2003). There are *criteria* that are supposed to be met in *identifying* personhood (presence of autonomy, rationality, consciousness) or the end of personhood (absence of heartbeat, brain signals) or the presence of life (a certain period of fetal development, a biological feature), but in themselves the concepts are either assumed, (arbitrarily) imposed, or in the case of state regulations, circularly defined, which may be a virtue: "(1) 'Dead body' means a human body or fetus, or a part of a dead human body or fetus, in a condition from which it may reasonably be concluded that death has occurred."[3] In other words, the discussions and the literature on the concept of persons and the criteria used to determine personhood are "mired in intractable dispute in a wide range of cases, including fetuses, newborns, the irreversibly comatose, God, extraterrestrials, and the great apes" (Beauchamp 1999, 319).There is neither social consensus on what a person is (and therefore what criteria to use to see whether it is present or not) nor a final, consistent, authoritative scientific, cognitive, legal, or philosophical position.

Effectively, technology has highlighted the tensions in the secular separation of organism and personhood. It is important to note that "organism" is not, at the point of death, a natural biological object simply because personhood or soul or the afterlife has been removed from it. It itself reemerges as a new social artifice, since its negation now is ascertained by picking a socially authorized but arbitrary point along a continuum of physiological viability, such that the remainder also requires a designation. Judgments regarding the decoupling of organism and personhood are socially embedded in language, institutions, modes of action, practices, rituals, and law. The way in which death, personhood, and corpsehood are defined, accepted, or declared, even by science, is a social process that describes not death itself but the range of conditions under which death may be declared and accepted—or not accepted.

Drawing on Collier and Lakoff's (2005) notion of a "regime of living" and Walter's (2005) concept of "deathwork," I use the notion of an immortality regime to refer to the set of norms, knowledge, practices, rules, authorities, and institutions through which the transition between states of being is evaluated and processed in society. The transition between states—life, death, ghost, cryopreserved figure, avatar, and so forth—is always fraught, for unlike animal death, human death has generally demanded some extra work, some form of judgment about when death has occurred (Wartofsky 1988), when the person has left the world, which is almost always followed by a social and cultural process of *transformation* that changes the *status* of the person—be it to a ghost, a cadaver, an ancestor, corpse, memory, or even to newer categories such as "persistent vegetative state" (Kaufman and Morgan 2005)—and that permits the *acknowledgment* of that change, while *managing* its consequences in terms, for example, of the redistribution of goods, responsibilities, and authority once held by that person (Goody 1962). "In all societies," write Beth Conklin and Lynn Morgan, "the complexities and contradictions in normative ideologies of personhood are heightened during the transitional moments of gestations, birth, and infancy, when personhood is imminent but not assured" (1996, 658). Such transitions mostly include more-complex clusters of states than simple binaries between life and death, and most often include some form of continuity beyond death, in a continuum of animation:deanimation:reanimation.[4] That is the function of an immortality regime in general: to render judgments about such transitions possible and stable.

A secular immortality regime, in our context, provides a necessary measure of what Anthony Giddens (1991) has called "ontological security." In the context of high modernity, in which lives are undetermined and contingent in relation to the future and largely influenced by "impersonal," "disembedded," and nonvisible systems (global finance, bureaucracy, global risks to the environment, etc.), "to be ontologically secure is to possess, on the level of the unconscious and practical consciousness, 'answers' to fundamental existential questions which all human life in some way addresses" (1991, 47). Giddens claims that ontological security may be attained without a secure, provable base for the "answers" as such. "Practical consciousness" and the structures of everyday life "provide modes of orientation which, on the level of practice, 'answer' the questions which could be raised about the frameworks of existence" (37)—so affects like trust and hope, generated by the culture, help anchor the existential dilemmas emotionally, without solving them cognitively.

As I explained, secular law and science have in fact not fared well in trying to define and determine "fundamental" categories such as death, life, and personhood; indeed, many scientists, medical professionals, and philosophers are increasingly agreed that any materially objective determination of death is a fiction (Youngner, Arnold, and Shapiro 1999; Crippen and Whetstine 2007), a problem that has been underlined by new technologies (Nwabueze 2002, 20; Waldby and Mitchell 2006). Political and social institutions, however, must pretend that there is a fact of the matter regarding such categories as death or personhood. Why? In order to bestow a measure of "ontological security" or, in the unusually candid words of the Warnock Commission, to "allay social anxiety." The Warnock Commission in the UK, in charge of assessing the rules about embryo research, emphatically stated that it ended up having to choose not the course that *was right* (which could not be found) but the one that *seemed right to most* and that could at the same time appear justifiable in concrete scientific terms. It chose the fourteen-day rule on embryo research because certain minor features become *visible* on the embryo at that point. Its solution, the committee itself said, was arbitrary. But it is thought to be in the best interest of the state and the public to officially designate a point, to settle the matter, especially when it comes to existential issues such as the "inviolability of life." Despite the arbitrariness, the commission openly said, "some precise decision must be taken in order to allay public anxiety" (Warnock Commission quoted in Franklin and Roberts 2006, 11). Similarly, the dissenting voice in *Roe v. Wade* clearly stated that the "viability" of a fetus is contingent and can be pushed back with the advent of more precise instruments and techniques. The precision—the picking out of a very specific fact to mark a concept—is not ontological but social. These decisions and the procedures that follow from them "normalize" an arbitrary designation by establishing them on scientific authority and then routinizing them through the institutional procedures of an immortality regime. But the original anxiety and the fragility of the whole arrangement is thereby also embedded in the system—in the institutions and authorities committed to implementing the arbitrary designation. When someone or some event then forces a confrontation with the arbitrariness, the immortality regime enters into crisis.

In short, regimes of medico-juridical reason in interaction with liberal publics require certainty, and the zones of indeterminacy constantly destabilize the assertions of such regimes and expose their limits. The remedy has come through medico-legal fictions. A variety of "legal fictions" associated with the human body are "normalized" by lending them authority through scientific commissions and then channeling them through the procedures of a secular regime. So in fact, and contra Giddens, it seems to me that ontological *in*security is built into or is very much one face of secular experience, riddled as it is with anxiety, dread, and other such secular affects.

When I say materialism is appealed to as epistemological bedrock, bearing a privileged relationship to the real, I am not arguing that materialism is the sole constitutive feature of the secular or that biotechnology and medicine are secular because they deal strictly with the material body, leaving the immaterial to religions. Rather, science and materialism are historically intertwined aspects of a constellation of ideas and conditions that make up the secular, part of whose history has been to grapple with what appears as the immaterial. So, closer attention to materialism and its relation to scientific practices is needed in a way that would incorporate both its historically contingent development in grappling with form and matter (Collingwood 1960), or person and body, as well as its conflicts with other aspects of the secular.

A secular immortality regime faces its own specific dilemmas. The conflict between rationalist notions of person based on aspects of consciousness such as reason, subjective feelings, autonomy, and intentionality, on the one hand, and the materialist notion that must find material correlates for socially sanctioned aspects of personhood and its cessation, on the other, takes place in what I have called the dualistic gap, producing a key tension that constitutes secular personhood. Even though the soul and the afterlife are removed from the reckoning in a secular court dispensing reward and punishment, the juridical person is not coterminous with the medical body, nor do the two easily coexist. Indeed, it is the endemic tension between them that can be understood as secular. That is the core of my argument: the secular body describes a particular way in which an unresolved relationship between mind and matter becomes problematized in the nexus of law and science. Secular institutions create bodies in tension, strung between rationalist and materialist conceptions of personhood—a tension embodied by the institutions themselves. Secular immortality regimes try to ease these issues, as in the case of the Warnock Commission. Immortalism, by contrast, makes them more pronounced.

DEANIMATION TIME

As soon as a cryonics member is in a state of critical care, he or she is placed on an emergency watch list. At any given time, Alcor or the Cryonics Institute may have several patients on a watch list. When there is a prognosis of nearing death, the cryonics case starts in earnest, signaling the initiation of a set of procedures that have been put into place over time, procedures that are both experimental and disruptive. The process takes the cryonics patient and the cryonics team through a vast landscape of liminal nodes as they negotiate their way through the immortality regime's established protocols, professional settings and networks, state codes and regulations, all designed to deal with death and the dead.

The body of Patient X33, a white male in his late seventies, first became visible in this medico-legal regime at 11:54 p.m. on April 18, when it was placed on life support at a hospital intensive care unit. The patient's next of kin were notified and, per the instructions on file and on the metal dogtag hanging from the patient's chest, so was the cryonics facility of which he is a member. The cryonics facility immediately verified the patient's life insurance status, which was confirmed to be active. Because of the man's ongoing condition, the facility already had a patient alert in place, but now a cryonics team is placed on standby outside the hospital, ready to move in with their special equipment and procedures in order to save his life—but only after the proper authorities pronounce his death.

The cryonics team, in touch with the hospital staff, monitored the patient's status wondering how intensively they should mobilize. Emergency response requires twenty-four-hour vigilance in order to be able to intervene immediately upon pronouncement, whenever it might occur. But death never comes in a controlled, predictable manner; the patient's conditions go up and down, infections stop and start again, medications can have unpredictable effects. If the response team is made up of volunteers, the standby is taxing on personal lives; if it is a professional team such as that set up by the company Suspended Animation, it can become expensive very quickly. In both cases, it is a tense and stressful period of time.

Two days after first admission, the patient had stabilized. Though he was still unconscious, the infection had been stopped and there was no heart attack. The response team and the cryonics facility went into stand-down.

The hospital is a complicated setting for cryonics. In the first two to three decades of cryonics, hospitals were generally uncooperative and obstructionist, according to veteran cryonicists. Hospitals would receive cryonics patients, with clear cryonics paperwork, but would not alert the cryonics facility; or they would not let the standby team in; or they would block the necessary paperwork; or refuse to help with the necessary procedures and so on. Things slowly started to change in the 1990s. In part, this had to do with patient advocacy, the rise of patient rights, and the acceptance of patient choice as the proper ethical standard; in part, it had to do with the growing knowledge within cryonics facilities about how to manage their relationship to the hospitals and present their case to the key personnel.

"But the hospital is not one uniform entity," a longtime, active cryonicist told me:

All of these things are renegotiated multiple times in the course of a standby. Because you have three shifts per day. And you have a nurse, a doctor, a charge nurse, hospital admins for different parts of the hospital. By the time you consider all the people who have a measure of authority you have a dozen fiefdoms that could potentially ruin your day. Let's say the head nurse, the charge nurse, and the doctor are all comfortable with this and then the case starts and a hospital administrator gets wind of it and decides for liability reasons to throw a hissy fit. So the hospital is not a single homogeneous entity. You never know who's going to decide to throw a hissy fit. The procedure is not an unobtrusive event. For several reasons. One is that there's a lot going on. You have a recently dead person. You are doing a ton of procedures, you have a tub full of ice water. You have an air compressor, ventilators running. It's very loud . . . and everyone wants to come and watch so all of a sudden you've got every nurse on the floor coming and sticking their head in. We have zero authority in that situation. And the patient is "dead." Legally patients have a right to privacy, but when a patient is "dead" there is no right to privacy, we have no mechanism for denying them access.

In a hospice setting the conditions are easier. The administration is leaner; the ideology of choice is paramount and so the patient's wishes are given priority; and the situation is geared toward guiding a patient to his or her death. Since they accept a patient as a dying customer, all the details are dealt with well beforehand. Hospices I spoke to may have had disagreements with the claims of cryonics, but nurses, doctors, and administrators there prided themselves on accommodating the wishes of the dying patient (choice being the paramount value in hospice ideology). Sometimes there are public-relations matters, however, wherein the hospice is wary of letting everyone or anyone be "witness" to a cryonics case. In one case, a hospice I spoke to had asked the cryonics "van" to be parked at the back and the patient to be wheeled out from the rear exit.

Two weeks after the first alert, Patient X33 was placed on standby again. Once more, the team was mobilized. A cryonics van was dispatched to the hospital. The patient was diagnosed with multiply antibiotic-resistant *Staphlococcus aureus* (MRSA), a highly contagious infection with no real cure. Unconscious but alive at the vortex of a pharmaceutico-machinic assemblage called life support, he was, in the eyes of Western secular law, already devoid of the most important attributes of personhood: awareness, reason, autonomy. Regardless of brain activity signals, then, the patient was already a personless body. Or to foreground the person, the patient became a "potential person" or a "quasi-person," unable to possess his own body, powerless to make decisions regarding it. That power was transferred legally to the man's siblings, who had power of attorney, and they decided to take him off medical support close to midnight.

A little after 6:30 a.m., Patient X33 "deanimated"—said another way, he was pronounced dead by the head nurse. The act of pronouncement by a properly authorized expert—previously the charge of religious officials— is the initial secular ritual signaling a key ontological transformation, with a simple gesture conferring on the body a new status, recodifying it as a corpse. No longer autonomous or self-regulating, the body becomes objectified in the law, and this reclassified body may be manipulated in ways that a person-full body may not (Crippen and Whetstine 2007)—including through cryonics.

After pronouncement, the cryonics team entered the hospital room with its special equipment to carry out initial stabilization—cooling the body slowly while oxygenating and medicating the patient to keep the cells alive during the process. At that point the body changes possession once again—belonging now to the cryonics facility. But in order that the transformation of life into death be complete, a series of actions, signatures, and confirmations are required. The death certificate, a form issued by the state, requires the certifying physician to state the time, date, place, and cause of death; the medical examiner must also sign the certificate to confirm the cause and manner of death (and in cases of suspicious death, to take over the medical investigation for legal purposes). Finally, the mortuary or funeral home assigned to the task of disposition and transportation of the remains must also sign, on release of the body. At what point in the process these signatures are officially required varies greatly. In some states, such as Florida, the regulations are more lax, such that the deceased body can be released and transported without the physician's signature, as long as the physician has made a statement to the effect that he or she will sign later. In other states, such as Michigan, the regulations are stricter and even require an additional "Burial-Transit Permit" for "transportation and any manner of disposition of a dead body including interment, storage and cremation." What's more, the transfer of bodies to cryonics organizations is regarded as a donation to scientific research and as such is regulated by the Uniform Anatomical Gift Act (UAGA), which treats the body essentially as a special kind of property, or thing. As a result, before leaving with the body, the cryonicists in this case had to await release papers from the hospital, which in turn had to pass the paperwork through their organ-donor-network bureaucracy.

The host of forms and regulations that determine who can pronounce death, who can handle the dead body, who can declare or question the cause of death, who can transport it, and *what* exactly may be transported, from where to where, and whose signatures are required causes perpetual worry to cryonicists, because all these forms can only be signed after official death. While for the medical profession the point of death marks a release from urgency, for the cryonics team, unable to orchestrate the event ideally, legal death is the beginning of urgency. For the former, the dead body cannot get deader. For cryonics teams, a person's life is about to be saved.

Over two hours after pronouncement, the hospital's donor network finally attached its approval and the hospital release papers were delivered. The patient's body was wheeled out in a specially made gurney, in which an ice bath circulates cold water and a chest thumper keeps the oxygen circulating. As he was further cooled in the back of the cryonics van, Patient X33 was to be driven to the next state, to the cryonics facility's headquarters several hours away, to undergo full perfusion and cryopreservation on the operating stage.

In this particular state, as in many others, a corpse can only be taken across state lines by a licensed mortician[5] and then only after obtaining signatures from the medical examiner and the state's vital statistics office. The cooperating mortician enlisted by cryonicists notices a problem: the body had been released but the death had not been certified by the attending physician, who had not been at the hospital. Without the attending physician's signature, the medical examiner won't sign off. The attending physician had left instructions with his answering service not to be paged.

"Probably off playing golf," a cryonicist quipped with disdain.

The team contacted the charge nurse, who finally located the doctor and faxed over the paperwork. The doctor signed and returned it. She then sent the paperwork over to the funeral director, who got the medical examiner's signature. This left just one last step, the deposition at the vital statistics office. By then it was 3 p.m. The state's vital statistics office closes at 3 p.m.! There would be no cross-state transport today! Not only that, it was Friday. It would be three days before they could cross state lines with that body, that still-viable patient whose life was in their charge. Three days before they could get it to the operating stage at headquarters in order to perfuse it and take it all the way down to cold and still temperatures! How much damage would simple old ice prevent in the meantime? How much more dead would the person be?

The body at that node occupied two legal classes. It was a corpse; but it was also an assemblage of organs, a status resulting from the organ donation forms. A corpse, presumably still carrying some vestiges of identity and dignity, may not be transported across state lines without the medical examiner's authorization; but body parts or anatomical donations may. One body, multiple laws.

As it happened, Patient X33 was a neuro patient, that is, he had taken the option to only freeze his head. A man's head is considered an organ, though as far as the cryonicists and the patient were concerned, the man's head is the man's life, is the man himself, still holding a person, a future ahead of it. As a result, the cryonics organization decided to carry out what it calls a "cephalic isolation," and thus, while the rest of the body, indisputably personless, stayed instate to be cremated by the next of kin, later that evening the essence of the patient arrived legally at headquarters, where the cryonics team would work through the night to fully perfuse and cool the head down. The experts cannulated the arteries, clamped the veins, drilled bore holes into the skull, inserted probes, and measured the pressure in the pumps every five minutes, his brain temperature, peripheral oxygenation readings, acoustic sensing of structural preservation—in short tracking his semivital stats. But they also kept looking at him, at his face and head, and saying, "He's looking good, he's looking good." Over the next two days, they would slowly take him all the way down to liquid nitrogen temperatures and transfer him to a cryostat, where he would join the pantheon of over two hundred cryopreserved patients in the United States.

THE MATTER OF PERSONHOOD

Cryonics forces the question: What is left of Patient X33 in suspension? A frozen corpse? A potential life? Aspects of personal identity? Or just a thing? When I asked one of the team members on the case, he asserted: "The person's identity is preserved in a retrievable fashion." Meaning that even when consciousness is suspended, aspects of personhood are still relevant, as long as they are imagined to be "retrievable." Retrievable identity here stands opposed to irreversible death. The team member continued:

> If a person is asleep they retain their personhood, if a person is in a coma they retain their personhood. But if a human is in a coma and has brain damage to the point where there is no hope of retrieving their identity then they are not a person. So to me the distinction between a person and a corpse is whether or not there is a reasonable hope of them regaining consciousness. . . . The question is, could this person be revived and become a conscious individual. If yes, then they retain their personhood.

If for cryonicists the bodiless head of Patient X33 stored in liquid nitrogen is still a potential person, it is not because his spirit or soul has been saved. To the contrary, the fundamental distinction between the living and the not-dead is made entirely through the continuities of the body—the body animate and the body in suspended animation. For the medico-legal regime, this is a frozen organ, a *thing* separated from what was previously a corpse. Notice, however, that cryonics, the UDDA, the different state regulations, the courts, and the UAGA are all secular, all rely in such matters on the authority of science and share the same brain-centered view of personhood, and yet they come up with separate fictions regarding its end. What can this schism, internal to the secular, tell us about secular positions and about the body emerging in their nexus?

The unspoken status in these discussions is that of the corpse. To compare cryonics patients to people in a coma or deep sleep is to make a distinction between the status of a cryonics patient and the status of a corpse, which is what the cryonics patient is to the general public and the law. For cryonicists, a person's identity, unlike a corpse's, is potentially retrievable. The problem is that unlike people in comas or in deep sleep, the "potential" has not yet been proven; that is, no one in a cryopreserved state has returned to life and thus there is an obvious empirical difference in the outcome between a cryonically suspended patient and others who temporarily lose consciousness. The potential is speculative.

I was struck by these thoughts as I was following Ben Best, then director of the Cryonics Institute, on his routine, daily chores. Best was going around the storage area, checking the level of liquid nitrogen in each cryostat, then moving outside to check the pressure on the large liquid nitrogen supply tank that supplies all the other ones through a network of hoses fastened to the ceiling. The liquid nitrogen in the cryostats "boils off" at a very slow rate and needs replenishment only every two weeks or so. But the tanks are checked every day to ensure that everything is functioning properly—to avoid low-likelihood problems such as leaks or valve failures.

Being there with Best through the day, I tried to think of those in the cryostats as potentially viable, physical life-forms *of this world,* which is what they are for cryonicists—though there are "no guarantees," as Best liked to repeat. Even in secularized settings, memorialization of the dead continues through various material practices, such as placing flowers, hair, or photographs on tombstones or keeping objects of memory or talking to and cleaning an urn; these are the private concerns and practices of family members who are extending the memory of the deceased in a deritualized secular setting (Hallam and Hockey 2001). The task of checking the liquid nitrogen valves began to strike me as different. If this was about maximizing the chances of survival, then it felt like something more than caring for the lawn, different from extending memories of the past or carrying out some sort of communication with the realm inhabited by the dead. It was more like a nurse doing his rounds of coma patients, checking the vital stats. Best said he did not know most of the people in the cryostats, but he had an intense sense that he was potentially caring for their future presence in this world.

When I later asked Best about his responsibilities, he said:

> In terms of all the things that need to be done and understanding the technology and the goals and the commitment, and all those kinds of things taken together: there's so many things that could happen, so many things that could go wrong, so many things I worry about, and sometimes the worry can get quite strong and I can get myself quite upset about things that can go wrong and things that could happen.

Cryonicists have argued that socially, conceptually, and legally we admit of partial persons: people with limited rights, people with limited functions, people who are wards of the state or have given up their decision making to others. Thus, in practice, personhood has gradations. And here, regardless of the concept of personhood used to determine the criteria (autonomy, rationality, etc.), it is clear that law and society work with the idea of personhood as a social function. That is, *personhood* is at least in part a matter of whether society, or a group, treats an entity *as* a person, or to take it farther, treats them *as if* they were a person (vs. *as if* they were dead). Personhood in this sense may be considered a *status* attributed to those who meet criteria for *social membership* (Conklin and Morgan 1996), with the understanding that social personhood is a gradation, not a binary. If personhood is a status based at least partly on whether society, or a group, treats an entity *as if* it is a person, then one of the factors that influences such a social claim on the entity is the social intensity and depth of relationships which have accumulated around it, the intensity by which social identity is recognized. In different ways, philosophers like Rom Harre (1981) have argued that a first-person point of view or subjectivity is necessarily linked to recognizing and being recognized by others. Thus, an amnesiac with only momentary awareness of himself (i.e., with self-consciousness but not memory and therefore no real sense of autobiographical continuity) is viewed to be a person largely because he is recognized as such by others who treat him as the same person. This still leaves the question of whether, and under what conditions, an inanimate body, such as a cryopreserved patient, may come to be treated as even a partial person. After all, even if the law and society admit of gradations in personhood, they don't treat corpses as any kind of person.

Or do they?

Here again the secular premises of the law come into play. In secular regimes, personhood and biological life are separated, yet signs of personhood must always be read in the biology of the human body. But there is a particular aspect of this separation that makes it distinctly secular, insofar as it is revealed through the materiality of a body caught between rationalist and materialist determinations: in this body, the secular body, the separation of personhood and life can only be understood unidirectionally—that is, a body in which life and personhood are separable but only in such a way that, for the first time, *the body's life may outlast the person,* and not vice versa. *The person cannot outlast the body.* The body without the person is known as a corpse; and a corpse, not being or containing a person, or even a ghost, is a thing.

In any continuum between person and thing, the corpse occupies a strange place. To begin with, there is an ontological problem, with the body having gone from an intentional subject to an inanimate one. The everyday course of things is to personify objects, whereas the corpse requires or even embodies the reverse operation. The person was a being with desires and projections into the future, unable ever to be just what it is, in the way a thing is just what it is. Suddenly, now, the corpse is a human body that coincides with itself, that is its own image. One might say that the corpse is a kind of body, only more so. Its tautological nature is unintentionally captured in state definitions of corpses or remains in American law. For example, the Michigan Public Health Code (Act 368 of 1978, section 333.2803) defines a dead body as follows: "(1) 'Dead body' means a human body or fetus, or a part of a dead human body or fetus, in a condition from which it may reasonably be concluded that death has occurred." Similarly, Arizona (ARS 32-1301, which establishes the funeral board and its rules, and ARS 36-325, which deals with public safety) has this definition, using the term "human remains" rather than "dead body": "'Human remains' means a lifeless human body or parts of a human body that permit a reasonable inference that death occurred." The definition couldn't be any simpler: a dead body may be defined as a dead body.

An examination of the status of the corpse in law and society helps us see how—through what rituals, codes, instruments, techniques—a person is socially transformed into a thing, if in fact a thing is what a dead person gets turned into. I will do this mainly by examining legal attempts in the United States to define the corpse. Even as the corpse turns out to be a much more varied and resistant cultural entity than secular concepts of death would outwardly claim, so the law itself, unlike its core assumptions and requirements, ends up being much more flexible, insecure, and vague in practice, allowing room for ambiguous subjects and objects.

In a modern American setting, the act that simultaneously acknowledges and constitutes the change of status from person to corpse is the act of pronouncement. As described above, with one act—a nurse's phone call, a signature—all the rules pertaining to permissible procedures suddenly and radically change. It is the moment when the body is deemed no longer intentional, even *potentially* so. Hence the idea of irreversibility. At that moment, the body becomes objectified in the law. A cascade of new laws and rules—from inheritance and estate laws to rules regarding disposition—kick in, and the reclassified body may be manipulated in ways that a living person cannot (Crippen and Whetstine 2007). Bodies may now be burned, they may be embalmed and displayed, or they may go to centers for scientific research, where they might end up in a crash test or get plastified and displayed in a museum.

The history of slavery is both the source and the brutal embodiment of the legal and political contradictions of property and personhood regimes in the United States. White colonists and later white legislators had to perform elaborate legal acrobatics to render African persons into property in order to justify their forced capture and labor for a term of life and to restrict their freedom at birth in the territories. According to legal scholars (Welch 2004; Finkelman 2012), the key turning points in laws regarding slavery came about from the 1660s onward in order to restrict the free status of those born to African mothers, especially when mixed-race, and to capture and return escaped slaves. The laws, then, had to enshrine the racism and protect the socioeconomic advantage of white landowners by specifying certain property rights over persons. At the same time, as scholars like Orlando Patterson (1982) have pointed out, similar property rights in various forms also defined relations to other persons, such as children. The issue, for Patterson, is that slavery had to be understood not in legalistic terms but as a set of social relations which he went on to define in the case of slavery as social death.

Today, too, the legal contradictions of property regimes should be understood in terms of the tensions between socioeconomic priorities and abstract legal principles. In the wake of slavery, the operating principles of secular humanism and secular law have been that humans possess their bodies and all of its parts in an absolute, inalienable way (Campbell 1992). Yet this is contradicted by another set of principles, which would state that humans should not destroy their bodies or sell their body parts for valuable consideration or have judgments levied against them (Zee 2008). That has been the issue, for example, with the most famous of such recent cases, that of Henrietta Lacks. The HeLa cell line was developed from cancerous tissue taken without consent from the body of Henrietta Lacks, a black woman from the South who visited Johns Hopkins Hospital in 1951 for treatment. The cells were called "immortal" because they would apparently divide without dying quickly and thus could be cultivated indefinitely and in great quantities. Like other, similar cases, the HeLa cell line raised questions of ownership, rights, and consent regarding biological tissue taken from the bodies of patients. The contradictions in the socio-juridical regimes are revealing of the larger cultural and economic context of these debates and the tension between property and personhood.

The point is that in secular law the *living* body has been and continues to be a site of tension (pulled politically between personhood and thingness), for a person is said to fully possess her body and yet cannot do with it what she can do with other things she possesses. But what if the person is no longer there? When the person is not there to possess it, what is the body then? What "remains" of the body or of the person?

In the case of organ transplants, anthropologists have clearly detailed the ways in which body parts are commodified, by presenting organs as objects or mechanical devices independent of the identity of the original owner and by managing anonymity through the entire course of the transplant chain. Personal identity is stripped from the body. Margaret Lock writes: "Only when corpses could be conceptualized as neutral biological objects, as part of nature and therefore autonomous and without cultural baggage, was it possible to divest them of social, moral and religious worth" (2002, 39).[6] Rendering the corpse anonymous in the research and donation process means entering it into exchange outside its own circuits of recognition, divorced from any specific history, identity, or biography (Sharp 2001). In fact, it is precisely this anonymity that enables research.[7] When brain tissue is examined or a body is used in a crash test, what is under study is not a particular person but the abstract universality of the human body. In research, not only is the corpse's identity unnecessary, but its individuality would be counterproductive: what matters is how the individual translates into or represents the general human. In the case of crossing state lines, anonymity is also at stake. As we saw in the case of Patient X33, crossing state lines with parts of the body, unlike with a full corpse, is not a problem; it is not subject to the same regulations and does not elicit the same scrutiny. A full corpse requires a different set of procedures, presumably because it is thought to *embody a person* more completely than an organ, more than a head or a heart; a corpse retains a problematic residual personhood in the secular imagination.

If the erasure of recognition is a key factor in the stripping of personhood as status, what is left, according to the law, once personhood is removed? The answer is property, since one of the foundational aspects of modern Western law is the ontological separation between person and thing, or person and property, the "two great legal categories for imposing order on the animate and inanimate world" (Naffine 1999, 104; Pottage 2004). The strict, modern distinction between property and person is not universal, however. Anthropological theories of personhood have analyzed ways in which persons are partible and become fused with histories of objects through which social relations are established (Battaglia 1990; Strathern 1988). Or take Roman law, which treated its categories more fluidly such that depending on the circumstances the same individual could be a person or property. Grandparents, for example, were persons but also property insofar as they could be "inherited," which itself entailed a certain responsibility (Maine 1963; Thomas 2004).

Admitting what was once a person into the category of property has caused some problems for Anglo-American law, which has slipped the corpse back and forth in a hazy territory between person and property. On the one hand, the dead body, unlike a live one, may be possessed for purposes of disposition and use of body parts; on the other, unlike other properties that may be possessed, a dead body cannot, in principle, be used in a commercial exchange (hence body and organ *donations* that in fact conceal the lucrative market world of organ transplantation); dead bodies cannot be sold or even solicited for, say, by a funeral home. In the Michigan code, for instance, solicitation for the purposes of obtaining a dead human body, before or after death, is punishable.[8]

U.S. law has tended to classify the corpse as "quasi-property" (Gilligan and Steuve 1995; Kester 2007)—a legal fiction made up by the courts to protect the survivors or next of kin without turning the corpse into full property (Nwabueze 2002). The decision often quoted in this regard is a Rhode Island case dating back to 1872. The pertinent passage states: "Although as we have said the body is not property in the usually recognized sense of the word, yet we may consider it as a sort of quasi-property to which certain persons may have rights, as they have duties to perform towards it, arising out of our common humanity. But the person having charge of it cannot be considered as the owner of it in any sense whatever; he holds it only as a sacred trust for the benefit of all who may, from family or friendship, have an interest in it."[9]

The terms "common humanity" and "sacred trust" are invoked by a secular court to make the corpse something more than an object without turning it over to the realm of religion, a concession validating the investment of the corpse with significance beyond its material existence (sacred trust) but also claiming it for the realm of human affairs (persons who have an interest in it). Originally, common law saw absolutely no property in the dead body, and the body was given up to God and the church. Physicians only prognosticated death, that is, detected signs of its onset, and were not charged with determining its arrival. Diagnosing and declaring death, pronouncing it, finding criteria for it—for most of Western history such things were nonmedical functions that often fell to the family and religious authorities (Pernick 1988). As the church lost greater authority to state and professional institutions, especially in the domain of public health (Pernick 1988), it became necessary for the state to take over the dead body for purposes of disposition with an eye toward protection of public health; yet, as grief and mourning were privatized (Walter 1991; Laderman 2003), it also became necessary to devolve some rights to next of kin. The only rights that could be so devolved were rights of "possession." Thus, the corpse became a body that must be gotten rid of, but with respect.

And in the end, these vestiges of respect don't last long either, as the quasi-property rights in a corpse are temporary. They are not property rights as such but custodial rights, valid only up to and for the purposes of disposition. After disposition, be it a burial or a gift to science, that "sacred trust" evaporates. Thus, for example, in a 1973 case in Michigan,[10] a mother with paramount rights over the disposition of the body donated it to science. During the transport to the medical facility the body was, apparently, lost. The mother sued, but the courts held that she had no case, as she had already given up custody at the moment she decided to donate to science. That was her choice of disposition. So much for sacred trust.

This in-between state of the corpse, its quasi-state, has left some indeterminate legal caveats. Not qualifying the corpse as property has implications. For example, the will as a legal instrument was designed for the lawful redistribution of the estate of the deceased according to his or her wishes. Not being property, however, the person's dead body cannot be considered part of the estate, thereby raising the question of whether the disposition of remains may be specified by the deceased as part of his or her will. In the past, some courts have actually ruled against this: "It is quite well established . . . that, in the absence of statutory provisions, there is no property in a dead body; that it is not part of the estate of the deceased person; and that a man cannot by will dispose of that which after his death will be his corpse."[11] But if the corpse were treated as property such that it could be fully possessed by others, in whatever order of precedence, then the notion of the person's will would be dissolved. Quasi-property gives the law some wiggle room. The next of kin can take possession of a corpse, but at the same time an individual has a right after death to have his or her body (and assets) disposed of in the manner he or she wishes. This has not solved the problem cleanly, of course, and there is little uniformity across the states on laws regarding bodily remains and wills (Kester 2007)—disputes regarding disposition come to court frequently, sometimes at great public cost, and disputes over the will itself and the estate, of course, are everyday matters.

What survives or remains of the person after death, then, is not regarded as merely a material body, a pure thing; something more than matter survives the death of the person. Terms like "quasi-property," "dignity," and "sacred trust" invoked by a secular court make the corpse something more than an object (one that perhaps hides under it a quasi-person), without fully turning it over to the realm of religion. The wavering of secular positions and institutions regarding the corpse arises not because they cannot account materially for what is there—the thingness of the body—but because they cannot account for what is not there—what had previously made the material body something more than its thingness, that autonomous willful person who used to be the very subject of the law.

There is the issue of the will as well, that is, of intentions that survive the intentional subject. In the United States, a person is partly defined by his or her will when alive: by competence and autonomy to make decisions. If these are deemed to be lacking then usually the "will," that is, the decision-making power, is given over to a surrogate of some sort, but *without* the person fully losing the status of personhood. So in such a case the living person is reduced to a quasi-person. This raises a question: if what remains of a person after death is not just a corpse but also decisions or desires in the form of a will or a trust, then could we surmise that the designation of a corpse as "quasi-property" may be hiding behind it a "quasi-person"?

Finally, let me add that the corpse is not just one thing. As studies of organ donation make clear, the corpse does not stop its status transformations at the moment of death. Rather, it can be divided and subdivided, labeled and relabeled in different ways,[12] all of which may have different consequences for the people involved. Even in the matter of a complete corpse, the law has made a clear effort to distinguish between a dead body and remains. In a grave robbery case, for example, an Ohio court ruled that skeletons are not considered dead bodies. "A cadaver," it noted, "is not an everlasting thing. After undergoing an undefined degree of decomposition, it ceases to be a dead body in the eyes of the law" (case in Gilligan and Steuve 1995). Similarly, the State of Arizona currently has two different definitions for "human remains." The first, quoted earlier, is under the public health and safety act, while the second figures under historic preservation and reads as follows: "Human remains means any remains of a human being who died more than fifty years before the remains are discovered."[13] Clearly, not all remains are created equal, and in an individualistic society where generational continuity is not the norm it seems reasonable that the law would find a distinction between newly deads and long deads, especially when it leads to profitable real estate developments over old cemetery grounds. In part, this is at stake when Native and indigenous communities claim their burials. In Australia there are further classifications, so that remains are considered "prehistoric" if they are more than two hundred years old. This turns them into fossils over which indigenous people can claim no jurisdiction. Not surprisingly, then, such definitions also delineate jurisdictions; they outline domains of authority over certain objects, rights, and even people—both dead and alive. To the extent that persons are produced socially, their absence too is socially produced.

WHAT'S NOT A CEMETERY?

While no law has been written to apply specifically to cryonics, where state institutions, the law, and cryonics have intersected, confusion between the categories of life and death, person and corpse has arisen. One effect has been to confound the state and the law regarding their classification of cryonics. For example, Alcor is legally understood to operate in Arizona as a "procurement organization" to which bodies may be donated for "scientific research." Conversely, the Cryonics Institute in Michigan is regulated as a cemetery, though its claim is that it is actually preserving lives, not disposing of dead bodies. The regulatory process is revealing.

On August 26, 2003, the Michigan Department of Consumer and Industry Services (CIS)[14] issued a cease-and-desist order to the Cryonics Institute, based in Clinton Township, a small suburb north of Detroit. Looking into the matter, the State of Michigan had noticed that the Cryonics Institute was not licensed and therefore was operating "outside the scope of the law," subject to no oversight or regulation whatsoever. The trouble was that the state seemed uncertain about the exact nature of the institute's business, not because the activities had been kept under wraps—to the contrary, almost everything that needed to be known about the institute was quite clearly delineated on their website and in their charter. Robert Ettinger, who was involved in the dispute, explained the situation to me as follows:

> We operated for many years without any regulatory oversight and we operated very openly. Everybody knew about it. In fact, when we applied for our articles of incorporation, our charter from the state, they recognized that our purposes were unusual, and the attorney general gave

it a special review and they said OK; they didn't see any problem with it. And when we bought our first facility in Detroit, we did not go to the city for any special permission, any special license, or anything like that. It didn't seem to be necessary, and it wasn't. Then when we moved out here, and bought this new building, this much better and much larger building in Clinton Township, we went to the township administration first of all and sat down with them and told them exactly what we were about and what we were gonna do and made sure that they didn't have any objections, and they didn't.

According to Ettinger, the county and the state were aware of the institute's activity—"there were even newspapers articles from time to time"—and had expressed no objections. In addition, the institute was a registered nonprofit and was affiliated with a funeral parlor, so its activities were already under regulation in many ways. "It was all in the open and everybody knew about it and nobody was concerned," Ettinger continued. "But then that Ted Williams thing came along."

Briefly, "that Ted Williams thing" refers to a public dispute involving the Phoenix-based cryonics organization, Alcor, and the surviving children of the baseball star Ted Williams. The dispute became a cover story in *Sports Illustrated* and went nationwide, as the reporter suggested that Williams had not signed up for cryonics but that his will was manipulated by one of his sons. The story initiated an investigation in Arizona that echoed in Michigan.

The State of Michigan knew that the Cryonics Institute was operating in Michigan and had dealings with dead bodies, but when it was forced to face the matter it didn't know what to make of it. The CIS nevertheless took the public position that it was not even aware, until the Ted Williams scandal, that the institute was doing what it was doing. So it charged its Bureau of Commercial Services, which is "responsible for assuring that more than 335,000 licensees for 26 different commercial occupations meet the minimum standards for operating in Michigan" and that they "are operating legally and competently,"[15] to investigate the institute. According to a funeral director who has served on Michigan's Board of Funeral Directors, the investigation was actually initiated by the "local cemeterian's complaint to the board." Due to cases where escrow funds for cemetery upkeep and storage had been usurped by funeral directors, the state had recently regulated cemeteries and funeral homes more strictly, separating the codes and rules for each. Jim, a funeral director who had worked with the institute, explained: "In Michigan, cemeteries can't own funeral homes and funeral homes can't own cemeteries. That's just in three states. So cemeteries had different regulations than funeral homes. So there were all those sorts of problems. And then Ted Williams happened, and cemeteries said, OK, if you are going to control us, what about them?"

The CIS insisted that one way or another the institute had to come under regulation in order to protect consumers. Having insisted, however, it now had to make some determinations. "Our investigation revealed the Cryonics Institute is clearly operating as both a funeral establishment and cemetery without any state oversight," CIS director David Hollister concluded. So because it had to designate the Cryonics Institute as something, the State decided to call it "an unlicensed mortuary science establishment and a non-registered cemetery . . . maintaining a mausoleum for crypt entombment for more than 40 human bodies" (Michigan Department of Labor and Economic Growth 2003). It was not stated that "the crypts" were large insulated tanks full of liquid nitrogen, that the bodies were being kept at a temperature of -196°C, and that the "more than 40 human bodies" thus stored were considered by the institute and its members as not "entombed" but "cryopreserved."

The cease-and-desist order required that the Cryonics Institute facility provide "proper maintenance for the 40 bodies currently in its care" but also that it should not accept "new cadavers." Emphasizing "cadavers," it recommended that the institute become either licensed as a funeral establishment or registered as a cemetery. As mentioned, state regulations now do not allow one establishment to serve in both capacities, so this put the institute into a catch-22. A cemetery can store bodies but cannot "prepare" them, while a funeral home can prepare bodies but not store them. Thus, if the institute were designated a cemetery it would have to carry out its initial perfusion and cooling operations elsewhere. For a practice to which time is crucial and the procedures are not standardized, this could create problems.

In the official press release, Hollister said, "Our ultimate goal is to encourage the Cryonics Institute to take the necessary steps to come in compliance with state law." That meant taking the formal step of becoming registered and licensed as a cemetery. The institute and its members were not happy about the recommendation and what it implied. As far as they were concerned, the institute was not receiving cadavers, running a mausoleum or cemetery, or dealing in any way with dead bodies. The organization did not hide the fact that it had human bodies in its care and that the bodies had been declared *legally dead*; it's just that as far as the institute and its members were concerned, they were only *legally* dead and not *truly* dead, as they are in cemeteries.

Explaining the conflict with the State of Michigan in the institute's newsletter, Best wrote that the institute's purpose "is to attempt to save lives, not to store dead bodies" (2004). Consequently, in his view the Michigan Department of Labor and Economic Growth (DLEG) had no real jurisdiction over this matter. But then again it was not clear which office, commission, or bureau would or should have jurisdiction over "suspended" persons. Neither the law nor medical science makes room for such a liminal category. As far as the law is concerned, you are either dead or alive, a person or not.

This conflict arising from radically different views of life, death, and personhood would seem intractable, but by the end of the year the dispute had been settled. After discussions between the DLEG, the Cryonics Institute lawyer David Ettinger, Ben Best, and some of the institute's board members, the two sides reached a compromise, agreeing to license the Cryonics Institute as a cemetery under Michigan law. The institute was permitted to continue providing its services and procedures almost as before, but is now *legally designated a cemetery* and, per Michigan law, *must carry out initial procedures at a funeral home* rather than at its own facilities, as had been the case. That is to say, all preliminary perfusion and cooling that is meant to save a person's potential life would have to take place in a funeral home first.

The institute was happy enough that it could operate normally again and provide its services to members, but members of the institute still bristle at being labeled a cemetery and being regulated under laws pertaining to dead bodies. However, some members think that being regulated may be safer, because at least now they can't be made "illegal." Though they are operating against the grain of the law, they are at least inside it, and in fact, through the process of regulation, cryonics was given a sort of special status. It is striking to notice that the State of Michigan's final press release refers to the cryopreserved bodies as "patients." "Patients," reads the document, "will still be stored at CI's facility in liquid nitrogen." References to "cadavers," which had appeared in abundance a few months earlier, vanished altogether from the official press release. The cryopreserved bodies formerly designated as cadavers had been transformed into patients; yet the State of Michigan did not feel the need to explain why it was acceptable to it that "patients" should be frozen in liquid nitrogen tanks, or be regulated according to laws pertaining to corpses, or, indeed, kept in a cemetery.

A similar case took place in Arizona when, at the prompting of the State Board of Funeral Directors and Embalmers, state representative Bob Stump introduced HB 2637 on February 6, 2004. It suggested two principal amendments. First it required any entity that stores dead human bodies or remains for a period of more than five years to be licensed or registered under the Arizona State Board of Funeral Directors and Embalmers.[16] Second, under articles regulating the "Disposition of Human Bodies" and specifically the "Revised Uniform Anatomical Gift Act,"[17] it prohibited any entity that charges a fee for receiving dead human bodies or remains from participating in the Arizona Anatomical Gift Act (AAGA).[18] The bill was designed to box cryonics into the funeral industry and bar it from acting under the Anatomical Gift Act. While it was specifically aimed at cryonics, however, the bill would also have an obvious impact on organ procurement and research agencies relying on the UAGA. The organ donors lobby, concerned that research institutions might be affected, sided with the cryonicists.

Additionally, the proponents of the bill underestimated the intense support Alcor members could generate. As the bill was weaving its way through, Alcor sent a desperate email to all its members urging them to email or call the Arizona house representatives in order to respectfully make their case. Alcor members were incensed by news of this legislation, so calls and emails flooded the Arizona house in loud and accusatory tones. They called the legislation tantamount to murder. Eventually, support for the bill waned and it was withdrawn. Alcor would continue to operate the way it always had: under the UAGA, doing medical research, and not storing dead bodies.

What for the funeral board was a matter of regulation, for cryonicists was a threat to the *lives* of their members. For one thing, being regulated as part of the funeral industry would imply that licensed embalming personnel, rather than Alcor employees, would be the only ones legally authorized to treat the "body." In Alcor's view, this would greatly reduce the chances of a person's survival. For the funeral board and legislature, the members were already dead anyway, so the procedures were only important in terms of the idea of "respect for the dead" as well as from a commercial and public-safety standpoint. This discrepancy made the anger of the calls and emails appear irrational to noncryonicists. Barry Aarons, a lobbyist hired to represent Alcor, told me he had to apologize to the legislature for those "inappropriate" calls and pulled back the Alcor membership from the process. But to me he insisted that their reactions must be understood in light of their views about life and death.

> Remember that our members consider their cryopreserved state to be life. . . . And our people got so emotionally overwrought, I actually had to hold them back. I mean some of their emails and letters and stuff were really harsh. But you were dealing with their lives. I mean they're people who believed that being cryonically preserved is a ticket to immortality. By potentially doing away with Alcor, by potentially passing a bill that could threaten the existence of Alcor, you're threatening their immortality.

What for state institutions is framed as a normal regulatory after-death matter (i.e., regulating a body devoid of a person) is taken by cryonicists to be a misinformed and threatening intervention into their practice of saving individual persons.

As soon as the state takes the plunge into the regulatory process it commits itself to specific categorizations. From a cryonicist point of view, the correct category does not currently exist. However, that is equally true for the state. It is evident from the record that the state's administrators themselves come face-to-face with perplexing problems regarding the classification of cryonics *given* the claims of cryonicists. In part this is due to the way the process of secularization has brought matters pertaining to the dead into the regulatory regimes of the state. The Department of Health, through its related offices, is in charge of declaring and registering life and death; as soon as a death is recorded and registered, the Department of Health has fulfilled its duty. From here on, the corpse, its interests, the survivors, the businesses profiting from it are no longer within its purview. Declared dead and thereby stripped of personhood by definition, cadavers and corpses are regulated by departments dealing with consumer services. How else could the modern state manage the realm of the dead but through an understanding of the dead as objects of consumption, albeit special objects with special rules of consumption, stripped of identity, culturally neutralized?

In other ways, however, the indeterminacy of the cryonics third state—the state of suspension—causes regulatory anxiety with larger questions regarding the state's ability—perhaps even the general ability—to classify, to record, to fix, and to manage slippery but crucial matters such as identity, life, and death: in short, the ability to govern, in a biopolitical sense. A clear illustration of what I call regulatory anxiety comes from the courts themselves, in a case involving Alcor and the State of California. In the 1990s, when Alcor was still based in California, the state's Department of Health and Safety (DHS) attempted to bring misdemeanor charges against Alcor's work on the basis that cryonics did not constitute "scientific research" and thus Alcor was operating illegally, since storing bodies anywhere but in a licensed cemetery was against the law. Additionally, if Alcor's work did not qualify as "scientific research" then what it was doing with bodies could also be deemed illegal, since those bodies would have to be disposed of by one of the two then-existing categories: burial or cremation.

The DHS had to withdraw this charge, since determining what qualified as scientific research was not in its jurisdiction. Subsequently, it pursued the case from a different angle[19] and then eventually gave up when the California attorney general ruled in favor of Alcor.

The court records reveal some key concerns behind the DHS's persistence. Insofar as the agency's explicitly Foucauldian duty is to keep precise records of birth, marriage, and death in order to understand and manage public health, cryonics practices and concepts seem to have raised some unusual quandaries, blurring the clear boundaries the DHS needs for population management and record keeping, namely, notching one under death, another under cremation. Presenting its case in court, the agency raised what it called some "serious questions": "Should cryonically suspended people be considered 'dead' or should a separate category of 'suspended' people be created? How should such people be registered in official records? . . . [W]hat happens to the estate and the assets of the 'decedent' after the decedent is put in cryonic suspension? . . . [W]hat would happen to such estate and assets if and when cryonic suspension is successful and decedent is restored to life? Whose identity is the person to assume or be assigned and what of the record of the person's death?"[20]

The administrative anxiety of the immortality regime is an anxiety regarding the vulnerability of existence and selfhood which is revealed in the progression of the questions from the issue of "categorization" (Should there be a separate category? How should they be registered in "official" records?) to the subject of identity (Whose identity is to be assumed?) to the recording of the person's status (dead? reborn?). Cryonics' liminality, its ability to operate within the cracks, against the grain of the state's power, brings these out. The policing of the boundaries confirms the power of the state, which wants to maintain the stability of the immortality regime. At the same time, it confirms the indeterminacies and anxieties with which it proceeds.

PERSONHOOD DONE AND UNDONE

Today's laws and the tensions they embody arose historically and were shaped through contests between ecclesiastical and secular law and priorities over control of the dead body (Nwabueze 2002). They are shaped by the materialist, immanent frame, which must hold that after death there can be no surviving person. In secular conceptions, there is a *necessary* erasure of personhood *after* death; "the dead" are denied certain forms of status. Yet under the domain of the secular, the law does offer postmortem protections, albeit in contradictory ways. Although material parts of the ex-person, such as tissue, may become property or quasi-property to be gifted away, researched, or commercialized, other, seemingly nonmaterial aspects of the person are treated differently. Thus the person's will is protected, the dignity of the dead body is protected, various interests such as the person's reputation and image are protected, and so on. And while the trend to provide postmortem rights and protections is growing daily in the United States (Madoff 2010)—many such issues arise when families take cryonics facilities to court—the law specifically avoids stating that these protections are given to a person.

The question is begged: If it is not to some version of personhood or a person's interests, to what is the protection accruing? Sometimes the laws regarding the dead are given in such a way as to protect the "survivors" (a sort of admission to the effects of "relational personhood"), but not always. The law talks about the inherent dignity of the dead and respect for the dead body, but without considering it either a person or an object.

Cryonics, by contrast, is clear that whatever protections it demands or gives in the form of care are given to the person. By pushing the logic of materialism to its limits, cryonics brings into problematic focus not its own preservation of personhood but rather the operations of the medico-legal regime as it socially produces the absence of the person while transforming the body into a thing. For, one might ask, what has survived death for secular law? Cryonics avoids the ghost of survival after death by eliminating death itself. In his cryostat, Patient X33 is considered a patient not because he has survived death; Patient X33 is suspended before "irreversible death" overtakes him. As a result, cryonicists confront the opposite end of the mind–body problem. Not, Does personhood survive matter? but, Is matter sufficient for personhood?

Secular immortality regimes say that in the continuity between animation, deanimation, and reanimation, the final term in the series is moot—it is either metaphor or psychosis. Yet in practice the ghost of survival haunts the secular. To the extent that in the West the secular arose through the materialist elimination of the soul as a condition of the real, it committed itself to a determination of personhood based on physical laws rather than transcendent, idealist, or nonmaterial ones. To the extent that the secular arose through rationalist conceptions, it advocated for the separability of personhood from its physical container. The latter risks being accused of cultivating mysteries unless it manages to ground some of its assumptions in a materialist science; the former is grounded in the material but has trouble accounting for mental features that constitute the secular subject, such as autonomy, intentionality, and consciousness.

The relationship between person, life, and death is established via arbitrary signs on the biological or material body and is rendered stable in the public imaginary through the secular authority of the medico-legal immortality regime. Hidden beneath its medico-legal fictions, the tensions remain. The secular person, I have argued, is not reducible to the materialist body but produced through the tensions, through the indeterminacies of the mind–body problematic. So a secular body is a body caught between the fictions of a nondeterministic, rationalist conception of personhood and its arbitrary materialist determinations. On the one hand, personhood, as consciousness, is separated in the secular from bare, biological life; on the other, the markers of personhood are located in biology itself, for example, in brain signals or features of an embryo. The secular body, then, is the body produced by materialism in order to be "real," at the same time as it is the body that must escape it in order to be "conscious" or "autonomous."

Despite, or rather precisely because of, the contentious relationship it has with the mainstream medico-legal regime, the cryopreserved figure of Patient X33 emerges as the secular's ideal-typical escape artist. By reductive materialism, its person has been shrunk to the brain. And yet, in this reduction it shows materialism to be insufficient, since its continued existence as some-sort-of-person depends not on the contested terrain of consciousness or autonomy, which are for now usefully absent, nor on any biochemical markers, which have been equally usefully stilled, but on a small community's recognition of its potential for consciousness or autonomy.

5

Convergence

SECULAR SOLIPSISM AND
THE MIND OF THE COSMOS

Terasem, an organization that bills itself as a "transreligion" dedicated to "diversity, unity and joyful immortality,"[1] has its headquarters on what is called the "Space Coast" of western Florida. It was there that about twenty of us ran out of a conference on the legal rights of the robot Bina48 in order to witness a Space Shuttle launch, with the sun taking its early winter descent behind us and the ocean to our side spraying up caustic spores from red algae that scraped our throats and made us cough through the rest of the gathering. The founder of Terasem, Martine Rothblatt, who, it should come as no surprise, is a Space Shuttle fan, suspended the mock trial of the robot in order that the excited attendees might witness takeoff. So it was that I found myself on a beach with Marvin Minsky of MIT, William Sims Bainbridge of the National Science Foundation, futurist philosopher Max More, founder and president of the Society for Universal Immortalism Mike Perry, Michael Anissimov of what was then the Singularity Institute for Artificial Intelligence, future Immortality Institute director Justin Loew, and a number of bioethicists enthusiastically taking snapshots of a little flame in the distance thrusting a white metal body into outer space, a region of the universe favored by most of those present. Terasem promotes, in the words of Rothblatt, "a vigorous space program to get people off the Earth."

It wasn't until later, though, that I understood the greater significance of the location and the moment. Though certainly for those gathered with us there the spectacle of liftoff generated awe at American technological prowess, for Rothblatt this evoked something more. For it was there on that same stretch of sand that Rothblatt had had a deep revelatory experience of the connection between people and the cosmos. Prior to what I thought were Terasem's somewhat banal origins at the 2003 bar association meeting, where the organization was officially launched, there was a more exciting founding event dating back to 2002, to a day in March, at 6 a.m. to be precise, when Rothblatt had gone out to the beach for her morning meditation.

Now, as a trans inventor with a "spiritual" practice, Rothblatt is not the typical transhumanist, Singularitarian, immortalist mind uploader. Many people in these groups balk at the use of this sort of language and imagery. Earlier that day, for example, as we walked around the Terasem reception room holding our lunch plates, I happened to find myself next to Marvin Minsky scanning the bookshelves. When we got to a shelf of books on spirituality, he sneered. Bear in mind that Minsky was an advocate of cryonics, mind uploading, and physical immortality. But as a reductive materialist, he bristled at anything that suggested "spirituality." Then again, Rothblatt also is into mind uploading, celebrates the imminent arrival of superintelligence, and disavows "mystical" and "supernatural" explanations, to use her own words. Yet she also has a syncretic meditation room in her headquarters and talks about superordinary consciousness being downloaded into her mind during a meditation session.

I have since heard this story on several occasions, but she described the event in detail at a virtual conference called "The Turing Church," organized by Giulio Prisco (see below). Here is my transcript of Rothblatt describing her fateful morning as she sat in meditation facing the horizon past the ocean on the very same Space Coast where we all had been standing, looking up at the sky:

> I felt two presences around me, so I immediately opened my eyes and witnessed quite to my surprise the blast-off of the Space Shuttle on its way to the Hubble telescope. I then felt a presence immediately behind me and found myself eye to eye with a large lanternback sea turtle which was clearly trying to make its way back to the sea and I was in its way. This

was kind of startling because it was pretty much dark still. An epiphany poured into me from the triangular vector of the shuttle's light blasting across the sky, and the lanternback turtle's consciousness streaming into my own and I felt a very strong triangular force of energy between the shuttle, the sea turtle and myself. An epiphany surged through me that there were three principles or values which united all life, all reality, indeed the entire multiverse. These were that the *purpose* [my italics] of the multiverse, the solution to the multiverse, was to balance diversity with unity in the quest for joyful immortality. . . . I slid myself out of the way of the lanternback turtle and it slowly but without delay pushed itself towards the sea. As it did, the shuttle quickly arced out of my sight . . . and again this triangular energy between the turtle and the shuttle remained connected to me and downloaded to me the entire gist of the truths of Terasem. It was like I received this multi-gigabit download that still had to be processed. . . . I was totally blown away. I stood up and it was like I was walking on air. I walked a couple hundred feet to where my soulmate and partner Bina was resting, up above the sea shores. I shared with her what happened and we slid right into the most wonderful, most erotic lovemaking that anyone could imagine. It was through this erotic lovemaking that the truths of Terasem were downloaded to me, and an epiphany channeled by the shuttle and the turtle became written into my soul, just as assuredly as a file is written into a digital or magnetic medium. I spent the rest of 2002 trying to print out what was in my mind and soul from that March morning.

The continuities described by Rothblatt—between her, the turtle, the shuttle, the universe, all of it in the form of a multi-gigabyte informatic download—prompted her to found Terasem and begin two related techno-scientific projects: first, the production of Bina48, an advanced robot based on her legal wife, Bina Rothblatt, an African American woman; and second, a project called CyBeRev, for cybernetic beingness revival, which later morphed into a parallel experiment called Lifenaut.

CyBeRev and Lifenaut are online portals on which members are asked to create an informatic version of themselves through two different approaches. In the first, member information is based on an extensive

psychological profile form, or personality capture form, designed by immortalist, transhumanist, and social psychologist William Sims Bainbridge,
who is also program director at the NSF's Information and Intelligent
Systems. The personality capture form, with hundreds of questions, is
designed to create a comprehensive but updatable picture of a member's
beliefs, values, behaviors, attitudes, self-image, and so on. In addition,
members are encouraged to upload a range of digital files from photos and
social-media archives to personal writings, from Amazon shopping lists
to medical reports. Together, these are called a person's "mindfile," that
is, a cumulative database of information reflecting the person's unique
mind. All of this is run through algorithms and AI software (versions of
chatbots) that are meant eventually to generate independent, autonomous
persons stored on a digital platform for subsequent uploading and animation with the "consciousness software into a cellular regenerated or
bionanotechnological body by future medicine and technology" (Rothblatt
2012, 148). In the meantime, the mindfiles are "space-cast," transmitted as digital information into outer space via satellite from Terasem's
beachside headquarters, where Rothblatt received her download. At the
Terasem conference, Rothblatt explained, "So every Terasem joiner or
participant who has mindfiles with us has already achieved a certain level
of immortality by having aspects of their mindfiles already anywhere from
up to five to six light-years away from the Earth depending on when they
started uploading."

It might be tempting to dismiss all this as a typical New Age syncre-
tism of technoscience and spirituality—it certainly echoes some formal
qualities scholars have analytically associated with the New Age, such as
the reach for authenticity and origins, the revelation, the projection of
unity and inclusiveness, the appeals to energy and immortality (McGrath
1999; Heelas 1996; Griffin 1988). But the complexities of the account,
the specific transhumanist context in which it circulates, the subsequent
organizational and technological forms it has taken, and the way immortality emerges in the attempts to connect the self to the universe can bear
a lot more unpacking. I will contextualize this through an analysis of the
secular orientation toward the cosmos in which the universe is construed

as "dead" and discontinuous with human life and consciousness. Cosmological quests like Rothblatt's should not be understood as religious per se but rather as tensions that constitute aspects of the secular, in this case in relation to the gap between the mind and the universe. As in other aspects, immortalists have reshaped this gap in interesting ways, imagining "information" as the concept and the science through which the gap could be bridged, thus producing the possibility of both individual and cosmic continuity. In the process, they are also producing cosmic subjectivities. If the usual gloss on spirituality as an inward turn begs the question of the self—Where and into what are humans retreating?—a "mindfile" as an informatic practice instantiates the self on a continuum with the external world and generates cosmic subjectivities. Many of those who use their mindfile regularly tend to use it as a kind of memory device uploading the usual array of travel images and personal files. But the sense that these are somehow relata of their selves and that such relata are being beamed into space alters their orientation in the world. "It feels like my imagination has been extended to something in between fantasy and a nostalgia for something that hasn't happened yet (contact with other planets/ET's)," one mindfile user wrote to me in an email. "I think knowing that a part of me (my information) is free and on its journey through the cosmos makes me think that I am in some way expanded beyond my here-and-now experience of life."

This view depends on the assumption that both people and the universe are constituted by information. The informatic cosmos is not an immortalist quirk; rather, it reflects other emerging views in other fields responding to the same secular gap. While the practice of transmitting mindfiles via satellite may be as eccentric as it is esoteric, and admittedly some people might not want their mindfile floating out near a red dwarf, the assumptions that enable these actions are the same ones that have allowed cognitive science and neuroscience their ventures into the naturalization and digitization of the mind. Indeed, it is the coming together of disparate fields—convergence—around the notion of information that has energized the possibility of immortality through AI and turned immortalism into a computational or informatic assemblage.

TECHNIQUES OF THE COSMIC SELF

In honor of Nietzsche, the roboticist and computer engineer Dr. Ben Goertzel named one of his sons Zarathustra. Goertzel's penchant for Nietzsche arises out of a combination of an attraction to the *Übermensch* and the conviction that humans must—and indeed do—claim values for themselves, rather than deriving them from gods or transcendent realms. After that, it is hard to see anything else in common between Goertzel and the German philosopher. For all Nietzsche's nihilism, Goertzel's worldview is optimistic. He believes in joy and growth as the key values and assumes an underlying purpose for humans in the cosmos, even if humans at this point might not know exactly what it is. Goertzel is one of the ubiquitous figures in the immortalist and transhumanist circuits (where there has been a contentious and long engagement with Nietzsche's *Übermensch*). He was one of the chief scientists at the well-known company Hanson Robotics and has a number of his own AI ventures running, while teaching and advising institutions such as the Singularity University in Silicon Valley and the Brain-like Intelligent Systems Lab at Xiamen University in China. Son of hippie and Marxist parents, Goertzel grew up learning a lot about Buddhism while reading science fiction, both of which he says are really about the nature of the mind, the world, and the cosmos. His wife, Gwendalin Qi Aranya, is an ordained Zen priest and a reiki master.

Goertzel is also the author of *A Cosmist Manifesto,* a handbook in which he advances joy, growth, and choice "not only as personal goals, but also as goals for other sentient beings and for the Cosmos." He explicitly calls his *Cosmist Manifesto* a worldview and hopes that it can "provide concrete guidance to the issues we face in our lives." He views "cosmism" as a way to actively seek to grow by being constantly aware of the universe. The principles behind the *Manifesto* are what "an individual or group can adopt." In a version of what, after Foucault, one might call techniques of the cosmic self, Goertzel encourages cosmists to "ongoingly, actively seek to better understand the universe in its multiple aspects"; to abandon habitual thought patterns, even through meditation or psychoactive substances if necessary; to gain greater control over emotions because that's an important step toward transhuman evolution; and to realize "that the fate of one single species, on one single planet, is seen as insignificant next to the fate of the known universe." He explains that "thinking daily about the grand themes of the cosmos" is important because "these ideas have simple, practical, everyday meanings" (2010, 23).

When I asked Goertzel about his daily relationship to the cosmos, he said he tries to keep these cosmist points in mind through his day. He meditates daily on the roof of his home, and that puts him in a "cosmic mood" and develops "feedback with the larger cosmos." Over the years he has developed a daily "mind-state" whereby he keeps himself constantly primed so that any activity "encompasses/acknowledges its embedding in and feedback with the larger cosmos, and also its construction within a certain ('my') stream of experience as a self-organizing pattern system."[2]

Once when we sat to talk on an old church pew in the 92nd Street Y in New York, where the Singularity conference was being held, we ended up talking about Hegel and whether the evolution of the universe was dialectical. His own cosmist view, as well as the view of Singularitarians, describes a universe that has evolved from simple to complex information patterns, of which human intelligence is one but not the ultimate manifestation. When I asked him whether the futurist view of an informatic unfolding was committing the secular sin of an anthropocentric teleology, he simply said, "Well, it's not about us." We are not an end for the universe, he believes; however, like Ray Kurzweil, he thinks we share in its purpose. In order for this view to hold water, one has to assume a kind of panpsychic continuity throughout the universe of matter; that is, mind and intention have somehow to find their way into the material account of the world out there. Indeed, Goertzel asserted that "there is a meaningful sense in which we can talk about awareness in the universe."

That doesn't sound like a mainstream Western secular scientific view, and that is exactly what Goertzel's *Manifesto* is manifesting against. "Modern Western culture," he writes, "has led to a world-view in which most of the universe is viewed as somehow 'dead' while only certain particular systems are viewed as having 'awareness'" (2010, 41). The dead universe Goertzel fights against is striking because it is a direct formulation of Weberian disenchantment: human existence was meaningful when it was located within a larger cosmological order which imbued everything with some kind of spirit; science and rational calculation have done away with that, so that all the matter around us is now inert, everything in principle subject to mastery by calculation, which is why, as Weber puts it, the "natural sciences" make "the 'meaning' of the universe die out at its very roots" (1958, 144). Weber didn't think we could make up the loss; the doors of the church remained open, but there was no turning back. Men of science just had to live or struggle with the meaninglessness and get on with their work—this was the mature secular scientific position adopted, with a few exceptions, throughout the twentieth century. Yet Goertzel is very much a man of science.

Goertzel's cosmism may be a particular vision, but talk with immortalists often turned to these kinds of questions, cast in the language of the humanities, what one cryonicist characterized as "that eternal question philosophers have been asking, Why? Why all this?" What came as a surprise to me was the extent to which these larger questions were foregrounded, sometimes as more important motivations for seeking immortality than the everyday joys of a long life. A cryonics researcher told me: "There's the daily things: going out, seeing friends, having a drink . . . but there's bigger things there: why are things the way they are? And the only way I'm going to understand the universe is if I live long enough, be able to explore and expand my own intelligence." Here is another cryonicist's account of his motivations for seeking radical life extension:

> The impulse that led me in this direction is just a desire that must be natural to me: to understand the big picture, of what this life is really all about. All of us grow up with this mélange of traditions about what the world really is about, what family's about, what society's about, what life means. . . . I guess I was never really satisfied with a lot of those answers. . . . I've read a lot about cosmology and the big bang and the universe and what's the nature of matter and how do cosmologists understand, or try to understand, where all this came from and where it's going. Meaning of life type questions.

In his quest for answers, this cryonicist, like many others I spoke to, ended up alienated from his family on existential and religious grounds, but found a community in transhumanists and immortalists, where such questions are not sequestered but discussed enthusiastically in specific nonreligious ways that suited him.

Admittedly, transhumanist questions and answers rely on a different set of permutations, a kind of nonreligion that keeps pointing to and needing the very allusions it sequesters as religion. Thus, an explanation around the quest for immortality might include affects and motivations such as these:

> None of it makes sense to me, so I figure there has to be a higher meaning than just objects interacting with objects for no reason. . . . I believe there has to be a higher power of some sort that has impacted the nature and structure of the universe in some way . . . for all I know it could be another race of humans who have gone through transhumanism to the point where they have the powers that they can manipulate time and space to the extent that they could've invented this universe out of another universe. . . . I think science could get us to that point, the only question is how long will it take and will we survive that long.

Echoing him are Ian Marks's words: "Curiosity. About everything. How the universe really works. All the unanswered questions. . . . Why are we here? I guess I could find out by dying if there's anything to me besides this. I'm not betting on it."

Immortalism and its mode of suspension may sensibly figure as a response to the pressures of temporal finitude, of a life without an afterlife, but what's with this constant reckoning of life on a cosmic scale, connecting inner selves to outer space and the whole cosmos? Why and how does an imaginary of "not dying" assume this particular form? What are its tacit cultural grounds? What kind of cosmological identity is this?

William Sims Bainbridge, NSF officer and Rothblatt's CyBeRev colleague, has also been worrying about the secular deadness of the universe for a long time, though from a rather different angle. In his early phases he lamented that a secular worldview could not provide strong social motivation for an advanced civilization. In 1982 he penned a utopian manifesto titled "Religions for a Galactic Civilization," explaining that "the human condition is one of extreme absurdity unless fixed in a cosmic context to provide meaning" (1982, 200). It would take the kind of "transcendent motivation" only offered by religion to build such a civilization. Without a religious structure, a transhumanist, interplanetary civilization will not be possible. And that is important because a transhumanist, interplanetary civilization is the only hope for long-term human survival and the survival of intelligence in the universe. Bainbridge proposes creating religions suited for the purpose: "We need several really aggressive, attractive space religions, meeting the emotional needs of different segments of our population, driving traditional religions and retrograde cults from the field" (198).

I heard similar laments from many a transhumanist and cryonicist: if only they could provide the right illusions, or the right sense of transcendent meaning, or the final answers via science to those "perennial questions," they would have as many followers as religions do! That's the quintessential secular scientific bind: the only way to replace religion is to become a religion, that is, by not abnegating on these questions, as has been secular science's troubling and contradictory position, but by answering them. And any answer, as Bainbridge lamented, would sound like religion. I will return to this dilemma.

In a later book, Bainbridge (2007) took another look at his developing posthuman and scientific environment and decided that a fabricated religion was no longer needed. He had underestimated the power of science to posit a true cosmology. In that book, *Across the Secular Abyss,* information and informatic sciences were charged with the task of answering the secular dilemma of purpose and meaning, of crossing that abyss. Interestingly, Bainbridge's ideas resurrected his earlier manifesto, which had sat silent for over two decades, spawning a technocosmological group, the Order of the Cosmic Engineers (OCE), as a practical way to deal with this issue. The OCE, which later turned into the Turing Church, was founded by Giulio Prisco, an Italian transhumanist activist, former physicist, and current Singularitarian who acknowledged that Goertzel's cosmism and Bainbridge's ideas were an inspiration in this regard. Writing against the dead universe view, Prisco formulated the goal of the OCE as follows: "to permeate our universe with benign intelligence, building and spreading it from inner space to outer space and beyond."[3] Despite its attempt at responding to the secular dilemma, the group never gained a significant following.

Nevertheless, such a quest for cosmological reembedding remains a key part of immortalism and transhumanism, rather than a minor side effect. And these notions are not merely fantastical; they are instantiated already in assemblages of people, software, hardware, organizations, practices, and techniques, producing distinct material practices (such as space-casting mindfiles and storing personal data in underground vaults). What's more, the informatic continuum is part of a wider quest or struggle in mainstream secular culture and science, which have been eminently aware of the cosmological inadequacies and anxieties.

COSMOLOGICAL COLLAPSE AND COSMIC SUBJECTIVITY

At the cusp of modernity, himself suspended between science and theology, Blaise Pascal had a vision of humans suspended between the infinite and the empty: the "Nothing from which he was made, and the Infinite in which he is swallowed up" (Pascal 1958, II, 72). "When I consider the short duration of my life," Pascal wrote, "swallowed up in the eternity before and after, the little space which I fill, and even can see, engulfed in the infinite immensity of spaces of which I am ignorant, and which know me not, I am frightened, and am astonished. . . . I see only darkness everywhere. Shall I believe that I am nothing? Shall I believe that I am God?" (III, 205, 227).

If Pascalian phrases have been repeated so often, it is because they articulated an emerging anxiety that would become common over the course of the next few centuries. The emergence of this anxious relationship to the cosmos owes, in Pascal's words, much to Galileo and his telescopes, expanding, as Pascal wrote, the number of moons and planets beyond our consideration, and literally decentering the place of the human world in their midst.[4] The sense of insignificance predates Pascal, perhaps even preceding Galileo, beginning with the Protestant Reformation and the breakdown of the ascending earthly and heavenly hierarchies reaching up into the firmament—the Protestant sky, after all, had no interceding angels. But the Copernican-Galileon decentering and opening up of the cosmos is the clear precursor. The once-reigning Ptolemaic system was not different only in terms of the Earth's centrality in the solar system. It also carried a completely different image of the universe as a closed and contained system in equilibrium. It was composed of nine concentric spheres, all wrapped in a layer of light called Empyrean—and, in some accounts, the only thing outside was God. Other than that it was an interconnected, permeated, and continuous set. With or without its Judeo-Christian God, this was a total and contained cosmology. So the Copernican-Galilean model didn't just displace the relationship between Earth and the sun; it broke open the cosmos.

The infinite universe was born of this, and along with it the finite human, a being plagued by the suspicion of its own unrescuable and irredeemable insignificance, isolated on this strange planet. Awe at the immensity of it all and the tremendous sense of mystery had certainly struck generations of humans, but before Pascal no one had looked up at the heavens and expressed terror at its *emptiness*. From Pascal, one can trace the affect through Schopenhauer and Tolstoy to Weber and much of Western philosophy, which became more and more interiorized.

Today, in the general secular modern context, it might seem that contemplating the cosmos or establishing a cosmic "feedback" loop is not part of people's everyday life—notwithstanding the Goertzels and Priscos. Yet, looking around a little, it is not hard to come upon astronomers and physicists, SETI researchers, NASA employees, and private space engineers who are doing just that; then there are astrologers with scientific charts, some of them employed by investment banks, and there's that one friend we all have who forwards NASA's picture of the day to his list. There are awe-filled reactions to new pictures of the sun. There are satellite companies and satellite launchers and installers mapping space, and there is GPS and Google. There are meteorite collectors, and when a solar eclipse happens millions rush to *experience* it. Indications of water in the solar system make headline news, the search for intelligent interstellar life is on, and though most people don't take the accounts very seriously it is hard to deny that alien abduction tales are at least a site of affective contact with the world beyond the planet. And, of course, there are the Richards (chapter 4) and all those others who through childhood engagement with science fiction *feel* themselves inextricably entangled with a future in space, think their lives through Andromeda, and are bound to the cosmos in a virtual way.

The anthropologist Goetz Hoeppe (2012) offers a touching account of astronomers who are not simply crunching data and mapping events but feel something deep when they "witness" an event that happened far away a long time ago or are even "touched" by a particle that has literally traveled that distance and that time. David Valentine (2012) has argued that private space ventures are as much about the sociality of space as they are about capitalist colonization. Debbora Battaglia (2014a, 2014b) has documented the impact of satellite orbits and falling debris on institutions such as insurance as well as on geopolitical affairs and diplomacy, considering in the midst the possibilities of a "cosmic commons." The universe may have been rendered cold and empty, disenchanted and mindless, but that hasn't stopped many people from letting the cosmos affect them, from having a cosmic subjectivity of some sort, figuring some aspect of their existence in relation to aspects of the cosmos. If anything, it would seem, the cosmic relationship is increasing in intensity; there is a need to exit from the cosmic alienation of the human, its unique, bounded but troubling existence in the dead expanse of the universe.

But it isn't that easy. There is no real set of scientifically sanctioned or secularly authorized forms of carrying out an exchange, of *relating* to the cosmos. After all, everything else is gas and particles and electromagnetic waves, isn't it? No mind out there, not even life, no continuity! So any story about cosmological belonging, any affect aside from the Pascalian terror at the impersonal vastness of it and the human loneliness in it all, is deemed illegitimate, akin to alien abduction stories or some religious delusion. In fact, that is one of the ways in which religion is secularly interpellated: you recognize religion, secularists say, when you see answers to this need for cosmic connection (e.g., Geertz 1973). And so every attempt to address cosmic alienation ends up looking like religion, because accounts of purpose, telos, and meaning are not found in standard scientific models of the mindless universe.

All this is severely exacerbated by various scientific projections of the future in which not just Earth or the species but the whole universe will come to an end. Recall, for example, Richard's worry that "our galaxy is on a collision course with the Andromeda galaxy and in roughly 300 to 500 million they will collide." Finitude and nihilism came wrapped together through secular history, generating an explicit sort of anxiety about total annihilation without the possibility of redemption or rebirth. That secular eschatology, its sense of an end-time, surfaced through scientific predictions. In fact, the first wide-scale manifestation of such a science-based end-time scenario was probably heat death, originally postulated by a German physicist in 1854. Derived from the second law of thermodynamics, the theory of heat death suggested that all energy-dependent forms, especially life, will eventually cease to exist in the universe. Lord Kelvin's apocalyptic prophecy, having taken the form of a scientific law, was only driving home the larger point that the whole enterprise was, as every secularist had to concede, pointless. So heat death came to serve as shorthand for these anxieties. Charles Darwin articulated his own fears with these words related to immortality as consolation to cosmic alienation:

> Believing as I do that man in the distant future will be a far more perfect creature than he now is, it is an intolerable thought that he and all other sentient beings are doomed to complete annihilation after such long continued slow progress. To those who fully admit the immortality of the human soul, the destruction of our world will not appear so dreadful. (1859, Appendix A, 433)

The philosopher Bertrand Russell (1903) was not atypical when he worried that "all the labors of the ages, all the devotion, all the inspiration, all the noonday brightness of human genius, are destined to extinction in the vast death of the solar system, and that the whole temple of Man's achievement must inevitably be buried beneath the debris of a universe in ruins." Russell described "the world which Science presents for our belief" as becoming all "the more purposeless, more void of meaning." Yet, like Weber, he braces himself and urges us to go on doing our work. "Amid such a world, if anywhere, our ideals henceforward must find a home. . . . Only within the scaffolding of these truths, only on the firm foundation of unyielding despair, can the soul's habitation henceforth be safely built."

From Kelvin to H. G. Wells to Russell to Freud to many others, the tensions and anxieties brought on by heat death run "like a leitmotiv through turn-of-the-century cultural formations," as Katherine Hayles (1999, 101) put it. Other science-generated end-time scenarios would follow. In the second half of the twentieth century, the Nobel Prize–winning French biologist Jacques Monod, while maintaining that God had been utterly refuted by science, could only repeat a Pascalian terror: "The ancient covenant is in pieces: Man knows at last that he is alone in the universe's unfeeling immensity, out of which he emerged only by chance" (Monod 1972, 180). Similarly, the Nobel Prize–winning physicist Steven Weinberg pronounced, "The more the universe seems comprehensible, the more it also seems pointless" (quoted in Easterbrook 2002). As it was discovering the universe, secular modernity was also uncovering it as meaningless— or dead, in Goertzel's terms. Without a cosmological explanation, or at least a cosmological promise, is scientific knowledge purchased only at the price of absurdity? It is not an exaggeration to claim that such a tension describes and constitutes a secular scientific subjectivity.

Contemporary scientific knowledge and power were made possible by removing mind (or soul or God or animistic spirits or forms of consciousness) from the cosmos. The cosmos, therefore, was no longer the font of meaning, nor could it have intentions or purpose. So there appeared *a cosmos without a cosmology,* a universe that could be known but was meaningless. I use "cosmology" here to denote the ways in which humans and other creatures hold a place along with what at other times would have been called heavenly bodies in a larger order, in which humans along with the other entities and systems can co-constitute each other relationally, or in intra-actions (Barad 2007), and manage to derive a sense of meaning and value as contributors to cosmic life (Becker 1973, 5; C. Taylor 1991, 3). It is the collapse of that cosmological tent that has been called disenchantment, sowing alongside the knowledge and power of science a kind of cosmic anxiety, of knowing the universe but not knowing one's place in it. In the disenchanted secular scientific worldview, cosmic destiny and human destiny are rent apart, and at any rate both are absurd, since each will in its own way at its own rate end in annihilation. What is life, thought, meaning, in that cosmological gap, where humans are not only alone but, as mindful observers, also always apart? With no mind in the cosmos, how may humans feel any true sense of connection with it? How and why should such a relationship matter at all? If it's all going to end, what's the point?

Within this secular imaginary, awareness of finitude is haunted by a deep ontological insecurity, imbued with anxieties over the meaninglessness of the universe, of a foreknowledge of entropy, of the irreparable rift between the parochial human world of meaning and Pascal's cold and indifferent universe. It is not the loss of meaning alone, nor only the effect of instrumental rationality or of the elimination of the afterlife, that accounts for the condition of modern disenchantment. It is their combined and coiled movement, of science reaching into more and more spheres, from atoms up to multiverses, while falling categorically short of the metaphysical extremities. Thus those extremities are reinforced as voids, as a collapsed cosmology. One wonders to what extent this sense of cosmic solitude motivates grand scientific projects. It certainly has shaped secular and scientific discourses and practices around life, death, and survival in modernity.

As exemplified by Bainbridge's assertions, secularists are well aware of all this *as* a problem. It has been presented *as a problem* and has been a major preoccupation through the intellectual and political history of the twentieth century (Nagel 2012; Brassier 2007). The philosopher of finitude Quentin Meillasoux, in a recent attempt to treat this bind, presents a prescription for the secular malaise: "The question must be resolved since to claim that it is insoluble or devoid of meaning is still to legitimate its celebration" (2008, 72). In other words, as Ray Brassier paraphrases Meillasoux via Wittgenstein, as long as such questions remain "unanswerable, the door is left open for every variety of religious mystification" (2007, 73). Meillasoux's secular solution is "to free ourselves from the question" not by falling for the game of finding a "first cause" that could lend us meaning but by acknowledging "the latter's eternal absence" (2008, 72). Yet this is just the problem—the absence itself is generated as important by the very conditions under which Meillasoux asks us to downplay its importance; in that system of monistic knowledge it cannot be downplayed. That's because science has totalizing claims—namely, its quest for "a theory of everything," in physicist Steve Weinberg's exposition—and so its very foundation trembles when phenomena seem intractable to its operations, leaving gaps or zones of indeterminacy. The indeterminacy also presents tensions for secular psychology and governance, which depend on science as a key source of authority. The "secular consensus" (Nagel 2012) begins to crack at these very junctures where the relationship of mind and matter in the universe is at stake.

The modern secular era, just like the immortalists who are often considered its extreme exemplars, has been characterized as an era of human hubris and delusions of grandeur and mastery over the nonhuman realm. Yet it has equally been an era of self-doubt and mounting anxiety. Descartes is known for his pronouncement on mastering and possessing nature, taken as the founding dictum of modernity and modern science, but it is readily forgotten that his whole venture for mastery rested on a fragile bed of doubt—almost a paranoia—about the nature of knowledge and being. I emphasize the latter here, because all too often the question is taken as a purely epistemological issue, and thus the affect running through such existential doubts are ignored. Indeed, this became the paradox of science, that while it increased technical mastery over the local environment, it shrank human existence to an absurd accident in a purposeless universe itself due to expire sooner or later. This torque of control and anxiety, of mastery and mystery, knowledge and meaninglessness, sits coiled inside secular modernity.

It's in light of all this that the particular relationship between the immortalist desire for long life and their connection to cosmic purpose becomes better visible. On the obvious level, long life without meaning would amount to a long stretch of meaninglessness, and the only way to restore that meaning is to build it right back into the universe: if the relationship to death is simultaneously a relationship to a dead universe, then overcoming death requires also overcoming that meaninglessness, linking the purpose of a human self to the whole cosmos, inner to outer space.

INFORMATIC COSMOS

In the cosmological gap, then, where intractable discontinuities persist, information suggests the possibility of a continuum between a number of secular scientific binaries—mind–matter, animate–inanimate, human–nonhuman, natural–supernatural—without altering the immanent materialist ontology beyond the recognition of science or adhering to a biased contingent social process (such as capitalism). The question of mind is a cosmic matter. In the words of the Brain Preservation Foundation, informatic neural networks are "rapidly-improving carriers and extensions of both our internal and external selves."[5] The self is seen as an emergent property of informational complexity, where information and complexity serve to diffuse older terms like "animate," "consciousness," and "self-awareness." Rothblatt (2009) writes: "Consciousness arose from genetic mutations that directed neurons to be connected (or grown) in a way that empowered self-awareness. In other words, inanimate molecules are ordered by DNA to assemble into conscious-trending clumps of neurons. It would not seem less improbable that inanimate lines of code can be ordered by human intelligence to assemble into conscious-trending clumps of software programs. What must be the case, barring mystical explanations, is that consciousness is an epiphenomena [*sic*] of a good enough relational database."[6] She sums up the claim: "People are information."

"People are information" is a literal statement with echoes and variations throughout immortalist communities. One of the participants in the Terasem conference, Mike Perry, a soft-spoken, smart cryonicist, and president of the Society for Universal Immortalism, has written: "We do not accept the supernatural definition of the soul, and instead provide a rational, scientific definition of what constitutes our soul. . . . The memories we have, the thoughts we consider, the emotions we feel—these form the essence of who we are . . . are represented in the brain biologically as information . . . like software, it is the pattern of information that is important, not the medium in which it resides or is expressed."[7] The Brain Preservation Foundation suggests that "we recognize ourselves as not merely physical but also informational entities. We are each special collections of unique and valuable experiences and information, carried in our biology, society, and technology. By consciously stewarding our informational selves, by daily choosing which thoughts, memories, and behaviors we wish to focus on in our lives, and what information we want to represent us, both now and in the future, we can best continue to improve ourselves and our communities."[8]

As they try out new vocabularies and concepts to move beyond the limits of some of the older dualities but *trying not to* depart from materialist premises, immortalists and transhumanists end up deploying concepts such as information (or pattern) that perform a sort of codependency, such that the material and immaterial come to both oppose and need each other. Neither transcendent in the traditional sense nor quite material (insofar as it is a pattern independent of the contingent material basis of its manifestation), "information" works to allow for the evocation of a cosmic unity whereby the inner and outer, self and universe, are connected. That is how the Acceleration Studies Foundation, tasked with forecasting, predicting, and advocating for technological singularity, can boldly see the development of computing as an evolutionary acceleration that is "racing towards inner space," into our thoughts and feelings, connecting them to the material complexity of the universe itself—recalling the triangulation of information, consciousness, and matter experienced by Rothblatt on the beach, her cosmic subjectivity connecting the ancient past in the turtle to the future in the space shuttle.

What matters equally here, then, is that in this cosmology the universe itself is informatic, moving with purpose toward greater complexity and intelligence. Information, emergence, intelligence, complexity—these all point toward ways of putting mind and purpose back into the cosmos. A cosmology has been wrought out of the cosmological gap, and Ray Kurzweil has been its biggest promoter. By 2045 or thereabouts, he has predicted, the pace of technological change will reach such dizzying speeds and such expansive breadth that "human life will be irreversibly transformed" (Kurzweil 2005, 7) by that event horizon now commonly known as the Singularity. The Singularity will start when AI becomes powerful, speedy, and intelligent enough to build its own intelligent agents, leading to an unimaginably fast proliferation of intelligent agents, in an "intelligence explosion" that will be able to access and manipulate with flawless Bayesian logic an unimaginable amount of information and matter in the universe—that's what will turn general human intelligence into superintelligence (also see Bostrom 2014). At that point, humans will merge with these superintelligent agents, thereby simultaneously realizing a potential for immortality by transcending their biological heritage and becoming one with a universe now saturated with intelligence, self-awareness, and knowledge. That is the awakening of the universe.

Clearly interested in biology and survival, Kurzweil proudly declares that he takes more than one hundred supplements and that they have kept him healthy at a "biological age" closer to forty. He is also a member of the Alcor Life Extension Foundation, where his body will be cryogenically preserved in case things don't go exactly according to his predictions. It would be a pity for the "prophet of the Singularity" to miss the Singularity. His first book, *The Age of Intelligent Machines* (1990), charted the acceleration of information-related technologies as exponential curves rather than linear growth up to a point where machines would develop to be faster and smarter than human minds. His second book, *The Age of Spiritual Machines* (1999), attempted to predict the "nature of human life" under conditions that would see machine and human cognition blend in what he later called an "increasingly intimate collaboration between our biological heritage and a future that transcends biology" (Kurzweil 2005, 3). But it was only in his last book, *The Singularity Is Near,* that his talent for predictions took on a programmatic, even cosmological, form.

Before exploring further, it might be worth recalling that from the beginning information theory was also a response to theories and anxieties of entropy. Norbert Wiener, one of the great architects of cybernetics, outwardly expressed his worries. In 1953, Wiener wrote: "We are swimming upstream against a great torrent of disorganization, which tends to reduce everything to the heat death of equilibrium and sameness" (Wiener [1954] 1989, xiii; also see Gleick 2011, 237). He immediately connected this to existential philosophy: "Heat death in physics has a counterpart in the ethics of Kierkegaard, who pointed out that we live in a chaotic moral universe. In this, our main obligation is to establish arbitrary enclaves of order and system." In a coauthored paper (Rosenblueth, Wiener, and Bigelow 1943) aptly titled "Behavior, Purpose and Teleology," Wiener had tried to make good on this obligation, developing a theory of electronic feedback loops, examining both positive and negative feedback. The main concern was behavioristic—as much in electronic circuits and mechanical objects as in humans and animals—as opposed to functional. That is, the paper aimed to jettison an explanation based on intrinsic and internal organization of an entity in favor of an explanation based on its relationship to the environment as a consequence of both inputs and outputs, in an ongoing stream of effects.

The problem was that in order to even interpret any event as a behavior consequent to a sequence of inputs and outputs they had to have a sense of the end state of that behavior. Otherwise, the event, though it would still be a sort of behavior, would not be "purposeful"; instead it would be merely what they ended up calling "random"—the chaotic result of no input in particular. "Purposeful denotes that the act or behavior may be interpreted as directed to the attainment of a goal. . . . Purposeless behavior is that which is not interpreted as directed to a goal" (Rosenblueth, Wiener, and Bigelow 1943, 18). They characterized purpose mechanistically: "All purposeful behavior may be considered to require negative feedback" (24). In lay terms, purposeful behavior is that which sends output into the world until it receives the input that satisfies a particular *predefined* condition. For example, a missile is not purposeful in itself; it is so only when it seeks heat until it finds it. When this was coupled with information theory it yielded the idea that information as such was not useful, but information processing (input–output) was: it was a movement away from noise toward order. Information-*processing* entities were always already purposeful entities. It was the trope of purpose that pushed later cyberneticians toward the formulation of emergent systems and self-organization. So while information theory banished semantic "meaning" from communication, it smuggled teleological "meaning" back into science. Hayles (1999, 102–3) makes a similar point, asserting that mathematician, engineer, and Wiener's colleague Claude Shannon's move of retaining a connection between entropy and information was a "crucial crossing point, for this allowed entropy to be reconceptualised as the thermodynamic motor driving systems to self-organization rather than as the heat engine driving the world to universal heat-death."

Cybernetics managed to suggest that information theory actually does away with entropy: information theory, it was said, shows that order *increases* in the universe!

Information theory worked against cosmological anxiety and legitimized teleology within the bounds of science. Within a decade of the publication of Shannon's original engineering definition of information, the theory was being applied to all sorts of "other forms of communication" in biology, psychology, robotics, and physics (Borgmann 1999, 132). Information was as much a property of a cell as of a human mind, a human-made machine, or any and all of the very primordial constituents of the universe. Soon everything was seen as a yes-no machine, an information-processing and information-transmitting unit. In the essay where he coined the term "bit," as a unit of information, the physicist John Archibald Wheeler, collaborator of Bohr and Einstein, theorizes the universe as fundamentally, even primordially, an informatic device: "No element in the description of physics shows itself as closer to primordial than the elementary quantum phenomenon, that is, the elementary device-intermediated act of posing a yes-no physical question?" (1990, 3). He concludes: "All things physical are information-theoretic in origin" (22). Neuroscientists talk about information as the basis of mind, and consciousness is being redefined as a product of computation. If cosmologists talk about the universe being an information-processing entity, molecular biologists will talk about biological matter and form as gene expressions read from the information in the DNA or RNA and output as proteins, thus simultaneously explaining developmental growth and accounting for generational consistency (genes copying their information) without needing to make recourse to vitalistic or transcendent concepts. Biologists who had turned away from teleology as a reaction to vitalism—no modern biologists would want to be labeled a vitalist!—found it possible to talk about biological purpose through information without bringing up the specter of vitalism (Keller 1995).

Starting in the mid-1980s a small but growing number of physicists and biologists began to reclaim teleology in a larger sense, in part thanks to information theory as well as to the marriage of information theory and biology (e.g., Barrow and Tipler 1986; De Duve 1995). In most of these teleological theories of the cosmos, the emergence of life or human intelligence is used to prove the directionality of the universe. But the teleology does not end with the human in the anthropological sense; rather, like Goertzel's vision, it passes disanthropically through humans to other forms.

The purposefulness and meaningfulness of information as a foil to entropy has been directly and joyfully imported into the immortalist and futurist context. As I showed in chapter 2, this was part of Ev Cooper's motivation when he was reading Wiener. It was also inscribed in the beginnings of the transhumanist movement, when Max More started the journal and then later the institute that went by the name of Extropy—a concept deliberately chosen to denote greater order and purpose rather than a pointless deterioration. In my interview with him, I came to understand that entropy had affective and social parallels, a postwar gloom-and-doom pessimism, a sense of constraint imposed by the Club of Rome and the Limits to Growth movement, that had also to be overcome. More explained his thinking and the formation of extropianism: "For me it was a response to what was going on in the seventies, which was a very gloomy decade. Very high unemployment rate, inflation and fears of nuclear war . . . and I always rebelled against that. I was an optimist, I wanted something better."

The Extropy Institute was crucial in the development of immortalist and transhumanist ideas, but it was Kurzweil who put numbers and dates on the anti-entropic ethos, using empirical data and mathematical models to chart the progress of complexity and predict the development of human-level AI or the merger of human and machine cognition. The Singularity has received attention mostly for its predictive self-assurance and views about the advent of superintelligent machines. I will focus on its cosmology, for part of its attraction has been just that.

Kurzweil has no apparent misgivings about interpreting the "very purpose of the universe," which, he writes, "reflects the same purpose as our lives: to move toward greater intelligence and knowledge" (2005, 372). In books and lectures, that cosmic destiny is presented via predictions based on the analysis of large data sets and graphs covering *technological* paradigm shifts and canonical milestones throughout "the history of the world," laying out everything from the Cambrian explosion to the invention of agriculture to the development of computing power and the decoding of DNA to microcosmic analyses of changes in transistor size and price, and processing power over the last few decades. The graphs, it is argued, show that from time immemorial, technology has been obeying Moore's law[9] and developing at exponential rates.

These claims rest on two premises. The first is that the entire universe is constituted by readable and, more important, useful information. Information is simply regarded as patterns with varying degrees of complexity. Any arrangement of atoms may be considered information. The atomic structure of a giant rock may contain more total information than the genetic code of a human, but a rock's information can be represented by a few simple specifications, so most of its information is redundant. The more complex, efficient, and *purposive* an entity becomes, the more intelligent it may be considered. Based on the ideas of some philosophers of mind, intelligence is considered to be "the activity of suitably arranged matter" (Churchland 1988, 167).

Through this frame, Singularitarians chart the history of the universe in terms of the evolution of intelligence. Kurzweil breaks down this universal evolution into six epochs, presented entirely through the increasing organization of information into intelligence, such that information works here as a *deus en machina,* an ordering mechanism internal to the totality of the cosmos, from its tiny particulars to its gigantic generalities: from the appearance of physics and chemistry on Earth (information in atomic structures), to the development of biology (information in DNA), to the rise of the brain (information in neural patterns), the creation of technology (information in design), to the merger of technology and human intelligence (the Singularity), to the final epoch—called "The Universe Wakes Up" or "The Intelligent Destiny of the Cosmos."

We are currently on the cusp of the fifth epoch. "Although neither Utopian nor Dystopian, this epoch will transform the concepts that we rely on to give meaning to our lives, from our business models to the cycle of human life, including death itself" (Kurzweil 2005, 7). Of the sixth and final epoch, Kurzweil writes: "In the aftermath of the Singularity, intelligence, derived from its biological origins in human brains and its technological origins in human ingenuity, will begin to saturate the matter and energy in its midst. It will achieve this by reorganizing matter and energy to provide an optimal level of computation to spread out from its origins on Earth.... [T]he 'dumb' matter and mechanisms of the universe will be transformed into exquisitely sublime forms of intelligence" (14–21). He closes the section on the sixth epoch with the following declaration: "This is the ultimate destiny of the Singularity and of the universe." Darwinian evolution, based originally on random (nonteleological) biological development, is transformed and applied to all matter in the universe as it develops now with purpose from simple, "dumb" matter into complex, "intelligent" matter. Whatever the data, cosmological ambition is the very purpose of Kurzweil's futurology: "I have begun to reflect on the future of our civilization and its relationship to our place in the universe," he declares modestly at the beginning of the book (4).

The second premise of the Singularity is that all information, however complex, is equally subject to computational manipulation. Information makes atoms, genes, neurons equivalent to each other via "bits," and through this equivalence, by flattening the universe into a primordial constituent, it generates the assumption that the larger complex structures that are built on them—for example, the human mind, culture, the universe itself—are *thoroughly* subject to manipulation, editing, and reformation.

So in the symbolic order of the Singularity's cosmology, information is the essence of the universe, and superintelligence is its teleology or purpose. The constitutive essence becomes increasingly organized as intelligence, the unit of "intrinsic value," as Goertzel (2009) called it in a *Forbes* article. But intelligence does not require biology—just some atoms and energy (Churchland 1988)—and thus may be produced on nonbiological or non-carbon-based substrates. Kurzweil's is part of an increasingly common view among scientists and theorists who are spinning information as the concept with the most privileged relationship to reality. Information has eclipsed the organism (Doyle 2003) as the matter of biological analysis, and a certain kind of wet biology seems in a practical and ideological sense obsolete. Since all development and decision making in the physical world works through algorithmic functions, so the world must be constituted by such functions. The world, then, is not only said to be representable by computation but a result of it. Life, mind, and this thing we call human are only the most recent and complex instances of a computational unfolding; biology is merely its current platform, a background technology, and a limited one at that. We are a modality of information.

Indeed, our very purpose as human versions of information in this very purposeful evolution of the universe is, in the words of Goertzel, "to give rise to other types of minds." He writes: "Life isn't just about one's own self. Just as there's intrinsic value in helping other humans, there's intrinsic value in helping higher intelligence come into existence" (Goertzel 2009, 253). But there's nevertheless a central role for humans. What effectively is offered in the infusion of matter with intelligence, in the "waking up of the universe," is the possibility of restoring to our disenchanted universe a kind of *mind* and to humans a sense of *our* role in *its* unfolding. Its most intelligent agents so far, humans are not just the decoders of the natural world but the purveyors of the next epoch. Evolution's accident-based, lumbering pace of change, "its lack of abstract intelligence, its reliance on random mutations, its blindness and incrementalism" (Yudkowski 2008, 323), to quote the founder of the Singularity Institute for Artificial Intelligence, will be superseded by a speedy, superintelligent agent engineered by humans.

When I asked Goertzel to elaborate on the idea of being surpassed or superseded, he said:

> Not being a religious person there's no basic value that I adopt on faith or instinct, at least not on an explicit level. So I thought about the core values that seemed important to me. And, of course, it's complex, the definition of those concepts. But basically I isolated our joy, growth and freedom. If you begin with these, you want happiness, growth, and change, and autonomous beings and minds being able to choose. Then the value of intelligence comes out of those. Intelligence enables growth and creation of new patterns. It's a way of realizing those things. Intelligent creatures can exercise growth and choice better than rats. Growth goes beyond humans, beyond mushrooms and ants and such. Are ants obsolete or pigs obsolete? They exist and continue to do what they do, but they are not the most complex or most interesting creatures on the

planet. That's what I'm assuming is the fate in store for humans. I hope some humans continue to exist in their current form, but there's going to be other minds. If it really came down to it, which is not a realistic choice, but if it came down to it, I wouldn't hesitate to annihilate myself in favor of some amazing superbeing.

CONVERGENCE AND AMERICAN COSMOCULTURE

In the absence of superbeings, deities, and spirits, the most convincing means of filling the cosmos seems to have become manifest through information and computation. Far from emptying the universe of meaning and abandoning existential inquiry, here science is used to fill the cosmos up with culture and meaning again through a telos of complexification. When these ideas are forwarded by immortalists and Singularitarians and transhumanists, they are portrayed as science fiction geek fantasies emanating from a marginal cult of "kooks." In fact, both sociologically and ideologically they circulate in a much wider context where notions like an informatic cosmos have come to occupy an important place in scientific practice and theory more broadly, "giving information a central role in a scientifically informed ontology" (P. Davies and Gregersen 2014, 8). For example, Freeman Dyson (2011, 10–11), a renowned physicist, has written:

> The explosive growth of information in our human society is a part of the slower growth of ordered structures in the evolution of life as a whole. Life has for billions of years been evolving with organisms and ecosystems embodying increasing amounts of information. The evolution of life is part of the evolution of the universe, which also evolves with increasing amounts of information embodied in ordered structures, galaxies and stars and planetary systems. In the living and in the nonliving world, we see a growth of order, starting from the featureless and uniform gas of the early universe and producing the magnificent diversity of weird objects that we see in the sky and in the rain forest. Everywhere around us, wherever we look, we see evidence of increasing order and increasing information.

As one writer has observed, this means that "if we can theoretically grasp the structure of information it is technologically possible to capture the surface and anatomy of reality . . . and in this way information about, for, and as reality can be structured in bits with powerful results" (Borgman 1999, 140). For some, the possibility "that information can reveal the secret of reality" (Borgman 1999, 174) has already become part of research designs and goals. The MIT engineer Seth Lloyd aims to create quantum computers as a way of joining up with the universe's "innate information-processing power." In his book *Programming the Universe*, he writes: "The history of the universe is . . . a huge and ongoing quantum computation. The universe is a quantum computer" (2006, 3). The futurist James Gardner speaks of "the Software of Everything" on which the natural cosmos runs and which will give rise to a cosmic mind. The astrobiologist and physicist Paul Davies (2014) has suggested that information precedes the laws of physics in the explanatory account of contemporary science, where deterministic matter has largely been diffused by probabilistic calculability. The information-to-intelligence view of the cosmos is beginning to spawn an effort to formalize the informatic cosmology by drawing on a range of sciences and scientists, imbuing their universal story not just with the usual teleologies of secular scientific triumph but with a technological unfolding in which human toil in the hitherto indifferent universe is reinscribed with a quasi-transcendent purpose—but without abandoning a commitment to scientific principles. Examples include James Gardner's recent *The Intelligent Universe* and physicist Frank Tipler's *The Physics of Immortality,* inspired by Jesuit paleontologist Teilhard de Chardin's Omega Point.

Thus, while the Singularity takes the logic of information and computation to its cosmological end, explicitly claiming the metaphysical extremities, similar assumptions underlie the mainstream convergence of the information-based sciences[10] with which the Singularity "grew up." In fact, ideas about the informatic universe are influenced by immortalist as well as informatic futurist networks, and under the label "convergence" they are being used to explicitly shape a wider, mainstream "scientific worldview" regarding the evolution of the universe, in which the key figure is "the cosmic primacy of information" (Gardner 2007, 25). The development and popularization of the project of convergence can be traced straight back to Bainbridge. In 2001, Bainbridge and his colleague Mihail Roco organized a conference sponsored by the National Science Foundation and the Department of Commerce to explore the potential of the information sciences. Roco has been one of the key advisers on federal funding for nanotechnology (Khushf 2007). Bainbridge's links to transhumanism and immortalism are not generally mentioned in any of the related reports or literature.

The first conference, and the resulting report, popularized the concept of convergence, referring to the unification of nanotech, biotech, informatics, and cognitive science (NBIC). It also launched a coordinated coalition of commerce, state, and science in this direction. Along with the scientists and labs, there were representatives from IBM and Lucent as well as political representatives from the Bush administration, with the undersecretary of commerce and technology presenting a paper. The concept of convergence very quickly spread to Europe and Japan. Today, with a slightly thinner vision than the original, the concept of convergence is being used across disciplines, in global health for example, to suggest solutions to large-scale planetary problems (Dybul 2013)—but, importantly, still confined to the planet. The basic notion behind NBIC convergence holds that these sciences are inevitably and naturally echoing with synergy because they enable each other—biology, for example, is shot through with information technologies, and neuroscience is the combination of the two at the cognitive level. NBIC is debated in terms of the "systems" approach it is ushering in, the acceleration of innovation in all the fields, and the enhancement of human performance (Canton 2005; Nordmann 2007). Thus, the power yielded by this convergence is said to be crucial to the human future, to the betterment of the human condition.

Nordmann (2007) has outlined some of the differences between European and American approaches to NBIC convergence, observing that technological and economic inevitability leads the project in the United States, whereas social management does in Europe. What's equally noteworthy is that the European and Japanese discourses on convergence emphasize social utility without making large claims about destiny, the universe, and the victorious unification of the scientific worldview. So the interesting question is how and why a cosmology, and one that is explicitly ideology laden, is being laid down with NBIC, especially in the United States.

The NSF report restated the totalizing project of science in its own informatic terms, claiming that convergence is driven by the development of new paradigms that are, for the first time in human history, allowing for "a comprehensive understanding of the structure and behavior of matter from the nanoscale up to the most complex system yet discovered, the human brain" (Bainbridge and Roco 2003, 1). Such an understanding is enabled by the treatment of everything—mind as well as matter in the whole cosmos—*as* information. The essays in that and subsequent volumes (Bainbridge and Roco 2005) projected a radical rupture—on a quasi-Singularitarian scale—in the landscape of human civilization, promising technologies such as "supercomputers the size of a cell in every human body, promoting health and preventing disease," and "100,000 machines generating energy from solar cells that can all fit on the head of a pin" (Canton 2005, 33). Echoing Bainbridge's 1982 warnings about intergalactic civilizations, the conveners of the convergence conference issued a similar warning: without the innovations of NBIC "the future of civilization itself is in doubt" (Bainbridge and Roco 2005, 2).

Informed by these origins, the NBIC sciences, like the Singularity, assume that information is constitutive of the world, that computational logic can represent and manipulate the universe (Golumbia 2009) and that "the mind" can be taken beyond biology, beyond the wetness of its human platform. The most comprehensive example is a recent NASA-sponsored publication, *Cosmos and Culture* (Dick and Lupisella 2010). Including multidisciplinary contributions from anthropologists and philosophers to systems theorists and engineers, and edited by Steven Dick, astronomer and NASA's chief historian, with Mark Lupisella, a NASA scientist, the volume is an effort to tie sociobiological views of cultural evolution to the informatic cosmos. Informed by convergence, the authors invent the term "cosmocultural evolution" (Lupisella 2010) and declare that we are witnessing the "emergence of a novel scientific worldview that places life and intelligence at the center of the vast, seemingly impersonal physical processes of the cosmos" (Gardner 2010, 379). From the NSF to NASA, the ultimate dream of convergence is to make the cosmos cultural and culture cosmological. Dick (2000) writes: "Surely the modern cosmos may serve as a new focus of meaning; it already has for many, and the numbers are increasing. . . . As we learn more about our place in the universe, and as we physically move away from our home planet, our cosmic consciousness will only increase." In one sense, at least, it is not an exaggeration to call the cosmos already cultural, as the informatic futurists have done, for our stuff is out there, at minimum in the form of radio waves sent out by Rothblatt's satellites—and the other twenty thousand satellites in orbit around the planet. Indeed, as Brad Allenby reminds us, we have already modified Earth's characteristic radiation spectrum through leakages and emissions from our information technologies: "In the Anthropocene, perhaps the most fundamental physical manifestation of our planet in the universe, its radiation spectrum, carries our signature" (2011, 67).

We do not know exactly what sort of consequence this portends, but Dick, Bainbridge, and other convergence gurus have been working toward preparing the ground for this shift (Bainbridge 2007b). Dick concludes: "The more we know about *science,* the more we know culture and cosmos are connected, to such an extent that we can now see that the cosmos is inextricably intertwined with human destiny . . . impinging on (and arguably essential to) questions normally *reserved* for religion and philosophy" (Dick 2010, 25, emphases mine). It is clear that the NBIC's forays into understanding not just the natural world, including the human body, but also the cultural world, including the human mind, and more importantly universal purpose, in terms of computation and information, will assume the power to shape more and more lives while redrawing the lines and recalibrating the tensions that distinguish the religious and the secular. The question of which culture the cosmos will benefit from is evaded, but all the writings seem to point to the consolidation of American and technocivilizational power structures—a topic discussed in the next chapter. As others have noted (Noble 1977, 1999), this sort of technoprogressive teleology is an extension and reformulation of American notions of Manifest Destiny; recall how frequently the word "destiny" is used in immortalist and transhumanist tracts, in which the unique liberation from the past in American historicity leads to human fulfillment via the unfettered construction of the future. The link between empire and information, between the informatic cosmos and the colonization of space, will need a closer study.

In the meantime, critics hold that the concept of information is itself generally vague and its deployment across, and even within, disciplines is not consistent. Peter Godfrey-Smith (2007) has outlined the word's range of different uses in evolutionary and genetic biology, from a deterministic semantic model to a computational program that plays some role in the generation of an organism to a simple code that merely posits a causal relation on the level of DNA and the production of proteins but not on the level of organisms. For Susan Oyama (2000) "information" is the contemporary version of "form," a preexisting guiding principle that leads to the specific shaping of matter. Evelyn Fox Keller (1995) has astutely pointed out that Shannon's original use was crucially different from information as it got to be conceived and used by geneticists.

Eugene Thacker (2005a) has pointed out that information cannot be categorized as matter in itself nor quite as immaterial. It's precisely this ambiguity, Stefan Helmreich adds, that lets information "be 'about' anything . . . which is the reason it can connect entities such as humans, transgenic mice, and digital organisms" (2001, 136). With reference to Georges Canguilhem's notion that information was a concept used to bridge the animate and inanimate, and making use of Sarah Franklin's notion of transbiology (2006b), I suggest that information functions to bridge incommensurable parts of disciplines and enables convergence; it allows for shape shifting, or what immortalists call "substrate independence," where one identity or one mind can be reinstantiated on a different material platform—a biological mind, for example, reproduced functionally in digital form.

The promise may be overextended and the immateriality a Platonic fantasy of disembodiment, but the ways in which information and informatic means have been deployed to facilitate the interpenetration of material, biological, digital, and mental layers are not to be downplayed. It's actually staggering that any of the following are even possible: computers are used to alter genetic sequences or codes within DNA molecules, thus altering an organism's behavior; controlling a beetle's actions by sticking electrodes into its brain and nervous system to manage its behavior from a computer;[11] functioning deep-learning algorithms based on the neural structure of the brain itself and brain-computer interfaces that allow tetraplegic humans to control robotic arms in complex actions only through the use of their minds;[12] brain-to-brain communication through transcranial magnetic simulation;[13] nanoscale implantable digital neurons that can communicate with receivers outside the brain.[14] Spanning major research labs at universities as well as futurist ventures like Kurzweil's (deep learning at Google) and Elon Musk's (brain–computer interfaces and mini-brains with Neuralink) in addition to government agencies (DARPA committed $65 million to digital neural implants in 2017), these achievements go beyond bionic or cochlear implants, hi-tech aids or enhancers that tap predictably into the mechanics of the brain and the nervous system. To transmit intention from the mind to a machine seems like an important shift, for how can there be something continuous between mind and matter on an intentional level?

The AI imaginaries, practices, and strategies must be viewed not as techniques or analytics that are finally getting us closer to physicalizing the essential qualities of the mind, thereby making the mind continuous with the physical universe. Rather, specific conceptions of intelligence and mind (and their continuity) are being produced through the sort of operations that nonbiological instruments can carry out. They will produce new kinds of persons or subjects, those with expectations, experiences, and mental operations that are made through machines. In the case of child development, for example, Keller (2007) describes how social robots designed to work with children are first modeled on child behavior, and although the robots end up being inadequate performers, their performance is not corrected; rather, in the next stage the children's proper development is nevertheless measured against the robot's behavior. In other words, the children are asked to grow up to be more like the robot. In the example of mental control of robotic limbs, it only appears that thoughts are controlling machines. In fact, the brain is trained as any organ might be trained to behave in a particular way, that is, to respond to a person's "intentions" by producing specific physiological signals, like EEG-detectable electrical signals or brainwaves in a particular spectrum. It is those signals, not the intention, that can be picked up by a machine. The interaction between the person and the robotic interface becomes a skill, like wielding a tennis racket or a javelin, rather than the deep ontology of the mind. Since computation is currently imagined as the only possible means of transferring consciousness, and since we are so deeply embedded in the infrastructure of computation, computation becomes the model imposed back on the brain itself.[15] The computational mind is overdetermined.

Algorithms mediate the face-off between mind and machine; they are not minds, but they may be good at representing and predicting certain behaviors, but more than anything, they, like other tools, are actually very good at *making* new behaviors which now are normalized and made ubiquitous through the political economy of the tech sector and portable devices. The upshot is that the mind via the brain responds to machines and gets closer to becoming machinic—rather than the other way around.

When mind becomes reducible to an algorithmic process of selecting the best path to achieve a desired outcome, then the mind (or person or self or . . .) has been characterized as a particular entity with particular functions, specifically, one whose main operations fit the current informatic theories and technologies. This worries many psychologists and some programmers, such as the virtual-reality pioneer Jaron Lanier, who penned a warning manifesto called *You Are Not a Gadget: A Manifesto* (2010). But to transhumanists and immortalists it is an idealized occasion of self-transformation. Immortalists, Singularitarians, and many transhumanists I met downgrade "regular" human brains—as "meatspace reality," as "wet," "messy," and "primitive" products of evolution. So while waiting for real uploads, cultivating algorithmic subjectivity becomes a second-best option: referring to their mind's workings as "my mental algorithms," they train their brain to be "less wrong" by making lists and outlines of human cognitive bias and correcting each other; or they train their brain—online, in workshops, at conferences—to process Bayesian probabilities when considering the uncertain future. The mind (or person or self or . . .) is lived, performed and experienced as "informatic," as a good enough relational database, in Rothblatt's words.

A whole lot of mind (or person or self or . . .) may be left out of these disciplines of the brain, but these examples only make evident an underlying development: that our subjectivities—our attention span, our way of thinking, our calculations, our desires, our lyricism, our dysfunctions, our intuitions and impulses, the very wiring of our plastic brains—have adapted to the algorithmic digital environment. This algorithmic subjectivity is harnessed efficaciously in computational form, that is, in the form favored by Silicon Valley's computational capitalism. Here mind becomes concomitant with capital, pressed into service, mirroring the infrastructures of profit derivation, so that mind and matter become coextensive through the processes and materialities of the algorithimic technologies—*that* is the real neural net, where the internal and external are co-constituted, where computationalism is not mapping the mind or some interiority onto an external machinic network, but mapping digital networks and other exterior forms onto the brain. This aspect of the harnessing of mind to infrastructures of power and capital is ignored by the informatic prophets of convergence, for whom the information sciences appear as a kind of salvation for the aporias of science, the pitfalls of progress, the limits of earthly resources, and the finitudes of being human in an infinitely open universe.

COSMOPOETICS AND THE MIND OUT OF PLACE

The development of posthuman thinking in general has been about figuring out ways to make the nonhuman world part of a sociopolitical reckoning against the grain of secular humanism, and in that sense it is also a posthuman*ism*. If secular humanism gained its power by dropping destinies into human hands, that power came at the expense of cosmic alienation and Cartesian solipsism—that is, secular humanism could only exist through the separation of the human from the matter around it over which humans were claiming dominion. If accounts of human history and belonging have provided a sense of being beyond the individual, wrapping the immanent in a temporal transcendence, it has been meager recompense for the grander cosmic alienation in which consciousness or mind is removed from the rest of the universe and isolated only in the human. As Thomas Nagel writes, the mind–body problem is not a local problem, "it invades our understanding of the entire cosmos and its history" (2012, 3).

If the informatic self in the informatic cosmos opens up nonsecular possibilities of relating mind and cosmos, it is *not* a return to religion, despite the occasional though modified use of that term by some immortalists and futurists as well as their critics. One argument throughout this book has been that what tends to get interpellated as religion—including by transhumanists themselves—is itself a product of secular rules and views, secular gaps, tensions, and binds. The mind and the cosmos present one of the bigger (universe-sized!) zones of indeterminacy around which this dynamic is activated. Indeed, secular law itself at various junctures, and especially where it has tended to create the concept and legal category of religious exceptions, appeals to the perennial mysteries as one important reason for carving out the zone of religious exception. For example, in an important abortion decision based on religious exceptions, the U.S. Supreme Court writing about the Fourteenth Amendment, liberty, and procreation, makes this stunning statement: "At the heart of liberty is the right to define one's own concept of existence, of meaning, of the universe, and of the mystery of human life. Beliefs about these matters could not define the attributes of personhood were they formed under compulsion of the State."[16] In this separation of the state from individual religious exigencies, secular law and the secular state define the parameters of the religious in relation to the very gaps of secular scientific logic itself (of existence, of meaning, of the universe). It is usually there, in zones of indeterminacy, that something is called religion by secular authority and comes to recognize its own legitimacy through the category of religion.[17]

How can we work and think outside that binary instead of secularly replaying it over and over? Is every appeal to the cosmos or to "en-minded matter" religious or spiritual in nature? Or worse, merely symbolic? And why does this matter?

Part of why and how it matters is that it has become increasingly important and urgent in the Anthropocene to construct a politics that goes beyond the centrality of the human actor. Secular anxieties about cosmic insignificance have engendered humanist fantasies of cosmic grandeur rather than orienting us toward a humbler dissolution, what Rosi Braidotti calls "becoming-imperceptible" (2013, 136). In a residual secular humanist gesture, Braidotti takes this toward an embrace of death as a form of merging into "the radical immanence of the earth itself and its cosmic resonance." What matters in this dissolution, unlike Goertzel's grand masculine gesture of sacrifice, is the openness to the proliferation and varieties of immanence, rather than to a narrow capital-enabled teleological vision that goes from prehuman to human to superintelligent posthuman. What Braidotti and other especially feminist posthumanists point to is an important politics embedded in the cultivation of a humbler relationship to and continuity with the cosmos, one in which the dedramatization of human existence (Povinelli 2016) may become part of a wider and more symbiotic responsibility, where we can see something about existence and our continuity from beyond the human perspective with which secular humanists tend to start. But I think this requires more than the invocation of life or zoe as the "motor"; it requires, as I have suggested, an engagement with minds—something that raises too many flags for secularists such as Braidotti. When mind appears in matter itself, the mind is found to be out of place, to make an inversion of Mary Douglas's formulation, and that makes many a secularist uncomfortable.

In some ways, the cosmology (reformulated as an ontology) proposed by Amazonian perspectivism (Viveiros de Castro 1998) suggests that universe of minds I am invoking. In parallel, others have been reviving variations of the mind-in-the-universe in the form of new animism (G. Harvey 2005) and panpsychism (Skrbina 2005; Chalmers 2002b; Nagel 2012).[18] These are cosmologies that allow us to start from other minds, *if mind* is taken to be more equitably distributed through the world, if it is put back into the universe. The consequences of perspectivism and/or panpsychism for death and continuity have not been worked out as far as I know. But what if, taking from the immortalist and transhumanist playbook, we asked whether "information" may be useful as a concept to connect inner and outer worlds? Might we think about *informatic perspectivism* as a kind of technoanimistic ontology, whereby the multinatural world is not exactly natural and the cultural substrate is not quite cultural, but actually information-based? Or, better, resonance-based, as a new theory of evolution proposes (Damasco and Giuliani 2017)? These possibilities arise for me as postsecular speculations when I read the possibilities of transhumanism and informatic futurism, for turbocharged as it all is with computational technologies and Silicon Valley capitalism, it nevertheless is one way to include nonhuman and *nonorganic* entanglements in the experiment of collective becoming—even if that is not what informatic futurists have in mind.

"What makes us human is not ours," writes Isabelle Stengers (2005) in "The Cosmopolitical Proposal": "it is the relation we are able to entertain with something that is not our creation." Futurism's fetishization of a human-created computational superbeing is not exactly what Stengers proposes as a cosmopolitics. But even as she declares that "a cosmic order" can protect us from an "entrepreneurial version of politics" (2005, 995) she calibrates the understanding of cosmos in particular ways. Cosmopolitics, she clarifies, does not refer to "the universe as an object of science" or to "the universal" in the scientific sense. "The prefix makes present, helps resonate, the unknown affecting *our* questions." It is a call for an openness to the unknown, for "the cosmos corresponds to no condition, establishes no requirement. It creates the question of possible nonhierarchical modes of coexistence among the ensemble of inventions of nonequivalence, among the diverging values and obligations through which the entangled existences that compose it are affirmed" (Stengers 2011, 355). Of course, as she makes clear, the cosmos is not to be taken as an entity as such.[19] In a way, there is no material or spatial cosmos in cosmopolitics; it is a concept holding together some sense of larger forces and belongings to which corresponds an ethics of "the coexistence of disparate technical practices corresponding to distinct forms of reciprocal capture, characterized by different logical constraints and different syntaxes" (359). Stengers's idea of cosmopolitics demands not only that we acknowledge the voids (science already does that) but also that, given the voids, we should not close off other beings' ways of knowing, doing, and being—their ecology of practices, which are, after all, symmetrical in some sense with science's ecology of practices, used to authorize its worldview on often unstable grounds. By contrast, as I have argued, the kind of intelligence that is deemed to be super by the immortalists and transhumanists, and slated to take over our universe, seems in fact to represent a narrow part of the spectrum of possible beings and minds.

Despite itself, informatic cosmism allows for certain reorientations out of the secular humanist paradigm in which humans the toolmakers and tool users dominate the world. In the humanist view, human technology can and ought to help master the world because technology is the external production of human rationality, of the only rational subject in the universe. From the informatic perspective, humans *are* technologies; they are intensified processes of a computational universe, themselves coded, themselves embedded into systems, some not of their making. These ideas have only been potentiated through utilitarian neoliberal pathways, amped by the models of funding and profit available on the West Coast. This has translated into a particular, though not always coherent, vision of continuity that flip-flops between human and transhuman continuity, between human obsolescence and human mastery, humanism and posthumanism. Even where the human's tools will dismantle the human's house, the human is still centralized in nonsymbiotic ways, in ways that do not reembed it in the environment but alienate it further as the creature that must take apart the planet Jupiter and rearrange it as a spherical shell around the sun where the next generations could survive with plenty of energy to harness. This is an idea, by the way, forwarded by the physicist Freeman Dyson, inspired by a science fiction story, and taken up as a potential future form of survival by many contemporary convergence theorists and informatic futurists (Tegmark 2017, 205–7). The question that remains open is whether an informatic cosmology and its concomitant cosmopolitics may be useful in imagining a cosmopoetics that would be about symbiosis, not superiority, incorporating the organic as much as the algorithmic to cultivate symbiotic intelligence rather than superintelligence.

6

Progress and Despair

THE PERVERSE DIALECTICS OF IMMORTALITY
AS TECHNOCIVILIZING MISSION

At one of the frequent immortalist get-togethers in Phoenix, a prominent young member acknowledged one of the tech-sector donors by saying that he was on his "list of three favorite persons in the world." I jokingly asked another member, a SENS researcher next to me, if he too had a list of three favorite persons in the world.

He responded in a serious tone: "No, I have a list of seven billion."

When it comes to life and death there are no favorites, he suggested. Every life out there is equally worthy of being saved from the blight of death.

Immortalists have come to portray their endeavor principally in terms not of preventing death but of *saving lives*, explicitly calling their projects "lifesaving techniques" or putting them on the same level as "critical care medicine." They write about cryonics, for example, as an intervention that places "critically ill patients . . . in a state of long-term care" (de Wolf 2007, 16). They have contributed to the journal *Critical Care Medicine*. The Methuselah Foundation has added an online consulting site to give advice about a "wide range of illnesses and conditions that may negatively impact your lifespan or the life of a friend or family member," as well as a New Organ Alliance, which helps secure organs for members who need them.[1] The SENS Foundation once focused on anti-aging medicine and is definitively in the business of "keeping people healthy" through rejuvenation medicine. Immortalism, they say, is about alleviating suffering and avoiding the tyranny of premature death. Who can fault any of it?

What more noble sentiment can one expect, what more obvious benefit, more aligned with nature's own movements, than to prevent ill-health and extend life?

Yet it is precisely this universalism that may be faulted, as its abstraction from the sociopolitical reality of present lives fails to acknowledge and address the structural problems that beset these ventures. Social problems, it is widely thought among immortalists, are technological problems to be solved with algorithmic prowess. It's not that questions of inequality and access aren't raised at all; rather, they are easily dismissed through stories about technological progress. Many people point to the cell phone as an example: the early adopters were a few wealthy people; today, everyone has a cell phone. It will be the same with life-extension technologies.

The social facts about mortality and socioeconomic difference quickly make the technological optimism of immortalists and technofuturists ring false. As medical science has advanced and life-extending treatments have been developed, even as the average life expectancy has risen in some places, the poor and the structurally disadvantaged have received little of the benefits. There has always been a difference in life expectancy based on wealth and income, but the disparity is growing. In a 2016 study, researchers at the Brookings Institution found that between 1970 and 1990—in just two decades—the gap in life expectancy at the age of fifty between the bottom tenth of income earners and the top tenth of income earners increased dramatically for both sexes: from 3.5 to 10 years for women and from 5 to 12 years for men.[2] In other words, according to these numbers, advances in medicine, whatever they might be, don't help low-income Americans with their life-and-death chances. Similar or greater discrepancies have been found for education levels and race, such that, for example, between 1990 and 2008 the gap in life expectancy between white women with high education levels (16+ years) and those with the lowest education attainment (less than 12 years) grew from 1.9 years to 10.4 years (Olshansky et al. 2012).[3] In general, the most important gains—and losses!—in life expectancy have come not from biotech or medical innovations but from *social* arrangements such as access to basic health care, clean water, income, and nutrition (Marmot 2004).

Nevertheless, immortalists see great irrationality in people who refuse to accept the benefits of solving the problem of death by these techno-scientific means. "Why does everybody else want to die?" the young SENS researcher asked me rhetorically. "I just consider it irrational." The rational position, for immortalists, is to understand that life always wants to avoid death; it is one of the ways in which life has been traditionally defined, after all, as matter organized against death.

When people question the impulse to life extension, immortalists tend to follow up—as de Grey and many others did with me—by asking whether you have life insurance or whether you would want to be cured of cholera if you got it today. Do you want to be alive and healthy tomorrow? And if tomorrow, then how about the day after? And the one after that? Ad infinitum, indefinitely, forevermore.

Ben Best, director of the Cryonics Institute at the time I met him, defended cryonics and research in radical life extension on these grounds: "You'll see a lot of that sentiment, of how socially irresponsible it is, with all these people in dire circumstance, and so on. . . . And my response is: so we might as well stop all medicine; there's no reason why we should be doing heart transplants or medical research either."[4]

So secular humanists, liberals, and progressives who dismiss immortalist projects as selfish or distracting (i.e., not tending to the urgent social issues) and yet use medicine to extend their own lives are either hypocritical or caught in the throes of "deathism," an ideology that has erected death as necessary not only in a biological sense but also in a moral sense. De Grey has called it a "collective hypnosis."[5] In this way, immortalists suggest that their venture is the logical extension of biomedical imperatives—which is to cure people of diseases and to save lives, or put biopolitically, to "make live."

But there is another argument regarding biomedicine that is important here: it has come with and perpetuated its own inequalities and myths, its own evils even, notoriously in Nazi eugenic schemes or in the Tuskegee syphilis experiments on black citizens. Scholars and writers have exposed the ways in which the inequalities of race and class are reproduced in medicine, not just historically but in advanced medical practice today—

what Harriet Washington (2008) has called "medical apartheid." That, too, has been a part of the history of medicine and biopolitics—not just letting die, as per Foucault, but discarding lives.

The secular tension between the technological imperative to save lives and the humanist dictum to accept death is complicated, then, by a biopolitics that makes some people live and lets other people die and, in some cases, even makes them die, for reasons and in contexts that have consistently been about race and geopolitics. The question "Why should life be extended?" is impossible to detach from the social and political question "Whose life is being extended?" As Sylvia Wynter has argued (2003), the biocentric model of humanity—subject no longer of God's will—was always a racialized humanity, a biological hierarchy authorized by science's account of evolution and driven by narratives of progress as a secular temporality and a colonially inflected account of human history and destiny. But if, as I argued in the previous chapter, immortalism is to be understood as part of a series of technoscience imaginaries that are shifting the grounds of politics *away from life itself* and toward informatic, posthuman, and nonbiological existence, how does this affect biopolitics? How does progress today relate biopolitics to the future, and how does the rise of posthuman technologies change biopolitics? What can techno-scientific immortality tell us about the politics of life as it intersects with ideologies of technological evolution and progress? Is extending life and defeating death a matter of universal concern, as many immortalists argue, or is it a matter of selective power, a new configuration of biopolitics, of racial and geopolitical reconsolidation via technological rather than social arrangements?

I will explore some of these issues through a useful distinction made by Nicholas Rose (2007) (taken from Wittgenstein, later also developed by Helmreich [2009]) between life-forms, somewhat akin to zoe or life itself, and forms of life, which I take to mean the sociopolitical formations that sustain life-forms—the former being the abstraction of existence as such and the latter being the structures in which existence finds itself embedded. There is a tension between life-forms and forms of life, such that the liberal focus on individual life-forms as the locus of universal value (life itself) hides the concomitant fact that biopolitics demands the perpetuation and protection of specific, nonuniversal *forms* of life, which in the modern geopolitical and ideological hegemony gain legitimacy through the notion of progress. In general, we need to understand not so much the value of life but the differential valuation of lives as they appear within particular social formations. I will begin by considering the value of life itself (life-forms) under biopolitics in light of immortalist bids for indefinite survival; in the second half of the chapter I will argue that as technoscientific immortalism and its informatic cosmology move beyond biological life they nevertheless are wedded to the perpetuation of a particular form of life, a Western—even American—technological civilization and its direct successor, a transhuman civilization. Here the differential valuation of life is not only obscured through the appeal to the universal value of life as such, while ignoring the structures that perpetuate disparities in life chances; rather, moving past the "bios" of biopolitics, it rests on the ways in which the promise of immortality merges with the promise of technological progress. If in biopolitics the promise of the end point of medical care becomes also the promise of the end point of human activity as biological being, in technocivilization the promise of technology, which lies precisely in the value of transcending biological being, becomes its own reified nonbiological value. At that juncture, the ideological value of "life" (as existence) is indexed to the projected brightness of the technological future. Achieving that "bright future"—that form of life—becomes the self-sufficient goal.

BIOPOLITICS OF IMMORTALITY

While the value of long, healthy lives might seem easy to uphold, polls show that a majority of people tend to balk at the possibility of very long lives, let alone immortality (e.g., Pew Research 2013). As I have argued, this is one of the quandaries of the modern secular immortality regime, suspended between the fostering of life and the acceptance of death. It is the contradiction that immortalists grind away at.

In countering immortalist projects, many secular humanists and bioethicists have argued for the value of death—as if to confirm the immortalist sense that this is a deathist culture. In a well-known piece, the Hastings Center bioethicist Daniel Callahan (2003a, 65) argued that medicine shouldn't conduct a war against death (even if, as he admitted, attacking disease is a way of forestalling death): "Death is, after all, a fact of biological existence and, since humans are at least organic, biological creatures, it might seem evident that it needs to be accepted. Death is just there, built into us, waiting only for the conditions necessary for it to declare and express itself. Why, then, should medicine treat it as an enemy, particularly a medicine that works so hard to understand how the body works and how it relates to the rest of nature?" Many others have argued along the same lines, suggesting death is not only the proper order of nature but also good for humans because it is the real basis of interest and engagement in human lives, of beauty and love, of virtue and moral excellence. Without these we would lose our "humanity."[6] Given that death is not actually a fact of all biological existence, as previous chapters showed, these arguments assume a metaphysical position about death and humanity, arguing that death is ultimately necessary and that without it humans would be human no longer.

At the other end of their defense of death, all these critics also share a position with immortalists insofar as none questions the value of life itself. How could one question the value of life, after all? Without having end-of-life ethics or contemporary immortalism in sight, a number of philosophical arguments have long countered the notion that extended lives are *necessarily* good (see essays in Fischer 1993). A line of Epicurean arguments holds that since the person after death is no longer present to experience any harm, then death cannot be said to harm the person. Ergo the elimination of death is not necessarily a good in the sense that it forestalls harm to a person. David Benatar (2006) and the nascent anti-natalist movement argue that coming into existence is always a harm as it produces suffering, and so nonexistence is preferable. A more political form of the argument in the Anthropocene suggests that the disappearance of humans might be an overall good (see Farman 2014b).

Another line of thinking suggests that personal continuity matters more in social contexts where the bounded, autonomous individual is the model of the self, and much less so in a setting where personhood is understood and cultivated relationally (in the anthropological, not strictly philosophical sense). In a relational setting, *my* interests and values and desires are shared by others and may be—should be—continued by others, who could even do a better job of developing those interests, and so a skin-bounded self (Battaglia 1995b) is not what matters in survival. This does not mean that any person would simply sit back and get mauled by a jaguar or give in to a virus. But relationality does suggest that the meaning of life is not anchored so instrumentally in the biological survival of the individual organism. Reminding us that life has carried deeper meanings and histories than the biological, Gil Anidjar (2011) notes that many modern thinkers, such as Benjamin, Arendt, Foucault, and Agamben, have grappled with the question of the value or sacredness of life and its origins. What shape does this sacredness take? From whence did it arise? Their answers range: lodged in modern modes of knowledge production as much as in the history or the moral prerogatives of Christianity. Regardless, for them that value has not referred to the biological life of the organism but to something more.

But in the age of technoscience, life has come to be equated with, counted in terms of, its biological rather than biographical outlines, the life of the organism, not the social being. It is the quantifiable lifetime that is reified as value—moral, social, and economic. There is a pernicious and pervasive bio-utilitarianism that can break up life and its value into an accounting exercise wherein the CPAs of life present us with our mortal exchange rate in terms of life expectancy and insurance rates, for example. As much as this is an effect of capitalism and liberalism, it owes much to secularization as well, to the elimination of transcendence and the physiological confinements of the immanent. In the immanent frame, how can one deny life's biological importance when this life is all there is?

Following Foucault's work on biopolitics, a number of scholars have documented the ways in which life has increasingly become subject to measure and understood through numbers, in particular regarding measures of time linked to biological activity and health (Bouk 2015; Farman 2017b; Jain 2010, 2013; Rottenburg et al. 2015). In addressing the life of the population in order to make live, biopolitical governmentality relies on the valuation of life as such, that is, on rendering life measurable rather than livable, counting it, as scientists like Raymond Pearl did, but not being accountable for it (also see Stevenson 2014). What allows these numbers to generate meaning and become part of the temporal experience of biopolitical subjects is death, the horizon of mortality against which all these indexes are measured. The future has a definitive end that makes measure possible, that allows governmentality to address the life of the population in relation to an absolute horizon. While improving those measures of life for some, biopolitical governmentality also and necessarily writes off the lives of others—those who have come to be socially devalued.

The immortalist approach to saving lives is firmly ensconced within biopolitical reason. Immortalism's abstract imperative is to make live, where life is reified in terms of rational bureaucratic quantification and numerical abstraction; in practice, immortalism also embodies the second aspect of biopolitical reason, to let die. The following conversation with a SENS researcher, NN, will help illustrate my point. We were sitting in a garden café full of young white students taking advantage of the early days of the summer of 2008. The conversation turned to his motivations for life-extension research. To explain it, he pointed to everyone around us:

> NN: Well, because all these people seem to be enjoying their lives enough in order not to want to die in a couple of decades. Look at them; they seem happy.
>
> AF: They seem happy right now, there are times when they are fucking miserable. There are a lot of people who are, you know, on balance, more in a state of misery than happiness. There are those people.
>
> NN: Yes, and some of them do commit suicide eventually, and that's sad and I can do nothing about that.
>
> AF: Some of them die in ridiculously stupid little social situations like starvation. Those are socially created, right?
>
> NN: Well, let's say they got fixed in some places but not in others. Like the default is there if it takes effort to overcome it, and I guess some regions are in pretty bad shape and pretty much hopeless. Because of what's in place elsewhere.
>
> AF: Hopeless, well it needn't be hopeless. It's not any less hopeless than fighting aging.
>
> NN: Yeah, right, but it's clear that millions of people more will die from

starvation and there's nothing we can do about it, that's what I mean by hopeless.

AF: Why not? What do you mean, "there's nothing we can do about it"?

NN: Well, if every billionaire in the world decided to donate their fortune tomorrow to save Africa from starvation, they couldn't do it, there would still be millions dying. That's all I'm saying by hopeless.

AF: Well, they could live a lot longer if certain structures were in place that didn't make big agri companies hoard their grains in order to keep the prices the same before feeding it to people who are starving, for example.

NN: Right, OK, yeah.

AF: I mean there are definite things one can do, that save millions of lives from expiring prematurely, not for living forever.

NN: Absolutely, and even if we couldn't save them this second, the fortune of all the billionaires in the world could still save an infinite number of lives in the long term. In starving countries.

AF: Sure, so it's not "there's nothing we can do about it." That's—

NN: No, I'm not saying there's nothing we can do about it. I'm just saying we can't do it tomorrow and that's all I meant by hopeless.

AF: Right. We can't stop aging tomorrow either.

NN: Right. That makes things, both things very much—

AF: Hopeless?

NN: Right, sure, of course. Probably none of these people I just pointed to can be saved from death, and that is very very sad and I think deserves the word "hopeless."

The racial geopolitics of letting die stands in contrast to the techno-optimism of making live—the imaginary of Africa as that place of hopeless starvation versus the garden of American university students pursuing careers in the sciences and the humanities to make the world a better place. In Foucault's concept of biopolitics, making live and letting die figure as two sides of the same coin, the value of some lives related to the disposability of other lives. Though the relationship is not made fully clear, the point is that some lives become subject to death without concern as part of the bio-waste necessary to biopolitical governmentality, which designates certain populations as acceptable subjects of death through biological ideologies of exclusion such as racism. Racism here is understood by Foucault through a biopolitical lens, as targeting a group, not in a kind of Hobbesian competitive belligerence over resources for survival, but because they are perceived to threaten, in their impurity, the *health and resilience* of a population. The health of a population (linked to labor and the market) is a key factor around which politics gets organized, the state's role becoming that of immunizing its population from adverse risks, financial or viral. Nonetheless, the valuation of the lives of some over the lives of others, the slide between value and vulnerability, even via health and purity, is not just a contingent matter; it is part and parcel of the workings of power in biopolitics, as states and institutions organize life while displacing risk and violence onto less powerful populations, and establishing the set of measures through which the differential valuation of lives and bodies (and body parts) gets normalized. In biopolitics, race is the calculus through which the displacement of vulnerability is justified, through which power and governmentality designate whose life is expendable (even though, as Weheliye [2014] notes, in much of the writing on biopolitics race has been sidelined in favor of an analysis of life itself). [7]

The temporality of technoscientific immortality changes the calculus: the time lost after death also becomes part of the measure, and thus a different grammar takes hold, where the past and future hypotheticals collapse into each other and become the grounds for understanding the present. If death is not inevitable, then what becomes measurable is all the lifetimes that would have been or will have been saved. Take this example from the Maximum Life Foundation, a nonprofit dedicated to reversing aging through biotech and nanotech. Their website tells us that "if we can speed up developing dramatic life extending technologies by just a few years, we could ultimately save more lives than were lost in every war since the beginning of time."

> Consider. The human death toll in the Year 2001 from all 227 nations on Earth was nearly 55 million people. 52 million of these were "natural" deaths and 37 million were directly related to aging. . . .
>
> Consider. The world's greatest resource is the knowledge, skills and

wisdom acquired by 2 billion older world citizens. We waste 100,000 of those resources every DAY due to unnecessary suffering and premature death from aging causes.[8]

The notion of *premature* death, then, becomes overloaded and unanchored at the same time; it carries the weight of all those abstract lives unsaved, but it makes premature death a socially meaningless concept, as every death is now premature. Wherever you set its parameters, no matter what ceiling you create through bioengineering or AI, the untimeliness of death is a social and political matter, not in itself a biotechnological one, just as who has access to clean water is a social matter, not a technical one. Categories like premature or natural death are not simple categories subject to individualized biotech intervention. By contrast, take Ruth Wilson Gilmore's institutional and political notion of "premature death"; it is anchored in social and institutional processes in which power and inherited hierarchies produce the conditions that may be recognized, felt, and analyzed as premature death in relation to certain populations who as a whole, and in contrast to others, are exposed to such conditions. That, indeed, is her definition of racism: "group-differentiated vulnerability to premature death" (2006, 28).

TECHNOCIVILIZATIONAL FUTURES

The valuation of life-as-such in biopolitics added to immortalism's politics of potentiality trains the calculus on lifetimes lost and lives that could have been saved as the moral rhetoric that is meant to drive the project forward. At the outer reaches of immortalism it is no longer even a question of lifetimes lost but of forms of life not attained, or informatic and transhuman futures not reached. The future imagined and desired and actively worked on by immortalists is a future imagined through a particular set of imperatives, tools, and especially promises, in which lives are saved through specific technological, not social, means, a future in which the seeds of technology will have finally borne their nonbiological fruit. In that future these technologies will prepare them to tackle all sorts of problems, of which defeating death is only one. It is not extended life in the abstract that is the goal, but extended life for the sake of a specific transhuman and techno-utopian future. Or as one cryonicist sarcastically characterized it to me from the opposite end: "I was never interested in heaven, given clouds and harps and angels."

If that other singular future should not be the theological eternity featuring angels sitting on clouds playing harps, nor is it meant to be some watered-down version of longevity, featuring wrinkle-free suburbanites with good knees who continue to jog until the age of 101. As one of the speakers in a recent RAADfest conference said about futures: "There are many potential futures, but there's *one* we've got our eyes on." That other future is projected as a particular technoscientific destiny where radical longevity is not just any kind of long life but one that is organized and focused through a vision of technoscience based on the notion of *convergence*, one that will bring together nanotechnology, biotechnology, information technology, and cognitive science in the service of a very enhanced cosmic future in which the slow and primitive biology bequeathed to humans by evolution will have been superseded by the creative evolution unleashed by the human mind and technological prowess. Immortalism's "future" is and can only be a nanotechnologically enhanced future. To "get the chance to become posthuman" and witness those "foreseeable technologies," a World Transhumanist Association pamphlet concedes, "is the primary reason for the strong interest in life extension and cryonics among transhumanists."[9] On his personal page, the former Cryonics Institute director Ben Best has written: "For many people, the prospect of living in the future means much more than extended lifespan. They are excited by the possibility of space travel and of the transformation of human life."[10] Immortalism, then, saves posthuman lives.

Underlying this overall vision of progress from human to posthuman is a sense of the importance of the continuity of Western civilization, not to mention American might, as the spearhead of any such progress. I want to return to José Cordeiro, a transhumanist activist I introduced at a RAADfest conference in chapter 2, this time holding forth to a group at the Church of Perpetual Life (COPL) in Hollywood, Florida. Housed in a building originally constructed as a Masonic temple, COPL is owned by the Life Extension Foundation, a large fabricator of health supplements started in 1987 by Saul Kent and Bill Faloon, among the main early back-ers of cryonics and other immortalist ventures. In its very early years LEF survived a murky pool of legal problems, having become the target of several FDA investigations and raids. By the mid-1990s the charges against the foundation were dropped. By 2009, according to GuideStar reports, it had declared assets of over $25 million and was netting more than $3 million on revenue of more than $18 million. Since then it got embroiled in another dogfight with governmental agencies, this time the IRS, which has been contesting its nonprofit status. Among other things, the IRS claims that LEF has funneled money to companies—mainly linked to cryonics and longevity—whose board members include the founders of LEF.

The gathering at COPL is mainly white, though attendees otherwise vary in age, sex, income, and profession, from cryonics researchers to health enthusiasts to people who organize a "secular solstice." Not every-one in the pews of this transhumanist "church" is a transhumanist. "We hold anonymous surveys of people who come into the church," the calm, quiet gray-haired director, Van DeRee, told me. "You'd think most people involved in age reversal are atheists, but maybe slightly over a third are atheists, and the rest would count themselves as Buddhists, Christians, and we also have members that are transhumanists." The secular Durk-heimian point, though, is that they congregate here under the same roof to listen not to an ordained priest but to a physician's lecture on nutrients and calorie restriction or a researcher's talk on adult stem-cell transplants or a cryonicist talking about insurance policies.

In the main hall upstairs, a band sings a cover of "Forever Young" on two screens flanking the stage. Then Cordeiro holds forth on the bril-liant future ahead, extolling the virtues of a Kurzweilian accelerationism, evoking an era in which computer technologies will be so advanced and powerful that we won't even need that old cumbersome tool of language anymore; our minds will be mutually transparent: no need of media-tion through symbols, no struggles to put things into words. Cordeiro presents these scenarios under the fold of an explicitly civilizational and socio-evolutionary canopy, contrasting that techno-utopian future with forms of life that did not conform and were therefore already obsolete.

First came projected pictures of Amazonian indigenous communities meant to serve as an example of people who, as he said, "live like they did five thousand years ago." This was followed by a picture of an Amish carriage to present an example of Ludditism, people who can but won't take up the promises of technological advancement. "There are always people who are opposed," he said, even if technology will just go ahead and develop and accelerate and make things better.

That civilizational projects were historically tied to the expansionism advocated by the ideology of the "white man's burden"—the title of Rudyard Kipling's poem sent to Theodore Roosevelt to encourage the U.S. colonial takeover of the Philippines—seems lost on most transhumanist proponents (and, as I argue below, post-9/11 security discourse). Progress as an explicitly technocivilizational unfolding, leaving behind indigenous lifeways and bypassing the poor people who refuse it, has always been part of immortalist imaginaries, as it has been a key part of Silicon Valley projects. Keep in mind Bainbridge's early warnings at the NBIC conferences: without technological convergence, "the future of civilization itself is in doubt" (Bainbridge and Roco 2005, 2). At COPL, Bill Faloon often talks about how bad they had it in the old days, when there was no electricity. Using depictions of Scrooge, he once asked attendees to imagine a bygone era when people had fun during Christmas, but in fact it was so miserable that only the rich had candlelight. Rationally, we need to embrace progress or risk being left behind, in societies and in bodies that will fall apart and remain half capable and largely boring and subject to extinction. John Schloendorn, a biologist and SENS researcher I spoke to frequently back in Phoenix, has explored his commitment to the humanitarian ethos of saving *all lives* in an academic journal: "A person from one of today's first-world countries can probably satisfy more of her desires than a citizen of the Roman Empire, who could in turn satisfy more desires than a caveman from the last ice-age" (Schloendorn 2006, 194). I cannot confidently pronounce on the desires of cave men or any being from the last ice age (though I am sure the latest iPhone was not on the Christmas list), but I see the telos of Schloendorn's argument repeated over and over in this age. Neuroscientist Kenneth Hayworth not only believes in the virtues of progress through the transference of mind from the biological body into "designed synthetic" platforms but, like others, told me that this is why he wants life extension, in order to arrive at that specific future. In an article in *Scientific American*, Hayworth stated, "I refuse to accept that the human race will stop technological and scientific progress," adding that in the future there will be smarter, better "posthumans poised to explore and colonize the universe."[11] Ben Best, who once told me he cannot imagine being bored in the future, has written: "The progress of civilization has been toward more freedom, more justice & more democracy for more of mankind. This progress is likely to continue and accelerate."[12]

At the same time, for what purports to be an optimistic technofuturistic movement so convinced of the curve of human progress and even the purposefulness of the universe, immortalism and its related groups are unexpectedly gripped by doomsday scenarios, producing and reproducing them promiscuously. I raise this because I think the technological and civilizational telos underlying the immortalist (and indeed the humanist) future can be better understood in light of the incessant obsession with endings—in part, as I will argue, this has been the underlying anxiety and tension in all notions of progress, where progress has been mobilized equally as a bulwark to secular nihilism.

In most conversations, just as at most conferences, a strand seems always dedicated to the threat of total annihilation. I remember being surprised at several postconference discussions by the sheer number of "bad futures" (Harding and Stewart 1999) described and imagined. And whenever someone threw up an alarming new scenario, a Singularitarian, shouldering the burden the rest of the world refused to acknowledge, would say, "Oh, now I'm really beginning to worry." From peak oil to nuclear terrorism to ecological disaster, doom echoes through many immortalist narratives, personal concerns, and listservs. It has motivated groups and conferences organized around "existential threats," a term popularized through the writings of transhumanist philosopher Nick Bostrom.

Although various forms of messianic and secular end-times, mobilized differently by different movements over the past couple of centuries, have provided imaginaries of the end of the world, the global spread of computational technology took these concerns around a new bend.

Computational power and intelligence are at once the saviors of the future and harbingers of its demise. Among the most effective of computational doomsday activists was Eliezer Yudkowski, cofounder of the Singularity Institute for Artificial Intelligence (SIAI, now MIRI). As mentioned in chapter 3, Eli was excited by the prospect of superintelligent agents but was worried that such agents might end up, willfully or accidentally, destroying what we care about, namely, human lives. SIAI was set up to find ways of preventing that. A common, crude illustration of an accidental case would be a superintelligence optimized to produce paperclips that would then take all matter in its vicinity and rearrange the atomic structure to obtain a lot of excellent paperclips. It may not despise you in particular, but since your atomic arrangement does not correspond to its aims, it would take you apart and transform you into a paperclip. Yudkowski urged research toward the development of Friendly AI.

SIAI's work and Yudkowski's writings influenced a number of other prominent projects set up to curtail worrisome threats to human futures. For example, Skype cofounder Jaan Tallinn, who read Yudkowski's work online, contacted him and ended up funding the work of his institutes and helping to set up the Center for the Study of Existential Risk (at Cambridge) and the Future of Life Institute (at MIT). Conferences on civilizational collapse, inspired by Jared Diamond's book, were organized by Terasem, Martine Rothblatt's group that holds annual conferences on aspects of a technological future. And various blogs and listservs have had discussions on peak oil, climate threats, biological threats, and other factors that could cause a global crisis of survival.[13]

The Future of Humanity Institute, set up by Nick Bostrom at Oxford University, is dedicated to the study of large-scale future threats to human existence. Bostrom's view of existential risk, to which I will turn below, has spawned, among others, an organization called the Lifeboat Foundation, a nonprofit connected to various AI, nanotechnology, and immortalist projects and figures that is dedicated to researching "extinction risks" and developing more than a dozen "shields" against each, including BioShields, AsteroidShields, NanoShields, AlienShields, and more.[14] It boasts many big names on its numerous boards, though its activities seem somewhat opaque and the founder's intentions and background are equally murky. Based in Nevada, the founder, Eric Klien, has been known, on sideblogs, to denounce science and suggest that science is the biggest risk we face, while recruiting scientists to be on the board; he has also recruited the right-wing anti-Islamist activist Pamela Geller, sparking concern among some board members.[15] In an interview he told me that he felt the urgency to found Lifeboat after 9/11. But those were just planes, he noted; there are bigger risks, and he wants, he said, to avert "the triumph of evil" over "civilization." Though counterterrorism is listed as one of Lifeboat's concerns, Klien emphasized that the majority of those risks will not come from terrorists. He is most concerned about something else: the biggest risks are built into the potential technologies themselves. "The same technology of health," he said, "becomes the technology of destruction."

The rehearsal of collective endings has inspired commentators to call these movements apocalyptic and characterize them as a thinly veiled form of millennialism (Geraci 2010). This is particularly the case in relation to groups around the Singularity, since these clearly present a kind of inevitable rupture that has affinities with the Christian rapture. I am wary of these generalizations, in part because they are based on formal features, and I don't think the social and historical specificity of these forms can be abstracted and flattened into a category that itself has a complex two-thousand-year-old history of variation; in part I am wary because end-time imaginaries and utopian futures are part of a wider cultural imaginary in mainstream secular and scientific American culture.

Through the secular history of finitude, from heat death to extinction, described in previous chapters, an explicit sort of secular anxiety emerged about total annihilation without the possibility of redemption or rebirth, expressing a particular secular sense of an end-time, especially since the projections themselves surfaced through scientific predictions. Added to those there was, and is, the nuclear bomb, triggering widespread apocalypticism in the United States. That was as much a religious affair as a secular one, vigorously renewing the idea of the Armageddon on the one hand and on the other leading to various secular apocalyptic groups and media representations of the end-times (Lifton 1987; Wojcik 1997; Harding and Stewart 2003). Since the nineteenth century, and certainly since the bomb and now climate collapse, it is no longer an exclusively millenarian exoticism to imagine, even *live*, the end of the world.

Reflecting on this context, Robert Lifton (1987) linked the rise of nihilism to a response to realities of nuclear annihilation, a sense of a collective destiny in destruction. That was the opposite of what he called "modes of immortality." Modes of immortality counteract not the physiology of death but the existential consequences of death for the living. The symbolic function of immortality is not to ensure personal survival as such but something more: it is the production of "a sense of historical connection beyond individual life" that gives "a sense of significance to experience" and reflects "man's relatedness to all that comes before him and all that follows him." When the "symbolization of immortality" as a form of continuity is threatened, as it was in the aftermath of Hiroshima and Nagasaki where Lifton carried out his research, "then it threatens a level of psychic experience that defines our humanity" (1987, 155). The erasure of immortality leads to a "radical futurelessness," a loss of the desire to move forward into the future. His Japanese patients and interlocutors had a word for this: "munashi," which he translates as "emptiness." One might think of munashi as the affect related to the emptying out of the future, a break in the possibility of continuity. Lifton thought that under the shadow of the mushroom cloud the whole planet had a case of the munashis. It was, he argued, just this *sense* of a break in continuity that drove people to imagine space colonies, new religions, and returns to nature.

In the secular, the end was not an escape from the world after the world had ended, as in millennialism; the end was already in the world. As the billboards and cartoons portrayed it, the end *was* near—it was right there, around the corner, part of the landscape of despair commonly inhabited and rehearsed in nuclear shelters, in the media, in science tracts, in literature. In this sense, Lifton's characterization of despair reached beyond the devastations of the bomb, pointing to the underlying sense of an inevitable ending without redemption, that is, annihilation as a totalizing aspect of existence. For Lifton, immortality—even if symbolic—was a way to get around the sense of radical futurelessness as a common destiny.

Progress has been another utopian answer to this secular eschatology.

People *must* believe in progress, Sidney Pollard asserted, because "the only possible alternative to the belief in progress would be total despair" (quoted in Lasch 1991, 41–42). Progress and despair—as secular visions—are thus rolled together not so much as opposites but as codependent concepts and affects reproducing each other. The secular dialectic of futurity moves ahead with both, progress and nihilism rolled into one, because, after all, in the secular, any conjuring of the future contains within it the negation of that future, its ending; like biological life itself, every future is already entombed. That is why the brightness of the future determines the value of life. It's not so much that futurelessness is a worry because some catastrophic event might threaten the bright future; rather, it's a prior condition through which bright technological futures are imagined and promised and take hold of the imagination. The hopeful future is actually what is required to counter the underlying meaninglessness of the future, the secular endings that disrupt imaginaries of continuity.

Because final secular death is a collective destiny (masked by individual tragedy), a global and civilizational concern, immortality as a project becomes inseparable from the fears of civilizational collapse. Indeed, the dialectic of future and futurelessness was a key part of the emergence of the radical life-extension movement. The two original texts of cryonics— the first texts of the contemporary life-extension movement—surfaced at the same time in part because they were products of a zeitgeist, synthesizing the cultural and scientific advances into an idea that had older roots. This was the early 1960s, with America on the brink of great social and cultural upheaval, when radical change and revolution seemed in the offing on all fronts, from the political to the technological and scientific. It was a time when immense possibilities seemed open to the imagination, but also a time when existential threats seemed quite at the fore of national consciousness, pitting the optimism of technological progress against the pessimistic probability of total annihilation. There was a man on the moon, but down on Earth, other men, having emerged from a world war, lived under the shadow of the mushroom cloud.

Both senses of the future—the moon and the mushroom cloud, immense potential to explore and to destroy—were present in the minds and texts of the two first cryonics authors. In imagining the future, both

Cooper's and Ettinger's books raise the possibility of devastating future wars, though they swiftly counter with a more-or-less utopian view that connects progress and rationality to some version of coexistence and peace. Ettinger was a soldier in war, as were a handful of other early figures in the movement. The cultural context of the second half of the twentieth century was ripe for imagining existential threats and wide-scale destruction, and hence generating a heightened emphasis on survival, wherein physical immortality would be the redemptive achievement, the apotheosis of success.

Today, there is only more, the techtopian future contrasted with the overabundance of catastrophic scenarios, as demonstrated in the list of Lifeboat shields. American private space ventures, whose members overlap with the immortalist and transhumanist groups, are a case in point. As David Valentine has shown (2012), and my interlocutors confirm, the prominent narrative is that humans (or Americans in Silicon Valley, at any rate) are going to settle outer space as a way of avoiding extinction, because this planet is going to hell and humans must reinvent life elsewhere. By contrast, the dual threat of promise and destruction is not played out in the European space program in the same way.[16]

The American obsession with new beginnings and ruptures linked to technology (the "American technological sublime," in David Nye's words [1999]), the possibility of second lives and the continuity of civilizations in splendid isolation, is reflected in the title of Goertzel's coedited volume on transhumanist thought, *The End of the Beginning* (Goertzel and Goertzel 2015). In this sense, immortality echoes with much more significance than just individual longevity. If the various possibilities of annihilation and catastrophe, including nuclear holocaust and the Anthropocene, can stand in as recognizable figures of futurelessness, then immortality can serve as secular America's redemptive counterfigure of the future, the new frontier of escape and continuity.

Technology has been part of the vision of progress in general, and very strongly so in the American context (Noble 1999)—though it's worth noting M. Giulia Fabi's observation that African American utopian writings gave less credence to "the liberatory potential of technological progress" and molded their futures around collective ideological change and autonomy (2004, 46). Secular humanism held that with science, rationality, and proper social planning humans can and will make things better. Science was only one domain that would contribute to the betterment of society. Science was part of a triumvirate of S's that included society and state. The proper arrangement of these together would lead to progress. In laying out their plan, most political theorists, modern states, or even utopians had envisioned or predicted transformations through three facets of human modern life: through the rearrangement of *social units* such as the appropriate means of production, the proper (re)distribution of goods, the provision of education and knowledge to all, the rearrangement of kinship; through the power of *science*, in revealing a unified world subject to control and calculation, and helping shed the illusions of religion; and finally, through the rule of law and the guarantee of security by a just *state*. These are the three S's: society, state, and science.[17] In a sense, secularism rides on the set of promises tendered by science, state, and society, this triumvirate of nonreligious institutions. Now faith has waned in two out of three of them. For many people, certainly for immortalists and transhumanists, society and state have been emptied of their promise; indeed they have become the source of trouble. Even when the "economy" is switched for society, the triumphalist narrative of progress loses its sheen as inequalities in global neoliberal regimes grow and crises repeat more and more frequently and the economy's machines and machinations gouge the Earth to the brink of collapse. What is left to show for progress is the incessant development of scientific thought and technological things. The remaining vision of progress then can only be encapsulated in a technocivilization of the future—that is, we are now told that the great secular humanist promises of development, comfort, knowledge, and security for all humans are now mainly going to be achieved not through the rearrangement of social units but of technological ones.

The growing attraction of immortality projects, in this light, might be understood not as the mere extension of individual life but as a recalibration of eschatology oriented toward this technocivilizational telos of progress. What matters in the politics of transhumanist survivalism is not, as the rhetoric purports, saving humanity, but *producing* or evolving a particular sort of future being embedded in a particular form of the future, the transhumans of Western civilization who can leave behind not just limitations of the human but also the human mess its civilization has helped create. Thus, perversely, existential risk applies to a form of being that is not even yet in existence, whose existence is only anticipated and must be protected![18] In his key paper, "Existential Risk: Analyzing Human Extinction Scenarios and Related Hazards," Bostrom defines existential risk as "one where an adverse outcome would either annihilate Earth-originating intelligent life or permanently and drastically curtail its potential. An existential risk is one where humankind as a whole is imperiled. Existential disasters have major adverse consequences for the course of human civilization for all time to come."[19] Note that existential risks are not risks that could wipe out only a portion of humans, at least not at first glance. In Bostrom's taxonomy, genocide may be an evil, but it is not an existential risk of this sort. Neither is AIDS as we have encountered it. As he explains, "These types of disasters have occurred many times and our cultural attitudes towards risk have been shaped by trial-and-error in managing such hazards. But tragic as such events are to the people immediately affected, in the big picture of things—from the perspective of humankind as a whole—even the worst of these catastrophes are mere ripples on the surface of the great sea of life. They haven't significantly affected the total amount of human suffering or happiness or determined the long-term fate of our species." Over the course of human history, Bostrom contends, there have been very few existential risks to "our species," the closest being an asteroid impact. At least until the mid-twentieth century. And thus is the mushroom cloud invoked. The advent of nuclear weapons was the first instance where humanity became aware of an existential risk, the first time that "our species" had the subjective experience of contemplating and envisioning its own ultimate destruction. Since then such possibilities have proliferated: global warming does count, in the Bostrom taxonomy, as one of the existential risks, alongside nuclear holocaust, deliberate or accidental misuse of nanotechnology, asteroid impact, and, on a lower scale, a repressive global regime or misguided world government, a flawed superintelligence, and/or a takeover by a rogue mind upload. Of course, the classic question posed by anthropologists when they encounter claims about humanity as a whole is valid here: Which humans count for the future? Who is being subsumed by the invocation of "our" species? of civilization? Whose risk is actually being evaluated?

Indeed, in Bostrom's text the question of "humankind" suddenly gets transformed. What is most fascinating about Bostrom's taxonomy is revealed in his criteria for assessing risk, because of course risk is never abstract—it is a risk *to* someone or something. And here, reading more closely, we see risk assessed in relation not just to "our species" but to the future development or evolution of *posthuman intelligent* life. As explained, one version of existential risk is when the future entails a threat in which "Earth-originating intelligent life goes extinct in relatively sudden disaster." But because intelligent life is not limited to humans alone, there is another scenario that would count too, one in which "the potential of humankind to develop into posthumanity is permanently thwarted" *even if* "human life continues in some form."

The ultimate standard by which existential risk is assessed, then, is any threat to the future establishment of a posthuman cyborg techno-civilization. In this view, if humans—say, Cordeiro's atavistic humans of the Amazon or equally atavistic Muslim terrorists!—get in the way of posthuman development, then such humans become the existential threat, ironically painted as a threat to their own human *potential*. Potential matters because it embodies the teleological account of the universe, where the past as history reveals itself to be pouring out into a particular future, becoming a particular sort of mind, a particular form of life—the informatic form. Kurzweil also invokes the future in this way, suggesting that it would be immoral to not develop superintelligence. Again, these narratives are not limited to transhumanists and immortalists—much of technofuturism and a good portion of the neoliberal ethos is embroiled in the same rehearsal of technoprogressive myths of the future against the backdrop of futurelessness: before his death, the world-renowned physicist Stephen Hawking, for instance, had been repeating the transhumanist story of existential threats to humanity while happily expressing his gratitude to Virgin Galactic for providing him a free ticket onboard their spaceship to take him off the planet. The company, owned by billionaire Richard Branson, has built its "Spaceport America" in Las Cruces, New Mexico, within eyeshot of the American border wall.

The invocations of "humanity" and "civilization" belie the fact that what ends up mattering as civilization is de facto white and Western, and more specifically, American—I am aware of the problems with categories such as "the West" and "whiteness," so I don't put forward these observations lightly or arbitrarily. They reflect the emic discourses, subject positions, and imaginaries I have been witnessing and weaving together. Not exactly coterminous but deeply related to each other, the West and whiteness start from geographical and biological referents, respectively, but rather than essentializing them in those terms, I understand them here mainly as historical formations that are imbricated with notions of progress and superiority (biological, cultural) and thus also imply particular relations to the future. Both describe a subjectivity that has imagined itself at the technological, intellectual, cultural, and even biological vanguard of humanity. The claim and control over the future has also implied the negation of other futures. Transhumanist (even many posthuman) visions of the future are not exactly replete with, say, indigenous, Muslim, African, or African American figures. Indeed, the threat to future civilizations, as the transhumanist philosopher Julian Savulescu warned in a 2009 talk, is often said to come from "minorities" who are going to gain access to the means of destruction (read revenge).[20] Interestingly, a most haunting specter seems to be a deeply internal one: threats to reproductive continuity. One of the justifications I heard from many people is reflected in the words of one transhumanist: "Western European countries are way below [population] replacement rates, so the only way to save Western civilization is life extension" (also see More 2015).

In this context, it should make sense that, leaving an immortalist presentation on AI, a Latina friend would turn to me and quip, "I have seen the future, and it's white." AI, of course, is meant to be beyond race, which is where the quip acquires its additional edge. For AI may or may not end up racialized through its avatars, but the social structures through which it will propagate, and certainly the social structures in which biopolitics more broadly is embedded (Weheliye 2014), are clearly marked by race and geopolitics. Though the immortalist organizations don't collect data on race and origin—they collect data on belief and psychological profile—it is no news to state that all the meetings and conferences I have attended have been overwhelmingly white (and mostly male, though decreasingly so). Writers on Afrofuturism have astutely warned that structures of inequality get perpetuated through imaginaries of a technologically enhanced race-free, even identity-free, future that purports to bloom into glory for all but in fact reproduces the same inequalities as far as racialized populations are concerned (A. Nelson 2002). This has also been true more broadly for the tech sector in Silicon Valley, where positions of power have remained white, lagging behind other sectors.[21] Frighteningly, it seems that what deep-learning AI is learning is deep-seated race and gender bias picked up from the same folks who, well, have the biases and design the code (Caliskan, Bryson, and Narayanan 2017).

CIVILIZING MISSIONS

Immortalist or transhumanist visions of the future always appear more extreme than they are, but like objects in the mirror, they are closer than they appear. For one, they influence the ideas and ethics of the tech sector, as I've shown through the example of the NBIC convergence or the immortalist/transhumanist involvement in major tech ventures. Equally important, they share in a broader understanding of important secular humanist concepts such as progress and civilization, concepts that have gained a renewed charge. These concepts are mobilized to justify the contemporary neoliberal technological governmentality, where inequalities in the distribution of the resources of life and death have widened to extremes while its defenders continue to present technology and the market as the solutions that will save civilization, showcasing the long history of betterment attributed to liberalism (global hunger reduction indices! fewer homicides per capita! dropping poverty rates! more billionaires!). When I revisited Silicon Valley in 2014, Steven Pinker's celebratory book, *The Better Angels of Our Nature: Why Violence Has Declined* (2011), was on many tongues as the ultimate empirical proof that things are just fine, the Cassandras be damned—the results are in, Western secular scientific civilization leads to better humans. In 2018, Bill Gates declared Pinker's subsequent Pollyanna publication one of his favorite books of the year.[22]

Pinker has argued that the Hobbesian state and its successor, Western democratic liberalism, has made us better human beings, as evidenced by dropping per capita homicide rates and fewer deaths in wars through the end of the twentieth century. He mobilizes numbers and charts based on anthropological data on hunter-gatherer groups, archaeological evidence, as well as death counts in wars both historical and contemporary. The conclusion he draws from his numbers is that humanity has gotten progressively better at limiting violence and increasing security, and when you do those things you make better human beings whose peaceful side, their better angel, emerges. We are heading toward better humans in better times. Pinker's numbers, timeline, and methods have been challenged; he has been accused of misreading problems of scale and history, cherry-picking his stats, of using the worst-case scenarios of the past, especially in early states, to represent the full picture, of doing the same for hunter-gatherer groups, whose context at any rate he misrepresents, ignoring the conflicts brought on in their encounter with states (Ferguson 2013; Falk and Hildebolt 2017; Oka et al. 2017; Cirillo and Taleb 2016; Arquilla 2012). In relation to modern states, Pinker does count atrocities such as World War II, Hitler, and Stalin as part of the twentieth century, but then he makes his claims about the reduced violence of modernity *after* those events have passed, that is, on the last fifty to seventy years, a narrow slice on which to build a theory of history (see Cirillo and Taleb [2016] on time intervals between mass atrocities). In doing so, Pinker also characterizes those regimes as not quite modern; rather he designates them as utopian counter-Enlightenment regimes—that is, he conveniently picks what counts as the proper child of Western Enlightenment and scientific thought. As John Gray remarks, "You would never know, from reading Pinker, that Nazi 'scientific racism' was based in theories whose intellectual pedigree goes back to Enlightenment thinkers such as the prominent Victorian psychologist and eugenicist Francis Galton."[23]

This blatant rewriting of history is coursed through with a racialized discourse constantly defending the civilizing process. Pinker calls Western states "respectable countries" (2011, 259) among which "conquest" is no longer a goal. He talks about Vietnamese insurgents as fanatics with no value for life (like Islamists, he says) and that if they didn't have a "fanatical dedication to outlasting the enemy" there would have been even fewer casualties imposed on them by the French and the Americans (307)! That they were killed in such great numbers, then, is not the fault of the killers and their colonial ventures; the problem was that the Vietnamese just didn't want to submit properly.

Aside from the numerical manipulations and racial-cum-moral white-washing, this mode of reckoning—in which the number of lives taken stands in for levels of insecurity and of moral worth—is exactly the bio-political abstraction that ignores lived experience, the level of insecurity people (usually the vulnerable) live under when suffering from debilities in the aftermath of war (three million affected by U.S. chemical warfare in Vietnam, a statistic ignored entirely by Pinker) or when the killer is hovering permanently over them with a Taser or a drone, like hunters over their prey (see Chamayou 2013). At that point, no one needs to *kill* in order to make others feel killable. The state and the market do not produce an even distribution of security throughout all populations. In Pinker's index of violence, the massive and rising incarceration rates in the United States have no place (two million humans a day in cages); nor does the fact that under the awful name "criminal justice system" another twelve million circulate through a dense web of supervision, parole, ankle monitoring, and check-ins at any one time;[24] and it indicates even less that these numbers refer to a highly racialized population of black and brown people. Durkheim pointed out in his study of sociality, way before the punitive scales we see today, that economic and social disruptions of this sort can cause much more harm to the collective body than an isolated case of homicide (Durkheim 1966, 33); but such disruptions cannot appear in the biopolitical calculus of life. Finally, it is far from clear that higher moral status can be claimed for a world that produces more than enough food to feed *everyone* on Earth and yet lets over 800 million people go hungry every day[25]—and instead, looking at its own humanitarian sentiments and paltry food aid, counts a few more stomachs that have a few more grains of rice in them and pats itself on the back for its immense moral progress.

It is not by accident that Pinker's views echo in the tech sector and among transhumanists. Though he doesn't focus on technology as such, technology is part of the social assemblage that he outlines as progress. His more sociocentric view of progress only serves to confirm the good consequences of technological change, especially since in the techno-civilizational discourse the social and the political have at any rate been evacuated. In technocivilizational discourse, in the economy of innovation imagined by informatic futurists and Silicon Valley, advanced technology becomes the self-sufficient marker of progress and of civilization. After all, what better index than clear and evident improvements in human tools, that is, technology? It's the direct doctrine of the technolibertarian vision, wherein, as Peter Thiel has repeatedly said, "technology will replace politics."

I have outlined a particular history of this "replacement" through shifts in Enlightenment utopianism, where the future unfolds as a dialectic of progress and annihilation and where social and state-led solutions ceded place to the supremacy of science and thus the predominance of technofixes that ignore their own social and political embeddedness. Fred Turner (2006) has documented a related history, tracing some of the ways in which the counterculture of the 1960s slowly morphed into the cyberculture of the new millennium, transforming social priorities into technological solutions. Mainstream figures such as Steve Brand and Buckminster Fuller are very much the immediate predecessors, calling for "revolution by design" (Scott 2009) and pushing the notion that with properly designed technology and non-institutional participation and innovation (technolibertarianism) the messiness of politics and social arrangements can be all ironed out.

We need closer studies of the ways in which Silicon Valley has recalibrated the American empire as a technolibertarian venture in collusion with the state it allegedly deplores, and the ways in which the hype of Silicon Valley has been crucial to the post-9/11 economic, psychological, and social transformation of the United States.[26] At the turn of the century, two complexes were mobilized to forcefully resurrect the trope of progress and the civilizational narrative in the United States, at the same time renewing the time of empire (Stoler 2016): technology and terrorism, temporally pointing in opposite directions, the former leaping forward, the other holding progress back, the one promoting life, the other destroying it. September 11 and its wars were crucial to the new ideological thrust of progress: the actual backward barbarians were going to be bombed back into the stone age where they belonged, and America had to be made great again moving into the future. Not just America. Civilizations would be able to move on with their high culture, advanced moral capacities, and economically driven innovations. This is civilization as the spectacle of technology, the ongoing and far-from-random linking of 9/11 as the existential threat to human civilization and the overcoming of such a threat through the genius of the human mind wedded to the market.[27] As one well-circulated story on *Wired* put it: "The iPhone Is Bigger Than Trump." And regarding Google, the piece continues: "As horrible as 9/11 was, the fact that one single corporation might connect almost all of the world's population on a single service is, in the long run, actually bigger news,"[28] meaning that the end of the iPhone would be a bigger disaster than the end of politics. The cruel fact is that, indeed, through Silicon Valley companies, including one owned by Thiel, the replacement of politics by technology is exactly what happened in the previous U.S. election, as ads and news and votes were algorithmized. Technology has taken over politics, but it hasn't replaced it.

There is something highly perverse in sermons of salvation and betterment coming from people who have benefited from the exploitation of others and the destruction of lives and habitats, and who now, while lecturing about how things are getting better, are themselves preparing for the worst, laying plans to leave and abandon all in order to survive elsewhere. If things are working so well, why then are technofuturists preparing to escape to Mars? Or buying large tracts of land in New Zealand, as transhumanist tech billionaires have? The ruptures of immortalism and transhumanism emphasize the neoliberal politics of abandonment (Povinelli 2011), where a departure from the human mess includes such literal strategies as colonizing Mars, thereby leaving the devastations behind to those who can't make it out—all the while erasing the fact that the devastations themselves have been caused by civilization and technology. This is part of the perversity of progress—rather than being lifted out of poverty and misery, people are always being pushed into precarity in the name of progress.

Now technological development has come to be seen as the true telos—not the social and political telos but the telos of the universe itself;

in that context the interests of the beings and communities technology purports to serve supervene upon the interests of technology. AI and technofuturism thus offer their promise mainly to those who have discarded community as the impossible dream,[29] or as the actual nightmare, of utopias; even more, the promise of technology serves those for whom political community appears as a threat to their form of life.

FUTURES OUT OF PLACE

The question I am addressing here may be summarized by asking: Who is the secular immortal? Immortality, as I have argued throughout the book, is not just a bid for personal survival but a social project for imagining and building particular futures. What future is being imagined or promised when technoscientific immortality is being imagined or promised? What future forms of life are immortal life-forms embedded in? Who is saving whom, and from what kind of destruction? What futures are being displaced?

The crucial distinction between saving individual lives and treating social persons (what I think of as life-forms and forms of life) in the biomedical setting is examined by Lisa Stevenson in the context of the Canadian state's purported attempts to "save" the Inuit from tuberculosis. Because it had no interest in their lived lives and forms of life, the state carried out its project by rendering the Inuit into numbered, serialized bodies who would be examined, removed, possibly treated, and if lucky returned. Their names and actual lives did not matter to the biopolitical state, but the removal of tuberculosis from the charts did. The state took the Inuit away by ship to distant tuberculosis wards and gave them tags and numbers; if they died, there would be some report of the death, often without the body being returned or the name being called or the grave being marked; identities were confused, and many went literally missing as a result. Who a person was did not matter; what mattered was a material body that needed to be saved. Stevenson writes: "The very impulse to save lives actually effaced who they were, their histories, their cultures, their desires. They became so many animals to be saved from death" (2014, 28). The colonizer wants the colonized to participate in its lifesaving efforts, even as those efforts are part of the establishment of a regime of sovereignty that perpetuates the subjugation of (some of) the people it saves or, even worse, a sovereign regime that often prefers their death or encourages it (through police brutality, for example, and economic underdevelopment). Their participation in the regime is crucial to the regime, for without it the biopolitical state project would founder. Stevenson makes a similar point in relation to what was called a suicide epidemic among the Inuit, in which the state also felt it must get involved, even as it helped create the conditions for the suicides and in a sense could only see the Inuit through the possibility of their premature death. Stevenson (who worked as a suicide hotline volunteer) writes that suicide is both expected and prohibited, a situation that simultaneously "bans and anticipates" Inuit death (96). Suicide as a response to "the pain of the now" and the vision of "the future's wreckage" suggests that suicide may be also "a leap into another way of being in time, one that questions whether there is always a brighter future around the corner" (147).

A counterbiopolitics entails rejecting a racialized immortality regime indexed to the brightness of the technological future, refusing its demand for cooperation and statistical interpellation, and repudiating the over-valuation of individual life as imagined by biopolitical ideologies. This, of course, can take things precariously close to suicide, as Stevenson recognizes. If suicide appears for many religions as an affront to the creator, in biopolitics suicide as the ultimate devaluation of life has been posited as a problem of misplaced self-contempt, a psychological short-circuiting of the proper biological impulses that leads to increased glutamate levels in the brain—a madness. Stevenson analyzes the devaluation of bare life not as an individual pathology but as a function of the social and political context in which life, removed from its social and cosmological context, is already devalued, in which its future is already grim.

What is at stake is not just the length of lives—the biopolitical exten-sion of bare life, the one life, the one body given to us in the secular—but situated judgments about kinds of lives and kinds of futures. The forms of lives and futures idealized, projected, and aimed for by immortalists and transhumanists are not idealized by all. Indeed, they may represent a form of life that poses a danger to many, especially those in already structurally precarious situations (racially, geopolitically, by class, by status, physical ability) as well as those engaged in political struggles that aim against the wider socioeconomic currents. As Braden Allenby and Daniel Sarewitz poignantly say, America's soldiers, "with their smart weapons, their body armor, their night-vision goggles, their special diets, their training in and integration into remote robotic combat systems, and, we would suspect, their ingestion of neuropharmaceuticals such as modafinil to keep them alert even when deprived of sleep for 36 hours" (2011, 24),[30] are the most physically and cognitively enhanced beings in the world today—the most transhumanist, one might say. This is not mere accident. The projected transhumanist technologies often emerge from military research and are fed back into the military. Despite outward libertarian gestures against the state, and their reputation as liberals, Silicon Valley and high-powered immortalists are enmeshed with the American state and the military. Ray Kurzweil has worked closely with DARPA and NASA. Peter Thiel owns a policing and surveillance company called Palantir (closely linked to Cambridge Analytica) and has backed the Trump presidency without as much as a peep from the immortalist community. When I asked de Grey whether he or SENS had a position on this, since Thiel has funded SENS, his pithy response was that "neither I nor SENS Research Foundation has a position on such matters."[31] Politics has indeed been taken over by technology.

For those who live through structurally foreshortened temporalities— due to, say, U.S.-led wars in Syria, Afghanistan, Iraq, or structural poverty or a racist political economy that subjects certain lives to premature death, or economically and politically induced famines, or "health policies" or prejudices that made living with AIDS a deviant terminal condition and a trans life a precarious one (Halberstam 2005), or continued violation of bodies and resources that threaten indigenous populations (Simpson 2014), or probabilities that determine hi-tech drone strikes on villages in Afghanistan (Chamayou 2013)—the abstractions of life itself, the bypassing of current social conditions of premature death *in the name of future life,* appears an insult, another way to perpetuate and reinforce the structures and histories in which inequalities lie firmly embedded. For those whose current lives are precarious because their deaths are made insignificant, those whose deaths are without the possibility of signification, of having meaning and importance in the world, those whose deaths, in Judith Butler's words, are ungrievable, for those on whom in their current lives the door of the future is closed, technoscientific immortality is not just a fantasy, it's also an affront. It means the repetition of exclusions that, under the fetish of the universal value of life, create social suffering and inequities in relation to death, differentially disposing lives to precarity and foreshortened life spans (Butler 2012; Gilmore 2006).

To separate the extension of bare life from the social and structural forms in which life is embedded—life-forms from forms of life—is to engage in what Siv Visvanathan (2003) calls the "moral infantilism" endemic to the technoscientific hype of the twenty-first century. Hannah Arendt saw the implications as she made a distinction between know-how and thought, wherein as we become dependent on the former we also become "thoughtless creatures at the mercy of every gadget which is technically possible, no matter how murderous it is" ([1958] 1998, 3). "The tragedy," Visvanathan adds, "is that science and technology, with their euphoric myth of progress, innovation, and obsolescence, have no format for understanding defeated and marginalized communities" (2003, 172).

Immortalism and transhumanism have pretty much been rolled into the power and promise of American libertarian futurism as practiced in Silicon Valley and many a university lab across the nation. Thus the constant rehearsal of the technocivilizational myths made me see immortalists and related groups not just as people seeking long life but as the new mythmakers of Western progress. They are necessary to the wider hype and hope ethos of Silicon Valley and the tech sector in general, built as they are on the promise of designing future life in more efficient, more comfortable, smarter ways, where the infrastructure of the brain and of society become one, where individual and public, inner and outer, merge because they are essentially thought to be one already. That is the consequence of the epistemological claim and cosmological underpinning of the algorithmic ideology: that human minds, like everything else in the universe, run on algorithms. This informs all the ways in which the information of the mind and the body are operationalized for cognitive and bio-capitalism (Moulier Boutang 2012) and in which polities are mobilized around bio- and informatic security. Threatened by its own prior modes of production and reproduction, as well as by the advances of other regions (a looming China is regularly invoked), American technocivilization has now tasked itself with tending to the development and protection of the full potential and purpose of "our species" and civilization understood through a posthuman technological index.

What is valued over humanness in the informatic cosmology is the perpetuation of a *posthuman form of life*—in which the power, accuracy, and speed of technology become the utmost measures of worth, mainly because these are also supposed to lead to the rise of conscious beings who, as one famous website has it, are less wrong, act without bias, and make the proper decisions in the algorithmic tree of being. In the case of Bostrom and transhumanism, unlike in regular neoliberalism, the transhuman life-form promised for the future will be embedded (if it isn't already) in the technocivilizational form of life, so it is *that* form of life which is subject to existential risk and comes to require protection. Notions of humanity and racial, ethnic, and national formations will no doubt be transformed over the course of the twenty-first century by the wide application of bio- and informatic technologies, and with the spread of new economic tools and spatial ventures (Seasteading, lunar colonies) that attempt to circumvent the territorial state as a way of escaping its regulatory power. Still, it is hard to ignore the resurrection of civilizational discourse not just in transhumanism but equally in liberalism and neoliberalism. If new population categories are deemed to pose a threat to technological and civilizational advance, the populations implicated in these transformations have been de facto racially and geopolitically defined; that is, white Americans, or Western-educated urban denizens more generally, are the assumed subjects of that future as transhumanism takes the form of a technocivilizational crusade.

In the previous chapter I indicated that the informatic cosmology of mind and cosmos moves beyond the secular desouling of matter and the universe. Here I am suggesting that informatic technologies also run the risk of narrowing the possibilities of mind by glorifying and fetishizing algorithmic intelligence (Ziewitz 2016). For in the name of expanding human capacities, algorithmic modalities are narrowing the range of possibilities, *even for posthuman life-forms*. Take the notion of convergence. I have described convergence as the site of a dominant technoscientific re-unification of destinies and belongings, an imaginary of universal purpose through which amped-up visions of secular technological progress and the totalizing machinification of the future may be justified. As a result, it leaves everything else behind—the nonmachinic, organic, noninformatic capacities and modes of creation, from literature to worms to soil to tree roots, as well as mental states and bodily practices like trance and possession rituals that it cannot fit into its vision of informatic convergence. To take the example from Cordeiro's COPL talk, there is no openness to Amazonian cosmologies that might relate to local context through radically different modalities of sensing and apprehending; and it is certainly not a future incorporating the range of Amazonian beings, human and nonhuman, in its projections of technocivilization, since these are imagined as headed toward extinction or, at best, obsolescence. Informatic futurist models, like most AI itself, know a lot about gaming minds or shopping minds but not much about hunting minds or shamanic minds. AI knows a lot about decision-making trees and chess games, but not much about the subjectivity of love and pain or the synesthetic sensing of fear or deep faith or the commitment to social life that has made lives thicker, sometimes perhaps more effective, and often enough, of course, imbued with messiness, melancholy, and suffering. As the NBIC fields fit their agendas into that of convergence, they move away from entanglements with life-forms such as plants and jaguars and their modes of sensing, inhabiting, and producing effects (Kohn 2013); thus they also lose concern for a wide range of forms of life (e.g., the rainforest) through which plants and jaguars make their life in relation to humans. It is a narrowing rather than the expansiveness promised in immortalist and transhumanist imaginaries. It is exactly what Stengers would not include in her cosmopolitical ethics. Technoscientific convergence is a proposal for unification rather than proliferation, a proposal for the elimination of the cosmological gap by stringing together mind and cosmos only through particular technologies: via information, silicon, algorithms, and capital.

The reduction of consciousness and social life to algorithms and decision-making trees may be an effect, and possibly an efficient one, of large social systems and digital infrastructures, but decision-making algorithms are not total aspects of who and what humans are, or even how humans think and judge and act or even how they might use digital technologies; algorithms do not represent some anterior, universal code that runs human beings. Indeed, as I argued in the last chapter, AI helps produce such forms (informatic selves), making new minds, new ways of being and thinking, new subjects and publics. The spectrum of subjectivity, thought, and capacity called mind is very wide indeed and is constantly being modified. In animist and panpsychic thinking that spectrum would be as wide as the universe itself. The danger in following paths of convergence down to a unified point with an underlying singular explanatory unit (information) is that we *narrow* the spectrum of possibility of minds while pretending to broaden it.

The particular future projected in the current civilizational imaginary, in which the technologically enhanced and immortal posthuman embodies the only true and most viable potential of humanity, erases any number of possible futures for human or even posthuman transformation. And it does so by generating a particular range of catastrophic fears about the future, or futurelessness — fears that affect and motivate privileged, mostly white Americans who feel that they have a valuable life (and maybe a two-car garage, i.e., a lifestyle) to protect and a bright future to produce, possibly somewhere off-planet.

At the same time, in a counterpolitics perhaps, some of us (jaguars, shamans, artists, anthropologists?) might gladly wave farewell to those taking off in their private space cars for Planet B. Elizabeth Povinelli's exhortation echoes loudly here, next to Stevenson's evaluations of life: "We must dedramatize human life as we squarely take responsibility for what we are doing. This simultaneous dedramatization and responsibilization may allow for opening new questions. Rather than Life and Nonlife, we will ask what formations we are keeping in existence or extinguishing?" (2016, 28). Not just keeping or extinguishing, but also what future formations are we imagining, producing, enabling, or acceding to?

In experimental communities across the globe there already are other "possible futures," social niches seeding evolutionary alternatives for unpredictable future political ecologies (Farman 2016b); if there is a "must," it is to imagine and produce more of these experiments, not accede to futures as they get issued in technocivilizational centers, overdetermined by the economies and infrastructures that embed them. But doing so really means imagining futures outside of the narrow frame of progress and betterment and humanism. Posthumanism, as opposed to transhumanism, may provide such possibilities through the acknowledgment of human entanglements in much larger forces and diverse agencies, and proposing ethical responses to this. Yet posthumanists tend to hold on fast to their secular views and invoke vague forms of vitalism or immanence to justify a posthuman ethics. My arguments in this book imply that posthumanism demands desecularization,[32] not by resorting to a Luddite atavism but through alternative conceptions of technology and telos that produce different imaginaries of futures and future formations, ones that might have to take forms of panpsychism seriously. A different cosmology would spell out a different technological or techno-ecological future.

It is important to consider the advantages of breaking beyond "the human" of secular humanism. But, first, without the kinds of reactions or presumptions that would jettison the transhuman view on technology wholesale, and second, by exploring and assuming the greater possibilities of connectivity with the universe's existing technologies, including plants and minds and radio frequencies. To outline or explore these alternative futures is for, well, a future project, but this suggests that a radical posthumanism perhaps requires not the defense but the implosion, the negation, of what has been known as civilization, and a pluralism of techno-ecologies, with different grammars connecting past-present-future temporalities and animate-deanimate-reanimate continuums.

May the futures proliferate.

Notes

1. A brief note on methodology. I conducted the main part of my fieldwork from 2008 to 2010, between Arizona, Florida, and Michigan, with several return trips over the years. I have visited the three most important cryonics centers in the United States: the Alcor Life Extension Foundation in Arizona, the Cryonics Institute in Michigan, and Suspended Animation, previously in Florida (which does not store cryonics patients but only acts as an intermediary, providing logistical support, deathbed care, transportation, and research), in order to carry out participant observation. I worked at one organization for over a month, mainly assisting in packing medication kits. I participated in a number of different trainings and assisted in one cryopreservation procedure. I also briefly helped with setting up a lab and conducting experiments with molecular biologists connected to a biogerontology project. I have attended several cryonics information sessions, member gatherings, and annual meetings. I have observed (taking notes, interviewing, and carrying out all the other general participant observation duties) more than fifteen gatherings and conferences on transhumanism, life extension, and artificial intelligence. In addition, I was given access to internal polls and membership data at several of the organizations. I returned to the Bay Area and Silicon Valley over the summer of 2015 for my final engagements in the field. But, it is important to note, this was and is not an ethnography of Silicon Valley.

   I have conducted more than seventy interviews, some recorded digitally, others written down manually, with a wide range of actors. I interviewed my subjects in various locations: at immortalist conferences around the United States, at and around cryonics facilities, in the towns where they live and where I happened to be, in their labs, and on several occasions either by phone or online. Immortalists, and particularly cryonicists, inevitably carry out crucial interactions with non-immortalist organizations and actors. Because I was interested in how the life–death transition is managed through institutional regimes, I also tracked peripheral actors, such as funeral directors, hospice workers, lawyers, lobbyists, doctors, and insurance agents, as their offices intersected with the practices of immortalists.

A great deal of interaction and discussion took place online (including laments over and solutions for the problem of "community"). Thus, a number of key online, public forums constituted an important site of investigation. These included Cryonet, Cold Filter, and the Immortality Institute's forums. I tracked and coded significant discussions taking place online, according to my interests, specifically marking discussions on the following: personhood, attitudes toward death, views on future life, visions of the social future, personal life-extension practices and strategies, and attitudes toward immortality.

INTRODUCTION

1. Musk's project is being conducted under the aegis of the company Neuralink, founded in 2016.
2. Greenblatt's reaction to Musk's announcement took the form of a December 23, 2018, editorial in the *Guardian,* https://www.theguardian .com/commentisfree/2018/dec/23/elon-musk-neuralink-chip-brain -implants-humanity.
3. Kant, in his second preface to the *Critique of Pure Reason,* concedes that reason has its limits and that upon reaching that boundary it is best to "remove knowledge in order to make room for faith" ([1781] 1966, B, xxxix). He points to several areas where human knowledge falls short because it is finite: the real limits of space and time, the original cause, freedom from causality, and the existence of a necessary being. He posits that human reason, the nonempirical rational faculty, is driven to synthesize and complete knowledge; it does not abide gaps. So, quixotically, mysteriously, it tries nevertheless to find answers to unanswerable questions and as a result generates stopgap concepts (ideas) that can close the circle of inquiry. For Kant, these included such ideas as God or primary cause, free will, and the soul and its survival.
4. A number of scholars have made interesting distinctions between posthumanism and transhumanism, describing the latter as a kind of uber-Enlightenment position; e.g., Wolfe (2010); S. Fuller and Lipinska (2015).
5. There are clear problems with Taylor's geographical delineation (North America/Europe) and how it articulates with his conceptual

parameters—do the edges of "the immanent frame" look different if you take just the United States as its center? Why is the immanent frame not applicable to Nicaragua?

6. Harris Poll, December 15, 2009, https://theharrispoll.com/wp-content/uploads/2017/12/Harris_Poll_2009_12_15.pdf.

7. Also see Bauman (1992, 170); and for a different take in relation to Orientalism, see Anidjar (2006).

8. I use the term "immortalist imaginary" (Marcus 1995; C. Taylor 2003) as an analytic that allows me to tack between projected, promised, or expected futures, normative notions and cultural concepts, assumptions or tensions underlying those futures, and the practices, institutions, and sites through which those futures are being actively produced in the present.

9. Also see Stoler (2017) on political concepts and Ingold (1993) on concepts in evolutionary theory.

10. Names have been changed to preserve anonymity. I have retained real names only when the person was a public figure, such as director of an organization or a conference speaker, and I spoke to them in that capacity.

11. David Segal, "This Man Is Not a Cyborg. Yet," New York Times, June 1, 2013, http://www.nytimes.com/2013/06/02/business/dmitry-itskov-and-the-avatar-quest.html?pagewanted=all&_r=0.

12. Blankholm (2018) skips subjectification, preferring to think of a more collective entity, secular people.

13. Exceptions include Kohn (2013) and some new animists (G. Harvey 2005).

14. I owe this point to Murphy Halliburton. Also see Mykytyn (2010).

1. AFTER LIFE

1. *The Immortalist,* July 2006, http://www.cryonics.org/immortalist/july06/news.htm (last accessed July 11, 2009).

2. A small flurry of self-published books in the 1990s referred to immortality in their titles: *Becoming Immortal* (1995) by Wesley du Charme presented an overview from the perspective of nanotechnology with techniques that would make reanimation from cryonics viable, and the future world a place of expanding intelligence and cheaper and better production; *Immortality: How Science Is Extending Your Lifespan and Changing the World* (1998) by Ben Bova focused mainly on medical techniques and dietary regimens that have been extending life in modern times; also in

1998, Frank Tipler's *The Physics of Immortality* presented a discussion in theoretical physics making an information-based argument about the physical resurrection of all existing things in infinite universes.

3. See the introduction for the use of this term. Related to it were associations with categories like "kooky," "cultish," and "far-out," as well as the term's appropriation by what was casually referred to as the "snake oil" vendors putting out anti-aging products with dubious claims about longevity and health.

4. Clearly, since 2008 much has changed regarding what shows up on search engines.

5. *The Immortalist,* November 2006, http://www.cryonics.org/immortalist/november06/news.htm (last accessed July 11, 2009).

6. I am not including histories that outline religious doctrinal developments and changes, e.g., Brandon (1967) or T. Brennan (2002).

7. More generally, the trouble with the centrality of belief in social studies of religion has been pointed out in the past (Needham 1972; Cantwell Smith [1962] 1991).

8. Hume's arguments against immortality were about Christian notions and institutions, but as with many other secularist forays, it was universalized to cover all notions and practices around the world that looked like versions of a belief in the afterlife and the survival of the soul as seen from a secular European perspective.

9. For example, d'Holbach's book was reprinted and retranslated extensively.

10. Leuba's survey of scientists was duplicated in 1996 by Edward Larson and Larry Witham (Larson and Witham 1997). Sending the questionnaire to a list culled at random from the 1995 *American Men and Women of Science,* they found that 46 percent said they disbelieved in immortality, 15 percent said they doubted it, and 38 percent believed in personal immortality.

11. A spate of recent neurological and cognitive experiments, using brand-new technologies, have simply extended the Tylorian/Frazerian explanation of religion into cognitive theory: they have treated beliefs in the afterlife as an illusion detectable as cognitive malfunctions, such as disorientation or schizophrenia, which have elicited post facto explanations and doctrines that turned, say, a flash on the retina into a ghost because it is evolutionarily advantageous (Bering and Bjorklund 2004).

12. Hertz's notion of death as an occasion for social regeneration was fruitfully extended by others such as Bloch (1971), Metcalf and Huntington

(1991), as well as by Bloch and Parry (1982), and Bauman (1992) in a more comparative and theoretical vein.

13. As well as an interesting correspondence with Alfred Wallace included in the footnote to the published text. See Weismann (1891, 22n1).

14. "Science: Physical Immortality," *Time,* November 30, 1925, http://www.time.com/time/magazine/article/0,9171,736596,00.html.

15. Raymond Pearl to A. Siegrit, Correspondence, Raymond Pearl Papers, Alan Mason Chesney Medical Archives of the Johns Hopkins Medical Institutions.

16. Infusoria is an obsolete taxonomic category that encompassed small aquatic creatures, including algae and amoebae.

17. See, e.g., *Cryonics,* July 1981, http://www.alcor.org/cryonics/cryonics 8107.txt.

18. See correspondence reproduced in M. Darwin (1991a, 1991b).

19. This bylaw (Section 2.04, Denial of Membership and Discipline of Members) is reproduced in M. Darwin (1991a) and in Cryonics Institute director Ben Best's online FAQ at http://www.benbest.com/cryonics/CryoFAQ.html#_IXC_); more recent bylaws, which I obtained from the society, have kept the same wording (until 2017), but the text appears in a different section.

20. "Society for Cryobiology—The Vote Is In," *Cryonics,* January 1983.

21. Reproduced in M. Darwin (1991b).

22. By the 1980s some simple life-forms, such as nematode worms and tardigrades, had been shown to undergo suspension and reanimation (Triantaphyllou and McCabe 1989).

23. According to cryonicists who have attended recent meetings of the Society of Cryobiology, this hostility is waning, and to their surprise a few cryobiologists are even open to scientific discussions on the matter. But the bylaw still stands.

24. Vitrification occurs when a liquid goes directly to a solid state without forming ice crystals—i.e., it vitrifies, or becomes like glass.

2. IMMORTALISM

1. As a cloning activist, Randy also appears in Debbora Battaglia's (2005) ethnography of Raelians.

2. Nathaniel Rich, "Can a Jellyfish Unlock the Secret of Immortality?" *New York Times Magazine,* November 28, 2012.

3. A recent study found that less than 1 percent of those involved in investment decisions in Silicon Valley are black, and only 1.3 percent are Hispanic. Peter Schults, "Introducing The Information's Future List," October 6, 2015, https://www.theinformation.com/introducing -the-informations-future-list.

4. See, e.g., Peter Thiel's company, Palantir, which has multiple contracts with the U.S. military and Immigration and Customs Enforcement. Shane Harris, "Palantir Wins Bid to Build Army Intelligence System," March 27, 2019, https://www.latimes.com/business/technology/la-fi -tn-palantir-army-contract-raytheon-20190327-story.html.

5. Mike Perry, "For the Record," *Cryonics,* February 1992.

6. Perhaps this can be traced back to the very beginnings of Western science itself and Francis Bacon, one of the fathers of the modern scientific method. Aside from his experiments and his writings on science, in 1626 Bacon also published *New Atlantis,* a futuristic utopian work of fiction in which voyagers discover an ideal island (off the coast of Peru) where the central institution is a state-led house of scientific research dedicated to the betterment of human life (Bacon [1627] 2008).

7. It is also to the point to remember that H. G. Wells had studied physics, zoology, and geology at the Normal School of Science in London before he became a science fiction writer.

8. This has a bitter and contentious history, well summarized in Regis (2004).

9. Fred Turner's (2006) history follows a different utopian line, tracing the hippie movement's merger with Silicon Valley through Stewart Brand, the Whole Earth Catalogue, and the Long Now Foundation.

10. AEA Press Release for Cyberstates 2002, "Growth of Tech Employment Declines Sharply in 2001, AEA Report Says," Washington, D.C., June 26, 2002, http://www.aeanet.org/PressRoom/idmk_cs2002_US.asp.

11. For a different group of researchers, the anti-aging group, see Mykytyn (2010).

12. Yudkowski email to SL4 list on "How Kurzweil Lost the Singularity," http://sl4.weidai.com/archive/0206/4015.html.

13. Steven Thrasher, "Conservative Facebook Investor Funded Anti-ACORN Videographer," *Village Voice,* September 22, 2009, http://blogs .villagevoice.com/runninscared/2009/09/acorn_videomake.php.

14. See blog post by Singularity Utopia on *Medium,* https://medium .com/@2045singularity/white-supremacist-futurism-81be3fa7020d.

15. Based on raw membership data for 2008 provided by Alcor.

16. See http://www.transhumanism.org/index.php/WTA/more/2007 survey/.
17. Mormon transhumanists are a particularly active subgroup. See new work by Jon Bialecki (2017).
18. Immortality Institute, 2003 online poll on religious/nonreligious identity. Results forwarded through private communication.

3. SUSPENSION

1. Temperature makes a difference. Sperm loses motility (the power to move) at a much greater rate when stored at −70°C than at −196°C. (Trummer et al. 1998).
2. Results for experiments on "semen quality" can be found in Sancho et al. (2007) and J. Bailey, Morrier, and Cormier (2003).
3. Also see various centers for resuscitation science, e.g., at the University of Pennsylvania, http://www.med.upenn.edu/resuscitation.
4. It may be interesting to note that Schopenhauer's ideological solution to the vanity of existence is antinatalism, the notion that it would have been better to not have been born at all (see Benatar 2006); but I also wonder if, having had the terrible bad luck of having been born, Schopenhauer would not have opted for the poetic stasis of cryopreservation at some point on his downhill trajectory. For notice that the man stumbling downhill toward his own eternal nonexistence is marked by a tragic inability to pause on this unavoidable trajectory.
5. Weber's position is not Nietzschean, insofar as Nietzsche thought technology to be irrelevant to the main task of power and joy; in fact, even as he did not deny science's improved ability to explain things in the world, Nietzsche suggested that these were merely better mechanical descriptions *(Gay Science)* and possibly disruptive of the ideal of return to an unencumbered Dionysianism or science of joy *(Gay Science, Twilight of the Idols)*. That world had been destroyed by religious morality long before the ascendance of science.
6. In a different vein, E. P. Thompson's (1967) analysis of time and industrial capitalism is an empirical incursion into this very idea, documenting, among other things, the shift in the practices of production from being task oriented—i.e., make this product—to being time oriented—i.e., work from nine to five on a time sheet; the imposition of time discipline; and the ideology of "time-thrift" upon the old irregular work habits required by pre-industrial production.

7. Asad writes of "the multiple ways in which the body and the world have come to be severed and yoked together" (2016, 338).
8. The framed analog photos have since been updated to a digital screen.
9. From a Seasteading press release that can be found at https://www.seasteading.org/the-seasteading-institute-announces-its-second-annual-conference/.
10. Alex Lightman, "Things to Do with Your Body While You Wait for Immortality," *H+ Magazine,* June 19, 2009, http://www.hplusmagazine.com/articles/forever-young/things-do-your-body-while-you-wait-immortality.
11. The text, taken from an older pamphlet, can now be found at https://hpluspedia.org/wiki/Transhumanist_FAQ_Version_3.
12. Supplements are favored for several reasons. First, they are unregulated and do not require prescriptions; indeed, they are often claimed to be ahead of the mainstream FDA-approved drugs and are often viewed as an alternative to the manipulations and lies of the pharmaceutical industry. For example, an online handbook from the Life Extension Foundation has a list of accomplishments cited in this vein: "In 1986, the Foundation recommended the drug ribavirin to treat lethal viral infections. It took twelve years for the FDA to approve ribavirin as a treatment for Hepatitis C. Hundreds of thousands of Americans died because the FDA denied Americans access to this broad-spectrum anti-viral medication" (http://www.lifeextensionfoundation.org/accomplishments_01.htm). Second, they are historically linked to life extension, especially through one of the major suppliers, the Life Extension Foundation, which also publishes a magazine by that name. The founders are longtime cryonicists who contribute large sums of money to cryonics organizations and research.
13. Interview with Aubrey de Grey of the SENS Research Foundation, at the Fight Aging blog, July 2017, https://www.fightaging.org/archives/2017/07/an-interview-with-aubrey-de-grey-of-the-sens-research-foundation/.

4. DEANIMATION

1. Uniform Determination of Death Act, http://www.law.upenn.edu/bll/ulc/fnact99/1980s/udda80.htm.
2. Even if it is agreed that the end of the person *should* be equated with the end of measurable brain functions, instruments make differential readings regarding those functions. At the beginning, when the criteria for whole-brain death were initiated, EEG measurements satisfied many, showing that at some point there was no longer any activity in the brain.

More precise instruments and studies, however, showed this to be untrue, revealing a range of active regions. The criteria were then adjusted, not by including the active regions but by making a distinction between "significant" and "insignificant" cellular regions and claiming that the newly discovered active regions were "insignificant" (see Veatch 2005 for a review).

3. Michigan Public Health Code (Act 368 of 1978, 333.2803).

4. Lambek (2016, 643) describes this in contexts that validate the sociality of ghosts and possession events.

5. When bodies arrive by air at airports, they are marked as "human remains," designated for pickup by a funeral director, who will have to sign the airbill and refile a disposition permit with the host state's vital statistics office.

6. However, it seems to me that such corpses *do* come with cultural baggage, but refilled with a different set of social, moral, or religious attributes. A different mode of stripping personhood is perfectly dramatized in the case of the Sabarl as described by Battaglia (1990). Upon the death of the person a set of mortuary engagements both build up and eliminate the memory of the person through the deployment and destruction of memory objects. Battaglia shows how the symbolic and practical elimination of the corpse is a necessary act of forgetting, because the world of "pervasive memory" that the Sabarl inhabit and in which the deceased person still remains embedded would prevent the Sabarl from moving on and contradict the fact of the person's absence. This ritual practice of forgetting runs counter to cultures of memorialization in which remembrance through the personification of objects is precisely what is valued. Interestingly, Andrews (1986) has argued that allowing for the commodification of body parts will in fact reduce the objectification or dehumanization of the whole person.

7. In an interview with an employee of Body Worlds, responsible for the high-traffic exhibit of the same name, I was told that the first step in their body-donation program is to cut off all connection to the family and to the history of that particular person as soon as the body is received. Once they are entered into the process of plastification, the connection to identity is entirely severed. In some cases, donors have requested that their identity be revealed because "they want to be seen by a brother or son," but the company does not allow that.

8. See PA Act 299, Occupational Code, 1980.

9. Pierce v. Proprietors of Swan Point Cemetery, 10 RI 227.

10. Rauhe v. Langeland Chapel, 44 Mich App 371, 265 NW 2d 313.

11. Enos v. Snyder, 63 P. 170, 171 (Cal. 1900).
12. As Sharp (1995, 368) shows, organs themselves have a further set of hierarchical categories: "organs fall into two categories. The first are the 'quality of life' organs, which include the kidneys and pancreas. Dialysis and insulin therapies replace the functioning of these organs and can sustain patients almost indefinitely. The second are the 'life saving' organs, including the heart, lungs, and liver. Patients who need these are in such ill health that they will soon die if they do not receive replacement organs."
13. ARS 41-865.
14. The name and the structures within the State of Michigan have been changed several times since 2006.
15. Statements were from the Michigan Bureau of Commercial Services website at the time. The numbers have been updated under the new bureau, Licensing and Regulatory Affairs, https://www.michigan.gov/lara/0,4601,7-154-89334_72600_72602---,00.html.
16. Proposed under Title 32, the regulation of "Professions and Occupations," chapter 12, "Funeral Directors and Embalmers."
17. Under Title 35, the regulation of "State and Local Boards and Departments of Health," chapter 7, article 3.
18. Drafts of HB2637 are from personal archives.
19. The DHS claimed that Alcor was not a procurement organization because it was not licensed as one. As it turns out, Alcor was not licensed as one because the DHS refused to give it a permit. Since there were no criteria for what qualifies as a "procurement organization"—unlike medical or educational institutions, which have accreditation—the California courts recognized the circularity of the DHS argument and ruled in favor of Alcor. That ruling has been the basis of Alcor's claims to legal operation ever since.
20. Alcor v. Mitchel, 7 Cal. App. 4th 1287, 9 Cal. Rptr. 2d 572.

5. CONVERGENCE

1. See http://www.terasemweb.org/ and http://terasemfaith.net/beliefs/.
2. Personal communication with the author, July 26, 2017.
3. Archived website: http://turingchurch.6.90com/2012/01/02/order-of-cosmic-engineers/.
4. The concepts of the infinite and of nothingness have their own histories (Koyré 1957). The infinite was related to God's unboundedness.

In most traditions, God was outside of physical systems of space and time, so attributes like unbounded and eternal referred to a different order. The contemplation of the *physical* universe as infinite was a recent phenomenon, according to Koyré. Although it had been broached by the atomists, most did not accept the possibility. The infinity of the universe was not brought up seriously until Nicolas of Cusa and then most clearly by Giordano Bruno. Descartes and Kepler thought about it, though like many others they didn't strictly think of the universe as infinite but only as "interminate" (Koyré 1957, 8).

5. The mission of the foundation has been reworded and may be found at https://www.brainpreservation.org/mission/. The old mission, quoted above, is preserved at https://www.guidestar.org/profile/27-3585957 and in my personal archives.

6. Martine Rothblatt, Mindclone blog, December 23, 2009, http://mind clones.blogspot.com/search?updated-min=2009-01-01T00%3A00%3A00 -05%3A00&updated-max=2010-01-01T00%3A00%3A00-05%3A00&max -results=10.

7. I first saw this on the now-defunct website of the Society of Universal Immortalism (SUI), which developed out of a short-lived organization called the Transhumanist Church. Through the years I spoke to him, Perry was the president of the SUI. The origins of the statement may be in the Transhumanist Church, dating back to 2004. See the remaining archives at http://www.croftpress.com/david/archives/transhumanistchurch/ beliefs/.

8. Mission statement of the Brain Preservation Foundation, as of October 2019, https://www.brainpreservation.org/mission/.

9. Intel cofounder Gordon Moore, who claimed that transistor power is doubling every year. This "law" is now commonly applied to all electronic devices that use transistors and integrated circuits.

10. Though he strongly commits himself to interpretation via religious categories such as "visions of heaven," Geraci (2010) outlines a similar connection between general AI, the information-based sciences, and the Singularity.

11. Reported in the *MIT Technology Review,* https://www.technologyreview .com/s/411814/the-armys-remote-controlled-beetle/.

12. Reported in *Wired,* http://www.wired.co.uk/article/braingate.

13. Report of the experiment was published in PLOS One, November 5, 2014, https://doi.org/10.1371/journal.pone.0111332.

14. Reported in *Scientific American,* https://www.scientificamerican.com/ article/3-brain-technologies-to-watch-in-2018/.

15. A number of social scientists have analyzed the social repercussions of the neurocentric shift, where everything about the person or the self is understood, and then experienced, as a facet of one's neurological constitution. But in important ways, the informatic self, although appearing neurocentric, is different from concepts such as Nikolas Rose's neurochemical selves (2003), the bioinformatically dispersed body of Mol (2002) or Brown and Webster (2004), or Malabou's (2008) call to be conscious of a relationship to "a plastic brain," or Dumit's (2004) objective self-fashioning, whereby people have "entered into a relationship with their brain that is negotiated and social." In the information context, although the brain *is* the self, it is also—ideally—disposable. The true relationship is to the computational processing of information in the brain.
16. Planned Parenthood of Southeastern Pa. v. Casey, 505 U.S. 833 (1992).
17. For a more detailed analysis of spirituality in transhumanist discourse see Farman (2019).
18. Another tack taken in this direction is through reenchantments of a different sort. For a criticism of the inadequacies of this view see Farman (2012).
19. I owe this point to Debbora Battaglia.

6. PROGRESS AND DESPAIR

1. For the New Organ Alliance see https://www.mfoundation.org/news/2018/6/25/new-organ-alliance-update.
2. See https://www.brookings.edu/opinions/the-growing-life-expectancy-gap-between-rich-and-poor/. On inequality, health, and violence see Wilkinson (2004).
3. Also see U.S. Life Tables 2009, https://www.cdc.gov/nchs/data/nvsr/nvsr62/nvsr62_07.pdf.
4. For permutations of this argument around immortality and stem-cell research see Pasquale (2013).
5. Aubrey de Grey, "Why Fight Aging?" (presentation at the Understanding Aging conference, UCLA, June 2008). Also see Perry's (2004) serious engagement with the concept. Others have also written about deathism. See http://en.wikipedia.org/wiki/User:Bogger/Deathism#cite_note-0 and "A Letter to Casual Deathists," *Longevity Meme Newsletter,* October 22, 2007, https://www.fightaging.org/archives/2007/10/longevity-meme-newsletter-october-22-2007/.

6. See Jonas (1992); Callahan (2003a); Kass (2005); Fukuyama (1992); Gems (2003); Habermas (2003); McKibben (2003).

7. See Weheliye (2014) for a comprehensive critique of Foucault and bio-political minimizations of the centrality of race, racial formations, and what Weheliye calls "racial assemblages." Michelle Murphy's (2017) study of the economization of life shows a link between current regimes of racialized and sexed value and a eugenics past.

8. See Maximum Life Foundation statement at https://www.maxlife.org/about-us/message-from-our-founder/.

9. See "What Is Transhumanism" at https://whatistranshumanism.org/.

10. https://www.benbest.com/cryonics/cryiss.html.

11. Michael Shermer, "Can Our Minds Live Forever," Scientific American, February 1, 2016, http://www.scientificamerican.com/article/can-our-minds-live-forever/.

12. https://www.benbest.com/cryonics/cryiss.html.

13. Terasem, Fourth Annual Workshop on GeoEthical Nanotechnology, http://www.geonano2008.com/; also see Michael Anissimov, "Immortalist Utilitarianism," https://lifeboat.com/ex/immortalist.utilitarianism.

14. Lifeboat Foundation's projects can be seen at http://lifeboat.com/ex/programs.

15. For example, https://becominggaia.wordpress.com/2011/03/07/please-quitboycott-the-lifeboat-foundation/.

16. I owe this information to Tamara Alvarez and her forthcoming dissertation on the European space program.

17. Casanova (1994) parses it as science, state, and economy.

18. See V. Adams et al. (2009) on anticipation and technoscience.

19. See Bostrom's article on existential risk: http://www.nickbostrom.com/existential/risks.html.

20. Julian Savulescu, "Genetically Enhance Humanity or Face Extinction" (lecture at Sydney's Festival of Dangerous Ideas, vimeo, minute 25+, https://vimeo.com/7515623).

21. See, e.g., the reports in *Gawker* and *The Information*: http://gawker.com/white-people-control-all-the-power-in-silicon-valley-1735006694; https://www.theinformation.com/introducing-the-informations-future-list.

22. Shana Lebowitz. "Bill Gates Says His Two Favorite Books from 2018 'Explain That the World Is Getting Better,'" *Business Insider,* February 27, 2018, http://www.businessinsider.com/bill-gates-favorite-books-2018-2.

23. John Gray, "Steven Pinker Is Wrong about Violence and War," *The*

Guardian, March 13, 2015, https://www.theguardian.com/books/2015/mar/13/john-gray-steven-pinker-wrong-violence-war-declining.

24. Peter Wagner and Leah Sakal, "Mass Incarceration: The Whole Pie; A Prison Policy Initiative Briefing," March 12, 2014, https://www.prisonpolicy.org/reports/pie.html.

25. Food and Agricultural Organization of the United Nations, *The State of Food Security and Nutrition in the World, 2017,* http://www.fao.org/3/a-I7695e.pdf.

26. The literature on biosecurity (e.g., Dillon and Lobo-Guerrero 2008; Thacker 2005b) takes up this aspect of post-9/11 securitization of life from a biopolitical perspective, though not with an explicit exploration of Silicon Valley.

27. In *Homo Deus,* Harari observes that many videogames are "civilizational-style strategy games" (2015, 246).

28. Stephen Levy, "The iPhone Is Bigger than Donald Trump," *Wired,* November 9, 2016, https://www.wired.com/2016/11/the-iphone-is-bigger-than-donald-trump/#.1xmye4ysx.

29. This is a blatant reformulation of Jacques Rancière's final passages in his book on the utopian projects of Icarians as they attempted to set up their utopian community in the United States (2012).

30. In a transhumanist journal, Evans (2007) argued that transhumanism has not affected the military. But this is based on a study of the prevalence of "transhumanism" as a term in military journals. By contrast, the use of terms such as artificial intelligence and nanotechnology were prevalent.

31. Personal email exchange, November 17, 2017.

32. Not a postsecularism à la Habermas, which is merely a call to offer mainstream religious authorities a seat at the public dinner table.

Bibliography

Adam, Barbara. 1995. *Timewatch: The Social Analysis of Time.* Cambridge: Polity Press.

Adam, Barbara. 2004. *Time.* Cambridge: Polity Press.

Adams, James Truslow. 1931 (2012). *The Epic of America.* New Brunswick: Transaction.

Adams, Vincanne, Michelle Murphy, and Adele Clarke. 2009. "Anticipation: Technoscience, Life, Affect, Temporality." *Subjectivities* 28 (1): 246–65.

Agamben, Giorgio. 1998. *Homo Sacer: Sovereign Power and Bare Life.* Trans. Daniel Heller-Roazen. Stanford: Stanford University Press.

Agrama, Hussein. 2010. "Secularism, Sovereignty, Indeterminacy: Is Egypt a Secular or a Religious State?" *Comparative Studies in Society and History* 52 (3): 495–523.

Agrama, Hussein. 2012. *Questioning Secularism: Islam, Sovereignty, and the Rule of Law in Modern Egypt.* Chicago: University of Chicago Press.

Alexander, Brian. 2003. *Rapture: How Biotech Became the New Religion.* New York: Basic Books.

Alexander, Irving, and Arthur Adlerstein. 1965. "Affective Responses to the Concept of Death in a Population of Children and Early Adolescents." In *Death and Identity,* ed. Robert Fulton, 111–22. New York: Wiley.

Allenby, Braden, and Daniel Sarewitz. 2011. *The Techno-Human Condition.* MIT Press. Kindle edition.

Anderson, Benedict. (1983) 2003. *Imagined Communities: Reflections on the Origin and Spread of Nationalism.* New York: Verso.

Andrews, Lori. 1986. "My Body, My Property." *Hastings Center Report* 16 (5): 28–38.

Anidjar, Gil. 2006. "Secularism." *Critical Inquiry* 33 (1): 52–77.

Anidjar, Gil. 2011. "The Meaning of Life." *Critical Inquiry* 37 (4): 697–723.

Ansell, Charles. 1997. "Introduction: A Personal Memoir of Herman Feifel." In *Death and the Quest for Meaning: Essays in Honor of Herman Feifel,* ed. Stephen Strack, xix–xxv. Northvale, N.J.: Jason Aronson Inc.

Anspach, R. R. 1993. *Deciding Who Lives: Fateful Choices in the Intensive-Care Nursery.* Berkeley: University of California Press.

Arendt, Hannah. (1958) 1998. *The Human Condition.* Chicago: University of Chicago Press.

Arendt, Hannah. 1976. *The Origins of Totalitarianism.* New York: Harvest Books.

Arendt, Hannah. 1993. *Between Past and Future.* New York: Penguin.

Arias, E. 2004. "United States Life Tables, 2002." *National Vital Statistics Reports* 53, no. 6. Hyattsville, Md.: National Center for Health Statistics, U.S. Department of Health and Human Services.

Aries, Phillipe. 1975. *Western Attitudes toward Death: From the Middle Ages to the Present.* Trans. Patricia Ranum. Baltimore: Johns Hopkins University Press.

Armstrong, Karen. 2006. "Is Immortality Important?" *Harvard Divinity Bulletin* 34 (1).

Arquilla, John. 2012. "The Big Kill." *Foreign Policy,* December 3, 2012. http://foreignpolicy.com/2012/12/03/the-big-kill/.

Asad, Talal. 1993. *Genealogies of Religion: Discipline and Reasons of Power in Christianity and Islam.* Baltimore: Johns Hopkins University Press.

Asad, Talal. 2003. *Formations of the Secular: Christianity, Islam, Modernity.* Stanford: Stanford University Press.

Asad, Talal. 2016. "Thinking about the Secular Body, Pain, and Liberal Politics." In *Living and Dying in the Contemporary World: A Compendium,* ed. Veena Das and Clara Han, 337–53. Berkeley: University of California Press.

Bacon, Frances. (1627) 2008. *The New Atlantis.* Project Gutenberg e-book. http://www.gutenberg.org/dirs/2/4/3/2434/2434.txt.

Baechler, Jean. 1979. *Suicides.* Trans. Barry Cooper. New York: Basic Books.

Bailey, J., A. Morrier, and N. Cormier. 2003. "Semen Cryopreservation: Successes and Persistent Problems in Farm Species." *Canadian Journal of Animal Science* 83 (3):393–401.

Bailey, Ronald. 2005. *Liberation Biology: The Scientific and Moral Case for Biotechnology.* New York: Prometheus Books.

Bainbridge, William Sims. 1982. "Religions for a Galactic Civilization." In *Science Fiction and Space Futures,* ed. Eugene M. Emme, 187–201. San Diego: American Astronautical Society. http://mysite.verizon.net/wsbainbridge/dl/relgal.htm#5.

Bainbridge, William Sims. 2007a. *Across the Secular Abyss: From Faith to Wisdom.* New York: Lexington Books.

Bainbridge, William Sims. 2007b. *Nanoconvergence: The Unity of Nanoscience, Biotechnology, Information Technology, and Cognitive Science*. Boston: Prentice Hall.

Bainbridge, William Sims, and Mihail C. Roco, eds. 2003. *Converging Technologies for Improving Human Performance: Nanotechnology, Biotechnology, Information Technology and Cognitive Science*. Dordrecht: Kluwer.

Bainbridge, William Sims, and Mihail C. Roco, eds. 2005. *Managing Nano-Bio-Info-Cogno Innovations: Converging Technologies in Society*. Dordrecht: Springer.

Balboni, Tracy A., Lauren C. Vanderwerker, Susan D. Block, M. Elizabeth Paulk, Christopher S. Lanthan, John R. Peteet, and Holly G. Prigerson. 2007. "Religiousness and Spiritual Support among Advanced Cancer Patients and Associations with End-of-Life Treatment Preferences and Quality of Life." *Journal of Clinical Oncology* 25:555–60.

Barad, Karen. 2007. *Meeting the Universe Halfway: Quantum Physics and the Entanglement of Matter and Meaning*. Durham: Duke University Press.

Barna Research Group. 2003. "Americans Describe Their Views about Life after Death." October 21, 2003. https://www.barna.com/research/americans-describe-their-views-about-life-after-death/.

Barrow, John D., and Frank J. Tipler. 1986. *The Anthropic Cosmological Principle*. New York: Oxford University Press.

Bartlett, Thomas. 2005. "The Man Who Would Murder Death." *Chronicle of Higher Education* 52 (10): 14–28.

Bataille, Georges. 1997. "Sacrifice, the Festival and the Principles of the Sacred World." In *The Bataille Reader*, ed. Fred Botting and Scott Wilson, 210–20. New York: Blackwell.

Battaglia, Debbora. 1990. *On the Bones of the Serpent: Person, Memory, and Mortality in Sabarl Island*. Chicago: University of Chicago Press.

Battaglia, Debbora. 1995a. "Fear of Selfing in the American Cultural Imaginary, or 'You Are Never Alone with a Clone.'" *American Anthropologist* 97 (4): 672–78.

Battaglia, Debbora. 1995b. "Problematizing the Self." In *Rhetorics of Self-Making*, ed. Battaglia, 1–15. Berkeley: University of California Press.

Battaglia, Debbora. 2005. "For Those Who Are Not Afraid of the Future: Raelian Clonehood in the Public Sphere." In *E.T. Culture: Anthropology in Outer Spaces*, ed. Battaglia, 149–79. Durham: Duke University Press.

Battaglia, Debbora. 2014a. "Cosmic Open Sourcery: Anxious Thoughts on Artifacts of False Witness and the Hypermediacy Effect." Paper presented to the New York University Department of Media, Culture, and Communication, April 10, 2014.

Battaglia, Debbora. 2014b. "Cosmos as Commons: An Activation of Cosmic Diplomacy." *E-Flux Journal* 58.

Baudrillard, Jean. 1994. *America*. New York: Verso.

Bauman, Zygmunt. 1992. *Mortality, Immortality and Other Life Strategies.* Stanford: Stanford University Press.

Beauchamp, Tom L. 1999. "The Failure of Theories of Personhood." *Kennedy Institute of Ethics Journal* 9 (4): 309–24.

Beck, Ulrich. 1992. *Risk Society: Towards a New Modernity*. London: Sage.

Becker, Ernest. 1973. *The Denial of Death*. New York: Macmillan.

Beer, Gillian. 1996. *Open Fields: Science in Cultural Encounter*. New York: Oxford University Press.

Bellah, Robert, Richard Madsen, William M. Sullivan, Ann Swidler, Steven M. Tipton. 1985. *Habits of the Heart: Individualism and Commitment in American Life*. New York: Perennial Library.

Belsky, Daniel W., Avshalom Caspi, Renate Houts, Harvey J. Cohen, David L. Corcoran, Andrea Danese, HonaLee Harrington, Salomon Israel, Morgan E. Levine, Jonathan D. Schaefer, Karen Sugden, Ben Williams, Anatoli I. Yashin, Richie Poulton, and Terrie E. Moffitt. 2015. "Quantification of Biological Aging in Young Adults." *Proceedings of the National Academy of Sciences* 112 (30): e4104–e4110.

Benatar, David. 2006. *Better Never to Have Been: The Harm of Coming into Existence*. Oxford: Oxford University Press.

Bender, Courtney, and Omar McRoberts. 2012. "Mapping a Field: Why and How to Study Spirituality." New York: Social Science Research Council, October, SSRC Working Papers.

Benecke, Mark. 2002. *The Dream of Eternal Life*. New York: Columbia University Press.

Bennett, Jane. 2001. *The Enchantment of Modern Life: Attachments, Crossings, Ethics*. Princeton: Princeton University Press.

Berger, Peter, ed. 1999. *The Desecularization of the World: Resurgent Religion and World Politics*. Washington, D.C.: Ethics and Public Policy Center.

Bergson, Henri. (1910) 1950. *Time and Free Will*. Trans. R. L. Pogson. London: George Allen and Unwin.

Bergson, Henri. (1922) 1965. *Duration and Simultaneity*. Trans. Leon Jacobson. Indianapolis: Bobbs-Merrill.

Bergson, Henri. (1947) 2007. *The Creative Mind: An Introduction to Metaphysics*. Trans. Mabelle L. Andison. Mineola, N.Y.: Dover Publications.

Bering, Jesse M., and David Bjorklund. 2004. "The Natural Emergence of Reasoning about the Afterlife as a Developmental Regularity." *Developmental Psychology* 40 (2): 217–33.

Bernal, J. D. 1929. *The World, the Flesh and the Devil: An Enquiry into the Future of the Three Enemies of the Rational Soul.* New York: Dutton.

Bernard, Hugh. 1979. *The Law of Death and Disposal of the Dead.* New York: Oceana Publications.

Bernstein, Anya. 2015. "Freeze, Die, Come to Life: The Many Paths to Immortality in Post-Soviet Russia." *American Ethnologist* 42 (4): 766–81.

Bernstein, Anya. 2019. *The Future of Immortality: Remaking Life and Death in Contemporary Russia.* Princeton: Princeton University Press.

Best, Ben. 2004. "President's Report." *The Immortalist.* http://www.cryonics.org/immortalist/january04/news.htm.

Best, Benjamin P. 2008. "Scientific Justification of Cryonics Practice." *Rejuvenation Research* 11 (2): 493–503.

Bialecki, Jon. 2017. "After, and Before, Anthropos." *Platypus: Committee on the Anthropology of Science, Technology and Computing (CASTAC) Blog.* http://blog.castac.org/2017/04/after-and-before-anthropos/.

Binet, Alfred. 1890. "The Immortality of Infusoria." *The Monist* 1 (1): 21–37.

Binski, Paul. 1996. *Medieval Death: Ritual and Representation.* Ithaca: Cornell University Press.

Blankholm, Joseph. 2018. "Secularism and Secular People." *Public Culture* 30 (2): 245–68.

Bloch, Maurice. 1971. *Placing the Dead: Tombs, Ancestral Villages, and Kinship Organization in Madagascar.* New York: Seminar Press.

Bloch, Maurice, and Jonathan Perry. 1982. Introduction. In *Death and the Regeneration of Life,* ed. Bloch and Parry, 1–44. New York: Cambridge University Press.

Bluebond-Langner, Myra. 1978. *The Private Worlds of Dying Children.* Princeton: Princeton University Press.

Blumenberg, Hans. (1966) 1983. *The Legitimacy of the Modern Age.* Trans. Robert Wallace. Cambridge: MIT Press.

Boellstorf, Tom. 2008. *Coming of Age in Second Life.* Princeton: Princeton University Press.

Bolte, Robert G., Philip G. Black, Robert S. Bowers, J. Kent Thorne, and Howard M. Corneli. 1988. "The Use of Extracorporeal Rewarming in a Child Submerged for 66 Minutes." *Journal of the American Medical Association* 260 (3): 377–79.

Bonnechere, Pierre. 1998. "La notion d''acte collectif' dans le sacrifice humain grec." *Phoenix* 52 (3/4): 191–215.

Borgmann, Albert. 1999. *Holding On to Reality: The Nature of Information at the Turn of the Millennium.* Chicago: University of Chicago Press.

Bortolloti, Lisa. 2006. "Moral Rights and Human Culture." *Ethical Perspectives: Journal of the European Ethics Network* 13 (4): 603–20.

Bostrom, Nick. 2014. *Superintelligence: Paths, Dangers, Strategies.* New York: Oxford University Press.

Bouk, Dan. 2015. *How Our Days Became Numbered: Risk and the Rise of the Statistical Individual.* Chicago: University of Chicago Press.

Boustead, Greg. 2008. "The Biohacking Hobbyist." *Seed Magazine,* December 11, 2008. http://seedmagazine.com/content/article/the_ biohacking_hobbyist/.

Boyer, Paul. 1992. *When Time Shall Be No More: Prophecy Belief in Modern American Culture.* Cambridge: Harvard University Press.

Braidotti, Rosi. 2007. "Bio-Power and Necro-Politics (Biomacht und nekro-Politik. Uberlegungen zu einer Ethik der Nachhaltigkeit)." *Springerin, Hefte fur Gegenwartskunst* 13 (2): 18–23.

Braidotti, Rosi. 2013. *The Posthuman.* Malden, Mass.: Polity Press.

Brandon, S. G. F. 1967. *The Judgment of the Dead: The Idea of Life after Death in the Major Religions.* New York: Scribner.

Brassier, Ray. 2007. *Nihil Unbound: Enlightenment and Extinction.* New York: Palgrave Macmillan.

Bravo, Michael, and W. Gareth Rees. 2006. "Cryo-politics: Environmental Security and the Future of Arctic Navigation." *Brown Journal of World Affairs* 13 (1): 205–15.

Brennan, Susan, and Richard Delgado. 1981. "Death: Multiple Standards or a Single Standard?" *Southern California Law Review* 54:1323–55.

Brennan, Tad. 2002. "Immortality in Ancient Philosophy." In *Routledge Encyclopedia of Philosophy.* London: Routledge. http://www.rep.routledge.com/article/A133.

Bridge, Stephen. (2006) 2014. "Has Cryonics Taken the Wrong Path?" *Cryonics,* February 2014 (originally published on the Alcor News Blog, August 2006).

Brieskorn, Norbert. 2010. "On the Attempt to Recall a Relationship." In Jürgen Habermas et al., *An Awareness of What Is Missing: Faith and Reason in a Post-Secular Age,* 24–35. Malden, Mass.: Polity Press.

Brinton, Crane. 1965. "Utopia and Democracy." *Daedalus* 94 (2): 348–66.

Brooke, John Hedley. 2005. "Visions of Perfectability." *Journal of Evolution and Technology* 14 (2): 1–12. http://jetpress.org/volume14/brooke.html.

Brostowin, Patrick R. 1969. *John Adolphus Etzler: Scientific-Utopian during the 1830s and 1840s.* Unpublished PhD diss., New York University (University Microfilms).

Brown, Nik. 2003. "Hope against Hype." *Science Studies* 16 (2): 3–21.

Brown, Nik, and Andrew Webster. 2004. *New Medical Technologies and Society: Reordering Life*. Malden, Mass.: Polity Press.

Brown, Wendy. 2007. "Idealism, Materialism, Secularism?" Post on *The Immanent Frame*, a Social Science Research Council blog, October 22, 2007. http://blogs.ssrc.org/tif/2007/10/22/idealism-materialism -secularism/.

Brown, Wendy. 2010. "The Sacred, the Secular, and the Profane: Charles Taylor and Karl Marx." In *Varieties of Secularism in a Secular Age*, ed. Michael Warner, Jonathan VanAntwerpen, and Craig Calhoun, 83–104. Cambridge: Harvard University Press.

Burant, Christopher J. 2006. "Optimism/Pessimism as a Mediator of Social Structural Disparities Effects on Physical Health and Psychological Well-Being: A Longitudinal Study of Hospitalized Elders." Unpublished dissertation, Case Western Reserve University.

Butler, Judith. 2012. "Can One Lead a Good Life in a Bad Life?" *Radical Philosophy* 176 (November/December): 9–18.

Bynum, Caroline Walker. 1995. *The Resurrection of the Body in Western Christianity*. New York: Columbia University Press.

Calhoun, Craig, Mark Juergensmeyer, and Jonathan VanAntwerpen. 2011. Introduction. In *Rethinking Secularism*, ed. Calhoun, Juergensmeyer, and VanAntwerpen. Oxford University Press. Kindle edition.

Caliskan, Aylin, Joanna J. Bryson, and Arvind Narayanan. 2017. "Semantics Derived Automatically from Language Corpora Contain Human-Like Biases." *Science* 356 (6334): 183–86.

Calkins, Gary. 1914. *Biology*. New York: Henry Holt.

Callahan, Daniel. 2003a. "Visions of Eternity." *First Things* 133 (May): 28–35.

Callahan, Daniel. 2003b. *What Price Better Health?* Berkeley: University of California Press.

Campbell, Courtney. 1992. "Body, Self, and the Property Paradigm." *Hastings Center Report* 22 (5): 4–42.

Canales, Jimena. 2015. *The Physicist and the Philosopher: Einstein, Bergson, and the Debate That Changed Our Understanding of Time*. Princeton University Press. Kindle edition.

Canguilhem, Georges. 1988. *Ideology and Rationality in the History of the Life Sciences*. Trans. Arthur Goldhammer. Cambridge: MIT Press.

Canguilhem, Georges. 2000. *A Vital Rationalist: Selected Writings from Georges Canguilhem*. Ed. François Delaporte. Trans. Arthur Goldhammer. New York: Zone Books.

Canine, John D. 1996. *The Psychosocial Aspects of Death and Dying*. New York: McGraw Hill.

Canton, James. 2005. "NBIC Convergent Technologies and the Innovation Economy: Challenges and Opportunities for the 21st Century." In *Managing Nano-Bio-Info-Cogno Innovations: Converging Technologies in Society*, ed. William Sims Bainbridge and Mihail C. Roco, 33–45. Berlin: Springer.

Cantwell Smith, Wilfred. (1962) 1991. *The Meaning and End of Religion*. Minneapolis: Fortress Press.

Carrasco, David. 1999. *The City of Sacrifice: The Aztec Empire and the Role of Violence in Civilization*. Boston: Beacon Press.

Carrel, Alexis. 1931. "Physiological Time." *Science* 74:618–21.

Carrel, Alexis. (1935) 1961. *Man the Unknown*. London: Burns & Oates.

Casanova, José. 1994. *Public Religions in the Modern World*. Chicago: University of Chicago Press.

Casanova, José. 2011. "The Secular, Secularization, Secularisms." In *Rethinking Secularism*, ed. Craig Calhoun, Mark Juergensmeyer, and J. VanAntwerpen, 54–74. New York: Oxford University Press.

Cave, Stephen. 2012. *Immortality: The Quest to Live Forever and How It Drives Civilization*. New York: Crown.

Chakrabarty, Dipesh. 1997. "The Time of History and the Times of Gods." In *The Politics of Culture in the Shadow of Capital*, ed. Lisa Lowe and David Lloyd, 35–60. Durham: Duke University Press.

Chalmers, David J. 2002a. "Consciousness and Its Place in Nature." In *Philosophy of Mind: Classical and Contemporary Readings*, ed. Chalmers, 247–72. New York: Oxford University Press.

Chalmers, David J. 2002b. "The Puzzle of Conscious Experience." *Scientific American* 286 (4): 90–99.

Chamayou, Gregoire. 2013. *A Theory of the Drone*. Trans. Janet Lloyd. New York: New Press.

Christakis, N. A. 1999. *Death Foretold: Prophecy and Prognosis in Medical Care*. Chicago: University of Chicago Press.

Churchland, Paul. 1988. *Matter and Consciousness*. Cambridge: MIT Press.

Cirillo, Pasquale, and Nassim Nicholas Taleb. 2016. "On the Statistical Properties and Tail Risk of Violent Conflicts." *Physica A* 452:29–45.

Clark, David. 2002. "Between Hope and Acceptance: The Medicalisation of Dying." *British Medical Journal* 324 (7342): 905–7.

Clarke, Adele, Janet K. Shim, Laura Mamo, Jennifer Ruth Fosket, and Jennifer R. Fishman. 2003. "Biomedicalization: Technoscientific Transformations of Health, Illness, and U.S. Biomedicine." *American Sociological Review* 68 (2): 161–94.

Collier, Stephen J., and Andrew Lakoff. 2005. "On Regimes of Living." In *Global Assemblages: Technology, Politics, and Ethics as Anthropological Problems,* ed. Aihwa Ong and Stephen J. Collier, 22–39. Malden, Mass.: Blackwell.

Collingwood, R. G. 1960. *The Idea of Nature.* New York: Oxford University Press.

Conklin, Beth, and Lynn Morgan. 1996. "Babies, Bodies, and the Production of Personhood in North America and a Native Amazonian Society." *Ethos* 24 (4): 657–94.

Connor, Linda. 1995. "The Action of the Body on Society: Washing a Corpse in Bali." *Journal of the Royal Anthropological Institute* 1 (3): 537–59.

Cook, Ian, and Mike Crang. 2007. *Doing Ethnographies.* Thousand Oaks, Calif.: Sage.

Coole, Diana, and Samantha Frost, eds. 2010. *New Materialisms: Ontology, Agency and Politics.* Durham: Duke University Press.

Coon, Deborah. J. 2000. "Salvaging the Self in a World without Soul: William James's *The Principles of Psychology.*" *History of Psychology* 3 (2): 83–103.

Cooper, Evan [Nathan Duhring]. 1962. *Immortality: Physically, Scientifically, Now.* http://depressedmetabolism.com/pdfs/cooper_immortality.pdf.

Cooper, Melinda. 2006. "Resuscitations: Stem Cells and the Crisis of Old Age." *Body Society* 12:1–23.

Cortright, Joseph, and Heike Mayer. 2002. "Signs of Life: The Growth of Biotechnology Sectors in the US." *Report of Center on Urban and Metropolitan Policy.* Washington, D.C.: Brookings Institution.

Counts, Dorothy Ayers. 1980. "Fighting Back Is Not the Way: Suicide and the Women of Kaliai." *American Ethnologist* 7 (2): 332–51.

Counts, Dorothy Ayers, and David Counts. 2004. "The Good, the Bad, and the Unresolved Death in Kaliai." *Social Science and Medicine* 58 (5): 887–97.

Cranford, R. 1988. "The Persistent Vegetative State: The Medical Reality (Getting the Facts Straight)." *Hastings Center Report* 18 (1): 27–32.

Crippen, David W., and Leslie M. Whetstine. 2007. "Dark Angels—The Problem of Death in Intensive Care." *Critical Care* 11 (1): 202.

Critchley, Simon. 2004. *Very Little . . . Almost Nothing.* New York: Routledge.

Dalsgaard, Steffen. 2008. "Facework on Facebook: The Presentation of Self in Virtual Life and Its Role in the US Elections." *Anthropology Today* 24 (6): 8–12.

Damasco, Achille, and Alessandro Giuliani. 2017. "A Resonance Based Model of Biological Evolution." *Physica A* 471:750–56.

Danforth, Loring. 1982. *The Death Rituals of Rural Greece*. Princeton: Princeton University Press.

Danowski, Déborah, and Eduardo Viveiros de Castro. 2017. *The Ends of the World*. Trans. Rodrigo Nunes. Malden, Mass.: Polity Press.

Darwin, Charles. 1859. *The Origin of Species*. http://darwin-online.org.uk/content/frameset?itemID=F373&viewtype=text&pageseq=1.

Darwin, Michael. 1981. "A Report on the 18th Annual Meeting of the Society for Cryobiology, June 14–18, 1981, St. Louis, Missouri." *Cryonics,* July 1981. http://www.alcor.org/cryonics/cryonics8107.txt.

Darwin, Mike. 1983a. "Ev Cooper." *Cryonics,* no. 32 (March): 7–9. http://www.alcor.org/cryonics/cryonics8303.txt.

Darwin, Mike. 1983b. "Why We Are Cryonicists," *Cryonics,* no. 30 (January): 25–26. Republished with edits in the Alcor library, https://alcor.org/Library/html/whycryonics.html.

Darwin, Mike. 1991a. "Cold War: The Conflict between Cryonicists and Cryobiologists." *Cryonics* 12 (6): 4–16. http://www.alcor.org/cryonics/cryonics9106.txt.

Darwin, Mike. 1991b. "Cold War: The Conflict between Cryonicists and Cryobiologists. Part II." *Cryonics* 12 (7): 2–14. https://www.alcor.org/cryonics/cryonics9107.pdf.

Darwin, Mike. 2005. "A Thoughtful Analysis of Death in the ICU." In "Pro/Con Ethics Debate: When Is Dead Really Dead?" *Critical Care* 9 (6): 538–42.

Davies, Douglas. 2002. *Death, Ritual and Belief: The Rhetoric of Funerary Rites*. New York: Continuum.

Davies, Karen. 1994. "The Tensions between Process Time and Clock Time in Care-Work: The Example of Day Nurseries." *Time and Society* 3 (3): 277–303.

Davies, Paul. 2014. "Universe from Bit." In *Information and the Nature of Physical Reality: From Physics to Metaphysics,* ed. Paul Davies and Henrik Gregersen, 83–117. Cambridge: Cambridge University Press.

Davies, Paul, and Henrik Gregersen. 2014. "Introduction: Does Information Matter?" In *Information and the Nature of Physical Reality: From Physics to Metaphysics,* ed. Paul Davies and Henrik Gregersen, 1–14. Cambridge: Cambridge University Press.

Dedon, John P. 2006. "How to Take It with You." *Legal Times,* February 13, 2006.

De Duve, Christian. 1995. *Vital Dust: Life as a Cosmic Imperative*. New York: Basic Books.

Derrida, Jacques. 1985. *The Gift of Death*. Trans. David Wills. Chicago: University of Chicago Press.

de Wolf, Aschwin. 2007. "Using Low Temperatures to Care for the Critically Ill." *Cryonics* 28 (4): 16–18. https://alcor.org/CryonicsMagazine/cryonics2007.html.

d'Holbach, Paul-Henri, Baron. (1771) 1835. *The System of Nature; or, the Laws of the Moral and Physical World*. Trans. H. D. Robinson (1835). New York: G. W. and A. J. Matsell.

Dick, Steven. 2000. "Many Worlds: Cosmotheology." Metanexus Institute. https://www.metanexus.net/many-worlds-cosmotheology/.

Dick, Steven. 2010. "Cosmic Evolution: History, Culture, and Human Destiny." In *Cosmos and Culture: Cultural Evolution in a Cosmic Context*, ed. Steven J. Dick and Mark L. Lupisella, 25–62. Washington, D.C.: National Aeronautics and Space Administration.

Dick, Steven J., and Mark L. Lupisella, eds. 2010. *Cosmos and Culture: Cultural Evolution in a Cosmic Context*. Washington, D.C.: National Aeronautics and Space Administration.

Dillon, Michael, and Luis Lobo-Guerrero. 2008. "Biopolitics of Security in the 21st Century: An Introduction." *Review of International Studies* 34:265–92.

Downey, Gary Lee, and Juan Lucena. 1998. "Engineering Selves: Hiring In to a Contested Field of Education." In *Cyborgs and Citadels: Anthropological Interventions in Emerging Sciences and Technologies*, ed. Gary Lee Downey and Joseph Dumit, 117–42. Santa Fe, N.M.: School of American Research Press.

Downey, Geraldine, and Phyllis Moen. 1987. "Personal Efficacy, Income, and Family Traditions: A Longitudinal Study of Women Heading Households." *Journal of Health and Social Behavior* 28 (3): 320–33.

Doyle, Richard. 2003. *Wetwares: Experiments in Postvital Living*. Minneapolis: University of Minnesota Press.

Draper, John William. 1876. *The History of Conflict between Science and Religion*. New York: Appleton Press.

Drexler, Eric. 1981. "Molecular Engineering: An Approach to the Development of General Capabilities for Molecular Manipulation." *Proceedings of the National Academy of Sciences* 78 (9): 5275–78.

Drexler, Eric. (1986) 1990. *Engines of Creation*. New York: Anchor Books.

Dumit, Joseph. 2004. *Picturing Personhood*. Princeton: Princeton University Press.

Dupuy, Jean-Pierre. 2009. *On the Origins of Cognitive Science: The Mechanization of the Mind*. Cambridge: MIT Press.

Durkheim, Émile. 1966. *Suicide: A Study in Sociology*. Trans. John A. Spaulding and George Simpson. Glencoe, Ill.: Free Press.

Durkheim, Émile. 1995. *Elementary Forms of the Religious Life*. Trans. Karen E. Fields. New York: Free Press.

Dworkin, Ronald. 1994. *Life's Dominion: An Argument about Abortion. Euthanasia, and Individual Freedom*. New York: Vintage Books.

Dworkin, Ronald, Thomas Nagel, Robert Nozick, John Rawls, T. M. Scanlon, and Judith Jarvis Thomson. 1997. "Assisted Suicide: The Philosophers' Brief." *New York Review of Books* 44 (5): 41–47.

Dybul, Mark. 2013. "A Grand Convergence and a Historic Opportunity." *Lancet* 382 (9908): e38–e39.

Dyson, Freeman. 2011. "How We Know." *New York Review of Books* 58 (4): 8–12.

Easterbrook, Gregg. 2002. "The New Convergence." *Wired,* December 1, 2002. https://www.wired.com/2002/12/convergence-3/.

Easterlin, Richard A. 1974. "Does Economic Growth Improve the Human Lot? Some Empirical Evidence." In *Nations and Households in Economic Growth: Essays in Honor of Moses Abramovitz,* ed. Paul A. David and Melvin W. Reder, 89–126. New York: Academic Press.

Easterlin, Richard A. 2005. "Feeding the Illusion of Growth and Happiness: A Reply to Hagerty and Veenhoven." *Social Indicators Research* 74:429–43.

Elias, Norbert. (1985) 2001. *The Loneliness of the Dying*. Trans. Edmund Jephcott. New York: Continuum.

Elliot, Gil. 1972. *The Twentieth Century Book of the Dead*. New York: Ballantine Books.

Emanuel, Ezekiel, and Linda Emanuel. 1998. "The Promise of a Good Death." *Lancet* 351 (Suppl. II): 21–29.

Ettinger, Robert. 1965. *The Prospect of Immortality*. https://www.cryonics.org/images/uploads/misc/Prospect_Book.pdf.

Ettinger, Robert. 1989. *Man into Superman*. http://www.cryonics.org/book2.html.

Ettinger, Robert. 1994. "The Past, the Present, the Future, and Everything." *Cryonics* 15 (3): 27–32. https://www.alcor.org/cryonics/cryonics1994-3.pdf.

Etzler, John Adolphus. 1977. *The Collected Works of John Adolphus Etzler.* Delmar, N.Y.: Scholars' Facsimiles and Reprints.

Evans, Woody. 2007. "Singularity Warfare: A Bibliometric Survey of Militarized Transhumanism." *Journal of Evolution and Technology* 16 (1): 161–65.

Ewin, R. E. 2002. *Reasons and the Fear of Death.* Lanham, Md.: Rowman & Littlefield.

Fabi, M. Giulia. 2004. *Passing and the Rise of the African Novel.* Champaign: University of Illinois Press.

Fabian, Johannes. 1973. "How Others Die: Reflections on the Anthropology of Death." In *Death in the American Experience,* ed. Arian Mack, 177–201. New York: Schocken.

Fabian, Johannes. (1983) 2014. *Time and the Other: How Anthropology Makes Its Object.* New York: Columbia University Press.

Fahy, Gregory M., Brian Wowk, Jun Wu, John Phan, Chris Rasch, Alice Chang, and Eric Zendejas. 2004. "Cryopreservation of Organs by Vitrification: Perspectives and Recent Advances." *Cryobiology* 48:157–78.

Falk, Dean, and Charles Hildebolt. 2017. "Annual War Deaths in Small-Scale versus State Societies Scale with Population Size Rather than Violence." *Current Anthropology* 58 (6): 805–13.

Farberow, Norman. 1975. "Cultural History of Suicide." In *Suicide in Different Cultures,* ed. Farberow, 1–16. Baltimore: University Park Press.

Farman, Abou. 2010. "The Mode of Prediction." *Anthropology Now* 2 (3): 89–94.

Farman, Abou. 2012. "Re-enchantment Cosmologies: Mastery and Obsolescence in an Intelligent Universe." *Anthropological Quarterly* 85 (4): 1069–88.

Farman, Abou. 2013. "Speculative Matter: Secular Bodies, Minds and Persons." *Cultural Anthropology* 28 (4): 737–59.

Farman, Abou. 2014a. "The Informatic Self." In *Ecologies of Care: Innovations through Technologies, Collectives and the Senses,* ed. Gergely Mohacsi, 273–82. Osaka: Osaka University Press.

Farman, Abou. 2014b. "Misanthropology." The Committee on the Anthropology of Science, Technology, and Computing, The CASTAC Blog, December 9, 2014. http://blog.castac.org/2014/12/misanthropology/.

Farman, Abou. 2015. "Social Science, Socialist Scientists and the Future of Utopias." The Committee on the Anthropology of Science, Technology, and Computing, The CASTAC Blog, September 29, 2015. http://blog.castac.org/2015/09/socialist-scientists/.

Farman, Abou. 2016a. "Cryonics in the Cradle of Technocivilization." The Committee on the Anthropology of Science, Technology, and Computing, The CASTAC Blog, December 9, 2016. http://blog.castac.org/2016/12/cryonics/.

Farman, Abou. 2016b. "Utopian Niche-ings and the Politics of If and Otherwise." *Public Seminar,* November 8, 2016. http://www.public seminar.org/2016/11/utopian-niche-ings-and-the-politics-of-if-and-otherwise/.

Farman, Abou. 2017a. "Informatic Cosmologies and the Anxieties of Immanence." The Immanent Frame, Social Science Research Council blog, October 25, 2017. https://tif.ssrc.org/2017/10/25/informatic-cosmologies-and-the-anxieties-of-immanence/.

Farman, Abou. 2017b. "Terminality." *Social Text* 35 (2): 93–118.

Farman, Abou. 2019. "Mind Out of Place: Transhuman Spirituality." *Journal of the American Academy of Religion* 87 (1): 57–80.

Farrell, James. 1980. *Inventing the American Way of Death, 1830–1920.* Philadelphia: Temple University Press.

Faubion, James D. 2001. *The Shadows and Lights of Waco: Millennialism Today.* Princeton: Princeton University Press.

Feifel, Herman, ed. 1959. *The Meaning of Death.* New York: McGraw-Hill.

Feldman, Noah. 2002. "The Intellectual Origins of the Establishment Clause." *New York University Law Review* 77:346–428.

Feldman, Noah. 2005. *Divided by God: America's Church-State Problem— and What We Should Do about It.* New York: Farrar, Straus and Giroux.

Ferguson, Brian. 2013. "Pinker's List: Exaggerating Prehistoric War Mortality." In *War, Peace and Human Nature: The Convergence of Evolutionary and Cultural Views,* ed. Douglas Fry, 112–31. New York: Oxford University Press.

Fernando, Mayanthi L. 2014a. *The Republic Unsettled: Muslim French and the Contradictions of Secularism.* Durham: Duke University Press.

Fernando, Mayanthi L. 2014b. "Intimacy Surveilled: Religion, Sex, and Secular Cunning." *Signs* 39, no. 3 (2014): 685–708. doi:10.1086/674207.

Feuerbach, Ludwig. (1930) 1980. *Thoughts on Death and Immortality.* Trans. James Massey. Berkeley: University of California Press.

Finkelman, Paul. 2012. "Slavery in the United States: Persons or Property?" In *The Legal Understanding of Slavery: From the Historical to the Contemporary,* ed. Jean Allain, 105–34. Oxford: Oxford University Press.

Finkler, Kaja. 2000. *Experiencing the New Genetics: Family and Kinship on the Medical Frontier.* Philadelphia: University of Pennsylvania Press.

Firth, Raymond. 1961. "The Fate of the Soul." In *Anthropology of Folk Religion,* ed. Charles Leslie, 301-32. New York: Vintage Books.

Fischer, John Martin, ed. 1993. *The Metaphysics of Death.* Stanford: Stanford University Press.

Flanagan, Owen. 1991. *The Science of the Mind.* Cambridge: MIT Press.

Flanagan, Owen. 2002. *The Problem of the Soul: Two Visions of Mind and How to Reconcile Them.* New York: Basic Books.

FM 2030. 1989. *Are You a Transhuman? Monitoring and Stimulating Your Personal Rate of Growth in a Rapidly Changing World.* New York: Warner Books.

Foucault, Michel. (1970) 1994. *The Order of Things: An Archaeology of the Human Sciences.* New York: Vintage Books.

Foucault, Michel. (1975) 1999. *Discipline and Punish: The Birth of the Prison.* New York: Vintage.

Foucault, Michel. 1978. *The History of Sexuality, Vol. 1.* New York: Pantheon Books.

Foucault, Michel. 1988. "Technologies of the Self." In *Technologies of the Self,* ed. Luther H. Martin, Huck Gutman, and Patrick H. Hutton, 16-50. Amherst: University of Massachusetts Press.

Foucault, Michel. 2003. *"Society Must Be Defended": Lectures at the Collège de France, 1975-1976.* Trans. David Macey. London: Penguin.

Fox, Rene. 1977. "The Medicalization and Demedicalization of American Society." *Daedalus* 106:9-22.

Franklin, Sarah. 2003. "Ethical Bio Capital: New Strategies of Cell Culture." In *Remaking Life and Death: Toward an Anthropology of the Biosciences,* ed. Sarah Franklin and Margaret Lock, 97-128. Santa Fe: School of American Research Press.

Franklin, Sarah. 2006a. "Better by Design?" In *Better Humans? The Politics of Human Enhancement and Life Extension,* ed. Paul Miller and James Wilson, 86-94. London: Demos.

Franklin, Sarah. 2006b. "The Cyborg Embryo: Our Path to Transbiology." *Theory Culture & Society* 23:167-87.

Franklin, Sarah, and Margaret Lock. 2003. "Animation and Cessation: The Remaking of Life and Death." In *Remaking Life and Death: Toward an Anthropology of the Biosciences,* ed. Sarah Franklin and Margaret Lock, 3-22. Santa Fe: School of American Research Press.

Franklin, Sarah, and Celina Roberts. 2006. *Born and Made: An Ethnography of Preimplantation Genetic Diagnosis.* Princeton: Princeton University Press.

Frazer, James George. 1913. *The Belief in Immortality and the Worship of the Dead.* Vol. 1. London: Macmillan.

Frazer, James George. 1994. *The Golden Bough: A Study in Magic and Religion.* London: Oxford University Press.

Fredkin, Edward. 2000. *On the Soul.* http://www.digitalphilosophy.org/Home/Papers/OntheSoul/Chapter1/tabid/157/Default.aspx.

Freeman, Elizabeth. 2007. *Time Binds: Queer Temporalities, Queer Histories.* Durham: Duke University Press.

Freud, Sigmund. (1917) 1989. "Mourning and Melancholia." In *A General Selection from the Works of Sigmund Freud,* ed. John Rickman, 124–40. New York: Doubleday.

Freud, Sigmund. (1919) 2003. *The Uncanny.* Trans. David McLintock. New York: Penguin.

Freud, Sigmund. (1920) 1989. *Beyond the Pleasure Principle.* Trans. James Strachey. New York: Norton.

Freud, Sigmund. (1930) 1962. *Civilization and Its Discontents.* Trans. James Strachey. New York: Norton.

Frickel, Scott, Sahra Gibbon, Jeff Howard, Joanna Kempner, Gwen Ottinger, and David Hess. 2010. "Undone Science: Charting Social Movement and Civil Society Challenges to Research Agenda Setting." *Science, Technology and Human Values* 35 (4): 444–73.

Friedman, David. 2007. *The Immortalists: Charles Lindbergh, Dr. Alexis Carrel, and Their Daring Quest to Live Forever.* New York: Harper Perennial.

Fuller, Robert. 1986. *Americans and the Unconscious.* Oxford: Oxford University Press.

Fuller, Robert C. 2001. *Spiritual, but Not Religious: Understanding Unchurched America.* Oxford: Oxford University Press.

Fuller, Steve, and Veronika Lipinska. 2015. "Transhumanism." In *The Encyclopedia of Ethics, Science, Technology, and Engineering,* ed. J. B. Holbrook et al., 3:410–13. 2nd ed. Boston: Cengage Learning.

Fukuyama, Francis. 1992. *The End of History and the Last Man.* New York: Free Press.

Fukuyama, Francis. 2002. *Our Posthuman Future: Consequences of the Biotechnology Revolution.* New York: Farrar, Straus and Giroux.

Fussel, Paul. 1977. *The Great War and Modern Memory.* Oxford: Oxford University Press.

Gallup, George, and William Proctor. 1983. *Adventures in Immortality.* London: Souvenir Press.

Gardner, James. 2007. *The Intelligent Universe: AI, ET, and the Emerging Mind of the Cosmos.* Franklin Lakes, N.J.: New Page Books.

Gardner, James. 2010. "The Intelligent Universe." In *Cosmos and Culture: Cultural Evolution in a Cosmic Context,* ed. Steven J. Dick and Mark Lupisella, 361–80. Washington, D.C.: U.S. Government Printing Office.

Geertz, Clifford. 1973. "Chapter 4: Religion as a Cultural System." *The Interpretation of Cultures,* 87–125. New York: Basic Books.

Gell, Alfred. 1995. "Closure and Multiplication: An Essay on Polynesian Cosmology and Ritual." In *Cosmos and Society in Oceania,* ed. Daniel de Coppet and Andre Iteanu, 21–56. Oxford: Berg.

Gellner, Ernest. 1993. *Postmodernism, Reason, and Religion.* London: Routledge.

Gelya, Frank, Leslie Blackhall, Vicki Michel, Sheila T. Murphy, Stanley P. Azen, and Kyeyoung Park. 1998. "A Discourse of Relationships in Bioethics: Patient Autonomy and End-of-Life Decision Making among the Elderly Korean Americans." *Medical Anthropology Quarterly* 12 (4): 403–23.

Gems, David. 2003. "Is More Life Always Better? The New Biology of Aging and the Meaning of Life." *Hastings Center Report* 33 (4): 31–39.

Geraci, Robert. 2010. *Apocalyptic AI: Visions of Heaven in Robotics, Artificial Intelligence and Virtual Reality.* New York: Oxford University Press.

Germain, C. 1980. "Nursing the Dying: Implications of Kübler-Ross' Staging Theory." *Annals of the American Academy of Political and Social Science* 447:46–58.

Germain, Gilbert. 1993. *A Discourse on Disenchantment: Reflections on Politics and Technology.* Albany: SUNY Press.

Gervais, Karen. 1986. *Redefining Death.* New Haven: Yale University Press.

Gibbon, Sahra. 2007. *Breast Cancer Genes and the Gendering of Knowledge: Science and Citizenship in the Cultural Context of the "New" Genetics.* Basingstoke, UK: Palgrave Macmillan.

Giddens, Anthony. 1991. *Modernity and Self-Identity.* Stanford: Stanford University Press.

Gieryn, Thomas. 1983. "Boundary-Work and the Demarcation of Science from Non-science: Strains and Interests in Professional Interests of Scientists." *American Sociological Review* 48 (6): 781–95.

Gilder, George, and Jay Richards. 2002. "Are We Spiritual Machines? The Beginning of a Debate." In *Are We Spiritual Machines? Ray Kurzweil vs. the Critics of Strong AI,* ed. Jay Richards, 1–14. Seattle: Discovery Institute.

Gilligan, T. Scott, and Thomas Stueve. 1995. *Mortuary Law.* Cincinnati: Cincinnati Foundation for Mortuary Education.

Gilmore, Ruth Wilson. 2006. *Golden Gulag: Prisons, Surplus, Crisis, and*

Opposition in Globalizing California. Berkeley: University of California Press.

Girard, René. 1979. *Violence and the Sacred*. Trans. Patrick Gregory. Baltimore: Johns Hopkins University Press.

Glaser, Barney, and Anselm Strauss. 1965. *Awareness of Dying*. Chicago: Aldine.

Gleick, James. 2011. *The Information: A Theory, a History, a Flood*. New York: Pantheon Books.

Glenn, Linda MacDonald. 2003. "Biotechnology at the Margins of Personhood: An Evolving Legal Paradigm." *Journal of Evolution and Technology* 13 (1): 1–58. https://www.jetpress.org/volume13/glenn.html.

Godfrey-Smith, Peter. 2007. "Information in Biology." In *The Cambridge Companion to the Philosophy of Biology*, ed. David L. Hull and Michael Ruse, 103–19. Cambridge: Cambridge University Press.

Goertzel, Ben. 2009. "AI and What to Do about It." *Forbes.com*, June 22, 2009. http://www.forbes.com/2009/06/22/ai-financial-advice-opinions -contributors-artificial-intelligence-09-goertzel.html.

Goertzel, Ben. 2010. *A Cosmist Manifesto: Practical Philosophy for the Posthuman Age*. Humanity Plus Press under a Creative Commons License. https://goertzel.org/CosmistManifesto_July2010.pdf.

Goertzel, Ben, and Ted Goertzel, eds. 2015. *The End of the Beginning: Life, Society and Economy on the Brink of the Singularity*. Los Angeles: Humanity+ Press.

Gollner, Adam Leith. 2013. *The Book of Immortality: The Science, Belief, and Magic behind Living Forever*. New York: Scribner.

Golumbia, David. 2009. *The Cultural Logic of Computation*. Cambridge, Mass.: Harvard University Press.

Goodwin, Barbara. 1978. *Social Science and Utopia: Nineteenth-Century Models of Social Harmony*. Atlantic Highlands, N.J.: Humanities Press.

Goody, Jack. 1962. *Death, Property and the Ancestors*. Stanford: Stanford University Press.

Gorer, Geoffrey. 1965. *Death, Grief and Mourning*. Garden City, N.Y.: Doubleday.

Graham, Connor, Martin Gibbs, and Lanfranco Aceti. 2013. "Introduction to the Special Issue on the Death, Afterlife, and Immortality of Bodies and Data." *Information Society* 29:133–41.

Gray, John. 2012. *The Immortalization Commission: Science and the Strange Quest to Cheat Death*. New York: Farrar, Straus and Giroux.

Greeley, A. M., and M. Hout. 1999. "Americans' Increasing Belief in Life after Death: Religious Competition and Acculturation." *American Sociological Review* 64:813–35.

Greenfield, Susan. 1999. "Soul, Brain, and Mind." In *From Soul to Self*, ed. James Crabbe, 108–25. New York: Routledge.

Griaule, Marcel. 1970. *Conversations with Ogotemmeli*. New York: Oxford University Press.

Griffin, David Ray, ed. 1988. *Spirituality and Society: Postmodern Visions*. Albany: SUNY Press.

Griffin, David R. 2001. *Reenchantment without Supernaturalism: A Process Philosophy of Religion*. Ithaca: Cornell University Press.

Gortmaker, Steven, and Paul Wise. 1997. "The First Injustice: Socioeconomic Disparities, Health Services Technology, and Infant Mortality." *Annual Review of Sociology* 23:147–70.

Guyer, Jane. 2007. "Prophecy and the Near Future." *American Ethnologist* 34 (3): 409–21.

Habermas, Jürgen. 2002. *Rationality and Religion: Essays on Reason, God, and Modernity*. Cambridge: MIT Press.

Habermas, Jürgen. 2003. *The Future of Human Nature*. Cambridge: Polity Press.

Habermas, Jürgen. 2010. "An Awareness of What Is Missing." In *An Awareness of What Is Missing: Faith and Reason in a Post-Secular Age*, ed. Habermas et al., 15–23. Malden, Mass.: Polity Press.

Halberstam, Judith. 2005. *In a Queer Time and Place: Transgender Bodies, Subcultural Lives*. New York: NYU Press.

Hall, Stephen. 2003. *Merchants of Immortality*. New York: Houghton Mifflin.

Hallam, Elizabeth, and Jenny Hockey. 2001. *Death, Memory and Material Culture*. New York: Berg.

Hamdy, Sherine F. 2008. "When the State and Your Kidneys Fail: Political Etiologies in an Egyptian Dialysis Ward." *American Ethnologist* 35 (4): 553–69.

Hansot, Elizabeth. 1974. *Perfection and Progress: Two Modes of Utopian Thought*. Cambridge: MIT Press.

Harari, Yuval Noah. 2015. *Homo Deus: A Brief History of Tomorrow*. London: Harvill Secker.

Haraway, Donna J. 1991. "The Biopolitics of Postmodern Bodies: Constitutions of Self in Immune System Discourse." In *Simians, Cyborgs, and Women: The Reinvention of Nature*, 203–30. New York: Routledge.

Haraway, Donna J. 1992. "The Promises of Monsters: A Regenerative

Politics for Inappropriate/d Others." In *Cultural Studies,* ed. Lawrence Grossberg, Cary Nelson, and Paula A. Treichler, 295–337. New York: Routledge.

Haraway, Donna J. 1998. "Mice into Wormholes: A Comment on the Nature of No Nature." In *Cyborgs and Citadels: Anthropological Interventions in Emerging Sciences and Technologies,* ed. Gary L. Downey and Joseph Dumit, 209–44. Santa Fe: School of American Research Press.

Harding, Susan, and Kathleen Stewart. 1999. "Bad Endings: American Apocalypsis." *Annual Review of Anthropology* 28:285–310.

Harding, Susan F., and Kathleen Stewart. 2003. "Anxieties of Influence: Conspiracy Theory and Therapeutic Culture in Millennial America." In *Transparency and Conspiracy: Ethnographies of Suspicion in the New World Order,* ed. Harry West and Todd Sanders, 258–86. Durham: Duke University Press.

Harley, Brian, and Glenn Firebaugh. 1993. "Americans' Belief in an Afterlife: Trends over the Past Two Decades." *Journal for the Scientific Study of Religion* 32:269–78.

Harre, Rom. 1981. "Psychological Variety." In *Indigenous Psychologies: The Anthropology of the Self,* ed. Paul Heelas and Andrew Lock, 79–104. New York: Academic Press.

Harris, Grace G. 1989. "Concepts of Individual, Self, and Person in Description and Analysis." *American Anthropologist* 91:599–612.

Harris, Marvin. 1981. "Why the Cults Are Coming." In *America Now: The Anthropology of a Changing Culture,* 141–66. New York: Touchstone Books.

Harris, Sam. 2010. *The Moral Landscape: How Science Can Determine Human Values.* New York: Free Press.

Hart, Bethne, Peter Sainsbury, and Stephanie Short. 1998. "Whose Dying? A Sociological Critique of the 'Good Death.'" *Mortality* 3 (1): 65–77.

Harvey, David. 1989. *The Condition of Postmodernity: An Inquiry into the Origins of Cultural Change.* Oxford: Blackwell.

Harvey, Graham. 2005. *Animism: Respecting the Living World.* New York: Columbia University Press.

Haub, Carl. 2002. "How Many People Have Ever Lived on Earth?" *Population Today.* Population Reference Bureau. http://www.prb.org/articles/2002/howmanypeoplehaveeverlivedonearth.aspx.

Hayflick, Leonard. 1998. "How and Why We Age." *Experimental Gerontology* 33 (7/8): 639–53.

Hayflick, Leonard. 2000. "The Illusion of Cell Immortality." *British Journal of Cancer* 83 (7): 841–46.

Hayles, N. Katherine. 1999. *How We Became Posthuman: Virtual Bodies in Cybernetics, Literature, and Informatics*. Chicago: University of Chicago Press.

Heatherington, Tracy. 2012. "From Ecocide to Genetic Rescue." In *The Anthropology of Extinction: Essays on Culture and Species Death,* ed. Genese Marie Sodikoff, 39–66. Bloomington: Indiana University Press.

Hecht, Jennifer Michael. 2003. *The End of the Soul: Scientific Modernity, Atheism, and Anthropology in France*. New York: Columbia University Press.

Heelas, Paul. 1996. *The New Age Movement: The Celebration of the Self and the Sacralization of Modernity*. Cambridge, Mass.: Blackwell.

Helmreich, Stefan. 1998. *Silicon Second Nature: Culturing Artificial Life in a Digital World*. Berkeley: University of California Press.

Helmreich, Stefan. 2001. "Transubstantiating Fatherhood and Information Flow in Artificial Life." In *Relative Values: Reconfiguring Kinship Studies,* ed. Sarah Franklin and Susan McKinnon, 116–43. Durham: Duke University Press.

Helmreich, Stefan. 2009. *Alien Ocean: Anthropological Voyages in Microbial Seas*. Berkeley: University of California Press.

Helmreich, Stefan, and Eben Kirksey. 2010. "The Emergence of Multispecies Ethnography." *Cultural Anthropology* 25 (4): 545–76.

Henrich, J., S. Heine, and A. Norenzayan. 2010. "The Weirdest People in the World?" *Behavioral and Brain Sciences* 33 (2–3): 61–83.

Hertz, Robert. 1960. *Death and the Right Hand*. Trans. Rodney and Claudia Needham. Glencoe, Ill.: The Free Press.

Hess, David. 2007. *Alternative Pathways in Science and Industry: Activism, Innovation and the Environment in an Era of Globalization*. Cambridge: MIT Press.

Hick, John. (1976) 1994. *Death and Eternal Life*. San Francisco: Harper and Row.

Hoeppe, Götz. 2012. "Astronomers at the Observatory: Place, Visual Practice, Traces." *Anthropological Quarterly* 85 (4): 1141–60.

Hogle, Linda. 2003. "Life/Time Warranty: Rechargeable Cells and Extendable Lives." In *Remaking Life and Death: Toward an Anthropology of the Biosciences,* ed. Sarah Franklin and Margaret Lock, 61–97. Santa Fe: School of American Research Press.

Hogle, Linda. 2005. "Enhancement Technologies and the Body." *Annual Review of Anthropology* 34:695–716.

Horrobin, Steven. 2006. "Immortality, Human Nature, The Value of Life and the Value of Life Extension." *Bioethics* 20 (6): 279–92.

Horvath, Steve, Yafeng Zhang, Peter Langfelder, René S. Kahn, Marco

P. M. Boks, Kristal van Eijk, Leonard H. van den Berg, and Roel A. Ophoff. 2012. "Aging Effects on DNA Methylation Modules in Human Brain and Blood Tissue." *Genome Biology* 13:R97.

Hubert, Henri, and Marcel Mauss. 1964. *Sacrifice: Its Nature and Function.* Trans. W. D. Hall. Chicago: University of Chicago Press.

Hughes, James J. 2001. "The Future of Death: Cryonics and the Telos of Liberal Individualism." *Journal of Technology and Evolution* 6 (July). http://www.jetpress.org/volume6/death.htm.

Hughes, James. 2004. *Citizen Cyborg.* Cambridge: Westview Press.

Hughes, James. 2012. "The Politics of Transhumanism and the Techno-Millennial Imagination, 1626–2030." *Zygon* 47 (4): 757–76.

Hume, David. (1779) 1970. "Of the Immortality of the Soul." In *Dialogues concerning Natural Religion,* 26–30. Indianapolis: Bobbs-Merrill.

Immortality Institute. 2004. *The Scientific Conquest of Death: Essays on Infinite Lifespans.* Buenos Aires: Libros en Red.

Ingold, Tim. 1993. "Technology, Language, Intelligence: A Reconsideration of Basic Concepts." In *Tools, Language and Cognition,* ed. Kathleen R. Gibson and Tim Ingold, 449–72. New York: Cambridge University Press.

Jacoby, Susan. 2004. *Freethinkers: A History of American Secularism.* New York: Holt.

Jain, S. Lochlann. 2010. "The Mortality Effect." *Public Culture* 22 (1): 89–117.

Jain, S. Lochlann. 2013. *Malignant: How Cancer Becomes Us.* Berkeley: University of California Press.

Jasanoff, Sheila. 2018. *Can Science Make Sense of Life?* Medford, Mass.: Polity Press.

Jonas, Hans. 1992. "The Burden and Blessing of Mortality." *Hastings Center Report* 22 (1): 34–40.

Jonas, Hans. (1966) 2001. *The Phenomenon of Life: Toward a Philosophical Biology.* Evanston, Ill.: Northwestern University Press.

Jones, Robert P. 2007. *Liberalism's Troubled Search for Equality: Religion and Cultural Bias in the Oregon Physician-Assisted Suicide Debates.* Notre Dame: University of Notre Dame Press.

Joyce, Kelly, and Meika Loe. 2010. "A Sociological Approach to Ageing, Technology and Health." *Sociology of Health and Illness* 32 (2): 171–80.

Juengst, Eric. 1998. "What Does Enhancement Mean?" In *Enhancing Human Traits: Ethical and Social Implications,* ed. Erik Parens, 29–47. Washington, D.C.: Georgetown University Press.

Kafer, Alison. 2013. *Feminist Queer Crip.* Bloomington: Indiana University Press.

Kane, B. 1979. "Children's Concepts of Death." *Journal of Genetic Psychology* 134:141–53.

Kant, Immanuel. (1755) 2008. *Universal Natural History and Theory of Heavens*. E-text. http://records.viu.ca/~johnstoi/kant/kant2e.htm.

Kant. Immanuel. (1781) 1966. *Critique of Pure Reason*. Trans. Max Muller. New York: Anchor Books.

Kaplan, Esther. 2005. *With God on Their Side*. New York: New Press.

Karppi, Tero. 2013. "Death Proof: On the Biopolitics and Noopolitics of Memorializing Dead Facebook Profiles." *Culture Machine* 14. https://culturemachine.net/wp-content/uploads/2019/05/513-1161-1-PB.pdf.

Kass, Leon R. 1974. "Averting One's Eyes, or Facing the Music? On Dignity in Death." *Facing Death: The Hastings Center Studies* 2 (20): 67–80.

Kass, Leon R. 1993. "Is There a Right to Die?" *Hastings Center Report* 23 (1): 34–43.

Kass, Leon. 2003. "Ageless Bodies, Happy Souls: Biotechnology and the Pursuit of Perfection." *New Atlantis: A Journal of Technology and Society* 1:9–28.

Kass, Leon. 2005. *Life, Liberty and the Defense of Dignity: The Challenge of Bioethics*. San Francisco: Encounter Books.

Kastenbaum, Robert. 1997. "What Is the Future of Death?" In *Death and the Quest for Meaning: Essays in Honor of Herman Feifel,* ed. Stephen Strack, 361–82. Northvale, N.J.: Jason Aronson.

Kastenbaum, Robert, and Paul T. Costa Jr. 1977. "Psychological Perspectives on Death." *Annual Review of Psychology* 28:225–49.

Kateb, George. 2009. "Locke and the Political Origins of Secularism." *Social Research* 76 (4): 1001–34.

Kauffman, Stuart A. 2008. *Reinventing the Sacred: A New View of Science, Reason, and Religion*. Perseus Books Group. Kindle edition.

Kaufman, Sharon. 2000. "In the Shadow of 'Death with Dignity': Medicine and the Cultural Quandaries of the Vegetative State." *American Anthropologist* 102 (1): 69–83.

Kaufman, Sharon. 2005. *. . . and a Time to Die: How American Hospitals Shape the End of Life*. Chicago: University of Chicago Press.

Kaufman, Sharon, and Lynn Morgan. 2005. "The Anthropology of the Beginnings and Ends of Life." *Annual Review of Anthropology* 34:317–41.

Kaufman, Sharon, Ann J. Russ, and Janet K. Shim. 2006. "Aged Bodies and Kinship Matters: The Ethical Field of Kidney Transplant." *American Ethnologist* 33 (1): 81–99.

Keane, Webb. 2013. "Secularism as a Moral Narrative of Modernity." *Transit* 43:159–70.

Kearl, Michael. 1997. "You Never Have to Die: On Mormons, NDEs,

Cryonics, and the American Immortalist Ethos." In *The Unknown Country: Death in Australia, Britain and the USA,* ed. Kathy Charmaz, Glennys Howarth, and Allan Kellehear, 184–97. London: Macmillan.

Kellehear, Allan. 1984. "Are We a 'Death-Denying' Society? A Sociological Review." *Social Science and Medicine* 18 (9): 713–23.

Kellehear, Allan. 1990. *Dying of Cancer: The Final Year of Life.* Melbourne: Harwood Academic Publishers.

Keller, Evelyn Fox. 1995. *Refiguring Life.* New York: Columbia University Press.

Keller, Evelyn Fox. 2007. "Booting Up Baby." In *Genesis Redux: Essays in the History and Philosophy of Artificial Life,* ed. Jessica Riskin, 334–45. Chicago: University of Chicago Press. https://doi.org/10.7208/chicago/9780226720838.003.0016.

Kelly, Kevin. 2006. "Will Spiritual Robots Replace Humanity by 2100?" *The Technium,* March 15, 2006. http://www.kk.org/thetechnium/archives/2006/03/will_spiritual.php.

Kent, Saul. 1983. "The First Cryonicist." *Cryonics,* no. 32:9–12. https://alcor.org/cryonics/cryonics8303.txt.

Kester, Tracie. 2007. "Can the Dead Hand Control the Dead Body? The Case for a Uniform Bodily Remains Law." *Western New England Law Review* 9:571–618.

Keyes, Charles. 2002. "Weber and Anthropology." *Annual Review of Anthropology* 31:233–55.

Khushf, George. 2007. "The Ethics of NBIC Convergence." *Journal of Medicine and Philosophy* 32 (3): 185–96.

Kierkegaard, Søren. (1844) 1957. *The Concept of Dread.* Trans. Walter Lowrie. Princeton: Princeton University Press.

Kirksey, Eben. 2017. "The Utopia for the Golden Frog of Panama." In *Cryopolitics: Frozen Life in a Melting World,* ed. Joanna Radin and Emma Kowal, 304–34. Cambridge: MIT Press.

Klawiter, Maren. 2008. *The Biopolitics of Breast Cancer: Changing Cultures of Disease and Activism.* Minneapolis: University of Minnesota Press.

Kneese, Tamara. 2018. "Networked Heirlooms: The Affective and Financial Logics of Digital Estate Planning." *Cultural Studies* 33 (2): 297–324.

Knobe, Joshua, and Shaun Nichols. 2008. *Experimental Philosophy.* New York: Oxford University Press.

Knorr Cetina, Karin. 2005. "The Rise of a Culture of Life." *European Molecular Biology Organization Reports* 6:76–80.

Koenig, Barbara A. 1988. "The Technological Imperative in Medical

Practice: The Social Creation of a 'Routine' Practice." In *Biomedicine Examined,* ed. Margaret Lock and Deborah Gordon, 465–96. Boston: Kluwer.

Kohn, Eduardo. 2013. *How Forests Think: Toward an Anthropology beyond the Human.* Berkeley: University of California Press.

Koselleck, Reinhart. 2002. "Time and History." In *The Practice of Conceptual History: Timing History, Spacing Concepts,* trans. T. S. Presner et al., 100–114. Stanford: Stanford University Press.

Kowal, Emma, Joanna Radin, and Jenny Reardon. 2013. "Indigenous Body Parts, Mutating Temporalities, and the Half-Lives of Postcolonial Technoscience." *Social Studies of Science* 43:465–83.

Koyré, Alexandre. 1957. *From the Closed World to the Infinite Universe.* Baltimore: Johns Hopkins University Press.

Kravel-Tovi, Michal, and Yoram Bilu. 2008. "The Work of the Present: Constructing Messianic Temporality in the Wake of Failed Prophecy among Chabad Hasidim." *American Ethnologist* 35 (1): 64–80.

Kruger, Oliver. 2010. "The Suspension of Death: The Cryonic Utopia in the Context of the U.S. Funeral Culture." *Marburg Journal of Religion* 15 (1): 1–19.

Kübler-Ross, Elisabeth. 1969. *On Death and Dying.* New York: Macmillan.

Kubota, Shin. 2011. "Repeating Rejuvenation in *Turritopsis,* an Immortal Hydrozoan (Cnidaria, Hydrozoa)." *Biogeography* 13:101–3.

Kuhn, Robert Lawrence. 2007. "Why This Universe? Toward a Taxonomy of Possible Explanations." *Skeptic* 13 (2): 28–40.

Kumral, E., M. Yüksel, S. Büket, T. Yagdi, Y. Atay, and A. Güzelant. 2001. "Neurologic Complications after Deep Hypothermic Circulatory Arrest: Types, Predictors, and Timing." *Texas Heart Institute Journal* 28 (2): 83–88.

Kurzweil, Ray. 1990. *The Age of Intelligent Machines.* Cambridge, Mass.: MIT Press.

Kurzweil, Ray. 1999. *The Age of Spiritual Machines.* New York: Penguin Books.

Kurzweil, Ray. 2005. *The Singularity Is Near: When Humans Transcend Biology.* New York: Viking.

Kurzweil, Ray. 2007. Foreword. In James Gardner, *The Intelligent Universe: AI, ET, and the Emerging Mind of the Cosmos,* 11–16. Franklin Lakes, N.J.: New Page Books.

Lacquer, Thomas. 2015. *The Work of the Dead: A Cultural History of Mortal Remains.* Princeton: Princeton University Press.

Laderman, Gary. 2003. *Rest in Peace: A Cultural History of Death and the*

Funeral Home in Twentieth-Century America. New York: Oxford University Press.

Lakoff, Andrew. 2010. "Two Regimes of Global Health." *Humanity: An International Journal of Human Rights, Humanitarianism, and Development* 1 (1): 59–79.

Lambek, Michael. 2016. "After Life." In *Living and Dying in the Contemporary World*, ed. Veena Das and Clara Han, 629–47. Berkeley: University of California Press.

Lamont, Corliss. (1935) 1990. *The Illusion of Immortality*. 5th ed. New York: Continuum.

Landecker, Hannah. 2003. "On Beginning and Ending with Apoptosis." In *Remaking Life and Death: Toward an Anthropology of the Biosciences,* ed. Sarah Franklin and Margaret Lock, 23–60. Santa Fe: School of American Research Press.

Landecker, Hannah. 2005. "Living Differently in Time: Plasticity, Temporality and Cellular Biotechnologies." *Culture Machine* 7. https://culture machine.net/biopolitics/living-differently-in-time/.

Landecker, Hannah. 2007. *Culturing Life: How Cells Became Technologies*. Cambridge: Harvard University Press.

Lanier, Jaron. 2010. *You Are Not a Gadget: A Manifesto*. New York: Vintage Books.

Lanier, William L. 2005. "Medical Interventions at the End of Life: What Is Appropriate and Who Is Responsible?" *Mayo Clinic Proceedings* 80 (11): 1411–13.

Larson, Edward J., and Larry Witham. 1997. "Scientists Are Still Keeping the Faith." *Nature* 386:435–36.

Lasch, Christopher. 1991. *The True and Only Heaven: Progress and Its Critics*. Norton. Kindle edition.

Lash, Scott, and John Urry. 1994. *Economies of Signs and Space*. London: Sage.

Latour, Bruno. 1993. *We Have Never Been Modern*. Trans. C. Porter. Cambridge: Harvard University Press.

Lavi, Shai. 2009. *The Modern Art of Dying: A History of Euthanasia in the United States*. Princeton: Princeton University Press.

Law, Robin. 1985. "Human Sacrifice in Pre-colonial West Africa." *African Affairs* 84 (334): 53–87.

Leach, Edmund. 1961. "Cronus and Chronos." In *Rethinking Anthropology*, 124–32. London: Athlone Press.

Lee, Lois. 2015. *Recognizing the Non-Religious: Reimagining the Secular*. Oxford: Oxford University Press.

Le Goff, Jacques. 1984. *The Birth of Purgatory.* Trans. Arthur Goldhammer. Chicago: University of Chicago Press.

Leibo, S. P. 2005. "The Early History of Gamete Cryobiology." In *Life in the Frozen State,* ed. Barry J. Fuller, Nick Lane, and Erica E. Benson, 347–70. Taylor and Francis e-Library.

Lerner, Monroe. 1976. "When, Why and Where People Die." In *Death: Current Perspectives,* ed. Edwin Schneidman, 138–62. Palo Alto, Calif.: Mayfield.

Leuba, James. (1916) 1921. *The Belief in God and Immortality: A Psychological, Anthropological and Statistical Study.* Chicago: Open Court Publishing.

Lienhardt, Godfrey. 1961. "Burial Alive." In *Divinity and Experience: The Religion of the Dinka,* 298–320. Oxford: Clarendon Press.

Lifton, Robert Jay. 1975. "On Death and the Continuity of Life: A Psychohistorical Perspective." *Omega* 6:143–60.

Lifton, Robert Jay. 1987. *The Future of Immortality and Other Essays for a Nuclear Age.* New York: Basic Books.

Lifton, Robert Jay, and Greg Mitchell. 2000. *Who Owns Death?* New York: William Morrow.

Lizza, John P. 2006. *Persons, Humanity and the Definition of Death.* Baltimore: Johns Hopkins University Press.

Lloyd, Seth. 2006. *Programming the Universe.* New York: Knopf.

Lock, Margaret. 1996. "Death in Technological Time: Locating the End of Meaningful Life." *Medical Anthropology Quarterly* 10 (4): 575–600.

Lock, Margaret. 2002. *Twice Dead: Organ Transplants and the Reinvention of Death.* Berkeley: University of California Press.

Long, Susan O. 2004. "Cultural Scripts for a Good Death in Japan and the United States: Similarities and Differences." *Social Science and Medicine* 58:913–28.

Lopez, Donald S. 2008. *Buddhism and Science: A Guide for the Perplexed.* Chicago: University of Chicago Press.

Luciano, Dana. 2007. *Arranging Grief: Sacred Time and the Body in Nineteenth-Century America.* New York: NYU Press.

Luckmann, Thomas. 1991. "The Constitution of Human Life in Time." In *Chronotypes: The Construction of Time,* ed. John Bender and David Wellbery, 151–66. Stanford: Stanford University Press.

Luker, Kristen. 1984. *Abortion and the Politics of Motherhood.* Berkeley: University of California Press.

Lupisella, Mark L. 2010. "Cosmocultural Evolution." In *Cosmos and Culture: Cultural Evolution in a Cosmic Context,* ed. Steven J. Dick and

Mark L. Lupisella, 321–60. Washington, D.C.: National Aeronautics and Space Administration.

Madoff, Ray. 2010. *Immortality and the Law: The Rising Power of the American Dead*. New Haven: Yale University Press.

Mahler, Sarah J. 1995. *American Dreaming: Immigrant Life on the Margins*. Princeton: Princeton University Press.

Mahmood, Saba. 2016. *Religious Difference in a Secular Age: A Minority Report*. Princeton: Princeton University Press.

Maine, Henry. 1963. *Ancient Law: Its Connection with the Early History of Society and its Relation to Modern Ideas*. Boston: Beacon Press.

Makari, George. 2015. *Soul Machine: The Invention of the Modern Mind*. New York: Norton.

Malabou, Catherine. 2008. *What Should We Do with Our Brain?* Trans. Sebastian Rand. New York: Fordham University Press.

Malinowski, Bronisław. 1954. *Magic, Science and Religion, and Other Essays*. New York: Doubleday.

Manderson, Desmond, ed. 1999a. *Courting Death: The Law of Mortality*. London: Pluto Press.

Manderson, Desmond. 1999b. "Et Lex Perpetua: Dying Declarations and the Terror of Sussmayr." In *Courting Death: The Law of Mortality*, ed. Desmond Manderson, 34–52. London: Pluto Press.

Mannheim, Karl. 2002. *Ideology and Utopia*. Routledge: London.

Manuel, Frank, and Fritzie Manuel. 1972. "Sketch for a Natural History of Paradise." *Daedalus* 101:83–128.

Manuel, Frank, and Fritzie Manuel. 1979. *Utopian Thought in the Western World*. Cambridge: Harvard University Press.

Marcus, George, ed. 1995. *Technoscientific Imaginaries: Conversations, Profiles and Memoirs*. Chicago: University of Chicago Press.

Marcuse, Herbert. 1959. "The Ideology of Death." In *The Meaning of Death*, ed. Herman Feifel, 64–76. New York: McGraw-Hill.

Marcuse, Herbert. 1970. *The End of Utopia*. Boston: Beacon.

Marmot, Michael. 2004. *Status Syndrome: How Social Standing Affects Our Health and Longevity*. New York: Henry Holt.

Martin, Emily. 2000. "Mind-Body Problems." *American Ethnologist* 27 (3): 569–90.

Martin, Raymond, and John Barresi. 2004. *Naturalization of the Soul: Self and Identity in the Eighteenth Century*. New York: Routledge.

Martin, Raymond, and John Barresi. 2006. *The Rise and Fall of Soul and Self*. New York: Columbia University Press.

Masco, Joseph. 2008. "'Survival Is Your Business': Engineering Ruins and Affect in Nuclear America." *Cultural Anthropology* 23 (2): 361–98.

Mavelli, Luca. 2017. "Governing the Resilience of Neoliberalism through Biopolitics." *European Journal of International Relations* 23 (3): 489–512.

May, John, and Nigel Thrift. 2001. *Timespace: Geographies of Temporality.* New York: Routledge.

Mbembe, Achille. 2003. "Necropolitics." *Public Culture* 15 (1): 11–40.

McAdam, Doug, and Ronnelle Paulsen. 1993. "Specifying the Relation between Social Ties and Activism." *American Journal of Sociology* 99 (3): 640–67.

McGrath, Alister E. 1999. *Christian Spirituality.* Malden, Mass.: Blackwell.

McKibben, Bill. 2003. *Enough: Staying Human in an Engineered Age.* New York: Henry Holt.

McKibben, Bill. 2004. "Do We Want Science to Redesign Human Aging?" In *The Fight over the Future: A Collection of SAGE Crossroads Debates That Examine the Implications of Aging-Related Research,* ed. Sage Crossroads, 19–43. New York: iUniverse, Inc.

McLean, Ian, and A. B. Urken. 1992. "Did Jefferson or Madison Understand Condorcet's Theory of Social Choice?" *Public Choice* 73:445–57.

McNamara, B., C. Waddell, and M. Colvin. 1994. "The Institutionalization of the Good Death." *Social Science and Medicine* 39 (11): 1501–8.

Meese, James, Bjorn Nansen, Tamara Kohn, Michael Arnold, and Martin Gibbs. 2015. "Posthumous Personhood and the Affordances of Digital Media." *Mortality* 20 (4): 408–20.

Meillasoux, Quentin. 2008. *After Finitude: An Essay on the Necessity of Contingency.* New York: Continuum.

Mellor, Phillip. 1993. "Death in High Modernity: The Contemporary Presence and Absence of Death." In *The Sociology of Death,* ed. David Clark, 11–30. Oxford: Blackwell.

Merkle, Ralph. 1992. "The Technical Feasibility of Cryonics." *Medical Hypotheses* 39:6–16.

Metcalf, Peter. 1981. "Meaning and Materialism: The Ritual Economy of Death." *Man* 16 (4): 563–78.

Metcalf, Peter, and Richard Huntington. 1991. *Celebrations of Death: The Anthropology of Mortuary Ritual.* New York: Cambridge University Press.

Michigan Department of Labor and Economic Growth. 2003. "CIS Orders Cease and Desist for Cryonics Institute of Clinton Township for Operating As an Unregulated Funeral Home and Cemetery." Press release, August 26, 2003. http://www.michigan.gov/dleg/0,1607,7-154 -10573_11472-74066-,00.html.

Michigan Department of Labor and Economic Growth. 2004. "Cryonics Institute Now Licensed as Cemetery to Settle Dispute with State of

Michigan." Press release, January 7, 2004. http://www.michigan.gov/
dleg/0,1607,7-154-10573_11472-84042-,00.html.

Miller, Paul, and James Wilsdon, eds. 2006. *Better Humans? The Politics of Human Enhancement and Life Extension.* London: Demos.

Mitchell, Lisa, Peter Stephenson, Susan Cadell, and Mary Ellen McDonald. 2012. "Death and Grief On-line: Virtual Memorialization and Changing Concepts of Childhood Death and Parental Bereavement on the Internet." *Health Sociology Review* 21 (4): 413–31.

Mitford, Jessica. 1963. *The American Way of Death.* New York: Simon Schuster.

Mol, Annemarie. 2002. *The Body Multiple: Ontology in Medical Practice.* Durham: Duke University Press.

Monod, Jacques. 1972. *Chance and Necessity.* Trans. Austryn Wainhouse. New York: Vintage Books.

Montague, S. 1989. "To Eat for the Dead." In *Death Rituals and Life in the Societies of the Kula Ring,* ed. Frederick Damon and Roy Wagner, 23–45. DeKalb: Northern Illinois University Press.

Moore, Thomas. 1997. *The Re-enchantment of Everyday Life.* New York: HarperPerennial.

More, Max. 2015. "Longer Lives on the Brink of Global Population Contraction: A Report from 2040." In *The End of the Beginning: Life, Society and Economy on the Brink of the Singularity,* ed. Ben Goertzel and Ted Goertzel, 81–94. Los Angeles: Humanity+ Press.

Mortimer, Ernest. 1959. *Blaise Pascal: The Life and Work of a Realist.* London: Methuen.

Moser, Kath, Vladimir Shkolnikov, and David A. Leon. 2005. "World Mortality 1950–2000: Divergence Replaces Convergence from the Late 1980s." *Bulletin of the World Health Organization* 83 (3): 202–9.

Moulier Boutang, Yann. 2012. *Cognitive Capitalism.* Trans. Ed Emery. Cambridge: Polity Press.

Mulvenna, Marie. 1974. "Cryobiology Explained by Noted Priest-Scientist." *Georgia Bulletin,* May 30, 1974. http://www.georgiabulletin.org/local/1974/05/30/a/?s=cryobiology.

Munn, Nancy. 1992. "The Cultural Anthropology of Time: A Critical Essay." *Annual Review of Anthropology* 21: 93–123.

Murphy, Ann M. 2007. "Please Don't Bury Me Down in That Cold Cold Ground: The Need for Uniform Laws on the Disposition of Human Remains." *Elder Law Journal* 15:381–471.

Murphy, Michelle. 2017. *The Economization of Life.* Durham: Duke University Press.

Mykytyn, Courtney. 2008. "Medicalizing the Optimal: Anti-Aging Medicine and the Quandary of Intervention." *Journal of Aging Studies* 22:313–21.

Mykytyn, Courtney. 2010. "A History of the Future: the Emergence of Contemporary Anti-ageing Medicine." *Sociology of Health and Illness* 32 (2): 181–96.

Naffine, Ngaire. 1999. "But a Lump of Earth? The Legal Status of the Corpse." In *Courting Death: The Law of Mortality,* ed. Desmond Manderson, 95–110. London: Pluto Press.

Nagel, Thomas. 1974. "What Is It Like to Be a Bat?" *Philosophical Review* 83 (October): 435–50.

Nagel, Thomas. 1986. *The View from Nowhere.* New York: Oxford University Press.

Nagel, Thomas. 2010. *Secular Philosophy and the Religious Temperament: Essays 2002–2008.* New York: Oxford University Press.

Nagel, Thomas. 2012. *Mind and Cosmos: Why the Materialist Neo-Darwinian Conception of Nature Is Almost Certainly False.* Oxford: Oxford University Press.

Nagy, Maria. 1959. "The Child's View of Death." In *The Meaning of Death,* ed. Herman Feifel, 79–98. New York: McGraw-Hill.

Needham, Rodney. 1972. *Belief, Language, and Experience.* Chicago: University of Chicago Press.

Nelkin, Dorothy. 1998. "Do the Dead Have Interests? Policy Issues for Research after Life." *American Journal of Law and Medicine* 23 (2/3): 261–91.

Nelson, Alondra. 2002. "Introduction: Future Texts." *Social Text* 20 (2): 1–15.

Nelson, Robert. 1968. *We Froze the First Man.* New York: Dell.

Noble, David. 1977. *America by Design: Science, Technology, and the Rise of Corporate Capitalism.* New York: Knopf.

Noble, David. 1999. *The Religion of Technology.* New York: Penguin Books.

Nordmann, Alfred. 2007. "Knots and Strands: An Argument for Productive Disillusionment." *Journal of Medicine and Philosophy* 32 (3): 217–36.

Nuland, Sherwin. 1995. *How We Die.* New York: Vintage Books.

Nuland, Sherwin. 2005. "Do You Want to Live Forever?" *MIT Technology Review,* February 1, 2005.

Nuland, Sherwin. 2007. *The Art of Aging.* New York: Random House.

Nussbaum, Martha. 2006. *Frontiers of Justice: Disability, Nationality, Species Membership.* Cambridge: Belknap Press.

Nwabueze, Remigius. 2002. "Biotechnology and the New Property

Regime in Human Bodies and Body Parts." *Loyola of Los Angeles International & Comparative Law Review* 24:19–64.

Nydahl, Joe. 1977. Introduction. In *The Collected Works of John Adolphus Etzler,* xvii–xviii. Delmar, N.Y.: Scholars' Facsimiles and Reprints.

Nydahl, Joe. 1981. "From Millennium to Utopia Americana." In *America as Utopia,* ed. Kenneth Roemer, 254–91. New York: Burth Franklin.

Nye, David. 1999. *American Technological Sublime.* Cambridge: MIT Press.

Nye, David. 2006. *Technology Matters: Questions to Live With.* Cambridge: MIT Press.

Oka, Rahul C., Marc Kissel, Mark Golitko, Susan Guise Sheridan, Nam C. Kim, and Agustín Fuentes. 2017. "Population Is the Main Driver of War Group Size and Conflict Casualties." *Proceedings of the National Academy of Sciences* 114 (52): E11101–E11110.

Olshansky, S. Jay, Toni Antonucci, Lisa Berkman, Robert H. Binstock, Axel Boersch-Supan, John T. Cacioppo, Bruce A. Carnes, Laura L. Carstensen, Linda P. Fried, Dana P. Goldman, James Jackson, Martin Kohli, John Rother, Yuhui Zheng, and John Rowe. 2012. "Differences In Life Expectancy Due to Race and Educational Differences Are Widening, and Many May Not Catch Up." *Health Affairs* 31 (8): 1803–13.

Oyama, Susan. 2000. *The Ontogeny of Information: Developmental States and Evolution.* Durham: Duke University Press.

Paine, Thomas. [1776]. *Common Sense.* http://www.bartleby.com/133/.

Palgi, Phyllis, and Henry Abramovitch. 1984. "Death: A Cross-Cultural Perspective." *Annual Review of Anthropology* 13:385–417.

Palladino, Paolo. 2016. *Biopolitics and the Philosophy of Death.* New York: Bloomsbury.

Parisi, Luciana, and Tiziana Terranova. 2000. "Heat-Death: Emergence and Control in Genetic Engineering and Artificial Life." *Ctheory,* May 10, 2000. http://www.ctheory.net/articles.aspx?id=127.

Park, Hyung Wook. 2016. *Old Age New Science: Gerontologists and Their Biosocial Visions, 1900–1960.* Pittsburgh: University of Pittsburgh Press.

Parry, Bronwyn. 2004. "Technologies of Immortality: The Brain on Ice." *Studies in History and Philosophy of Biological and Biomedical Sciences* 35:391–413.

Pascal, Blaise. 1958. *Pensées.* Trans. W. F. Trotter. Intro. T. S. Eliot. New York: Dutton.

Pasquale, Frank. 2013. "Two Concepts of Immortality: Reframing Public Debate on Stem-Cell Research." *Yale Journal of Law and the Humanities* 14 (1): 73–119.

Patterson, Orlando. 1984. *Slavery and Social Death: A Comparative Study.* Cambridge, Mass.: Harvard University Press.

Pauly, Phil. 1987. *Controlling Life: Jacques Loeb and the Engineering Ideal in Biology*. New York: Oxford University Press.

Pearl, Raymond. 1922. *The Biology of Death*. Philadelphia: Lippincott.

Pearl, Raymond. 1926. *Alcohol and Longevity*. New York: Knopf.

Pearl, Raymond. 1928. *The Rate of Living: Being an Account of Some Experimental Studies on the Biology of Life Duration*. New York: Knopf.

Pernick, Martin. 1988. "Back from the Grave: Recurring Controversies over Defining and Diagnosing Death in History." In *Death: Beyond Whole-Brain Criteria,* ed. Richard Zaner, 17–74. Boston: Kluwer.

Perry, Mike. 1991. "First Suspension No 'Blue Sky' Event." *Cryonics* 12 (7): 11–13. https://www.alcor.org/cryonics/cryonics9107.pdf.

Perry, Mike. 2004. "Deconstructing Deathism: Answering a Recent Critique and Other Objections to Immortality." *Physical Immortality* 2 (4): 11–16.

Peters, Ted. 2008. "Transhumanism and the Posthuman Future: Will Technological Progress Get Us There?" Metanexus Institute. http://www.metanexus.net/magazine/tabid/68/id/10546/Default.aspx#_ednref2.

Petryna, Adriana. 2002. *Life Exposed: Biological Citizens after Chernobyl*. Princeton: Princeton University Press.

Pew Forum on Religion and Public Life. 2010. *Religion among the Millennials*. Washington, D.C.: Pew Research Center.

Pew Research Center's Religion and Public Life Project. 2013. *Living to 120 and Beyond: Americans' Views on Aging, Medical Advances and Radical Life Extension*. Washington, D.C.: Pew Research Center. https://www.pewforum.org/2013/08/06/living-to-120-and-beyond-americans-views-on-aging-medical-advances-and-radical-life-extension/.

Pickering, Andrew. 2009. *The Cybernetic Brain*. Chicago: University of Chicago Press.

Pinker, Steven. 2011. *The Better Angels of Our Nature: Why Violence Has Declined*. New York: Penguin Books.

Piot, Charles. 2010. *Nostalgia for the Future: West Africa after the Cold War*. Chicago: University of Chicago Press.

Plessner, Helmut. 1983. "On the Relation of Time to Death." In *Man and Time,* ed. Joseph Campbell, 233–63. Princeton: Princeton University Press.

Porath, Nathan. 2007. "Being Human in a Dualistically-Conceived Embodied World: Descartes' Dualism and Sakais' Universalist Concepts of (Altered) Consciousness, Inner-Knowledge and Self." In *Anthropology and Science: Epistemologies in Practice,* ed. Jeanette Edwards, Penny Harvey, and Peter Wade, 187–204. New York: Berg.

Pottage, Alain. 2004. "The Fabrication of Persons and Things." In *Law, Anthropology, and the Constitution of the Social: Making Persons and Things,* ed. Alain Pottage and Martha Mundy, 1–39. New York: Cambridge University Press.

Pottage, Alain, and Martha Mundy, eds. 2004. *Law, Anthropology, and the Constitution of the Social: Making Persons and Things.* New York: Cambridge University Press.

Povinelli, Elizabeth. 2011. *Economies of Abandonment: Social Belonging and Endurance in Late Liberalism.* Durham: Duke University Press.

Povinelli, Elizabeth. 2016. *Geontologies: A Requiem to Late Liberalism.* Durham: Duke University Press.

President's Commission for the Study of Ethical Problems in Medicine and Biomedical and Behavioral Research. 1981. *Defining Death: Medical, Legal and Ethical Issues in the Definition of Death.* Washington, D.C.: U.S. Government Printing Office.

President's Council on Bioethics. 2003. *Beyond Therapy: Biotechnology and the Pursuit of Happiness.* Washington, D.C.: U.S. Government Printing Office.

Price, Michael. 2017. "Why Human Society Isn't More—or Less—Violent than in the Past." *Science,* December 15, 2017. http://www.sciencemag.org/news/2017/12/why-human-society-isn-t-more-or-less-violent-past.

Proudfoot, Wayne. 1985. *Religious Experience.* Berkeley: University of California Press.

Puig de la Bellacasa, María. 2017. *Matters of Care: Speculative Ethics in More Than Human Worlds.* Minneapolis: University of Minnesota Press.

Quandt, Jean. 1973. "Religion and Social Thought: The Secularization of Postmillennialism." *American Quarterly* 25 (4): 390–409.

Rabinow, Paul. 1996. *Making PCR: A Story of Biotechnology.* Chicago: University of Chicago Press.

Rabinow, Paul. 1999. *French DNA: Trouble in Purgatory.* Chicago: University of Chicago Press.

Rabinow, Paul. 2003. *Anthropos Today.* Princeton: Princeton University Press.

Rabinow, Paul, and Carlo Caduff. 2006. "Life—After Canguilhem." *Theory, Culture and Society* 23 (2–3): 329–30.

Rabinow, Paul, and Nikolas Rose. 2006. "Biopower Today." *BioSocieties* 1:195–217.

Radcliffe-Brown, A. 1940. "On Social Structure." *Journal of the Royal Anthropological Institute of Great Britain and Ireland* 70 (1): 1–12.

Radcliffe-Brown, A. 1964. *The Andaman Islanders*. New York: Free Press.

Radin, Joanna. 2013. "Latent Life: Concepts and Practices of Human Tissue Preservation in the International Biological Program." *Social Studies of Science* 43:484–508.

Radin, Joanna. 2017. *Life on Ice: A History of New Uses for Cold Blood*. Chicago: University of Chicago Press.

Radin, Joanna, and Emma Kowal, eds. 2017. *Cryopolitics: Frozen Life in a Melting World*. Cambridge: MIT Press.

Raman, Sujatha, and Richard Tutton. 2010. "Life, Science and Biopower." *Science, Technology, and Human Values* 35 (5): 711–34.

Ramsden, E. 2002. "Carving Up Population Science: Eugenics, Demography and the Controversy over the 'Biological Law' of Population Growth." *Social Studies of Science* 32 (5–6): 857–99.

Rancière, Jacques. 2012. *Proletarian Nights: The Workers' Dream in Nineteenth Century France*. Trans. John Drury. New York: Verso.

Rapp, Rayna. 1997. "Real-time Fetus." In *Cyborgs and Citadels: Anthropological Interventions in Emerging Sciences and Technologies*. ed. Gary L. Downey and Joseph Dumit, 31–48. Santa Fe: School of American Research Press.

Rapp, Rayna. 2003. "Cell Life and Death, Child Life and Death." In *Remaking Life and Death: Toward an Anthropology of the Biosciences*, ed. Sarah Franklin and Margaret Lock, 129–64. Santa Fe: School of American Research Press.

Rawls, John. 1997. "The Idea of Public Reason Revisited." *University of Chicago Law Review* 64 (3): 765–807.

Redfield, Peter. 2012. "Secular Humanitarianism and the Value of Life." In *What Matters? Ethnographies of Value in a Not So Secular Age*, ed. Courtney Bender and Ann Taves, 144–78. New York: Columbia University Press.

Regis, Ed. 1990. *Great Mambo Chicken and the Transhuman Condition*. New York: Addison-Wesley.

Regis, Ed. 2004. "The Incredible Shrinking Man." *Wired* 12 (10). https://www.wired.com/2004/10/drexler/.

Renan, Ernest. 1996 . "What Is a Nation?" In *Becoming National: A Reader*, ed. Geoff Eley and Ronald Grigor Suny, 41–55. New York: Oxford University Press.

Rice, N. L., Rev. (1871) 2005. *Immortality of the Soul and Destiny of the Wicked*. Ann Arbor: University of Michigan University Library.

Richards, Jay, ed. 2002. *Are We Spiritual Machines? Ray Kurzweil vs. the Critics of Strong AI*. Seattle: Discovery Institute.

Richardson, Robert. 2005. Introduction. In Baron Paul Henri Thiery d'Holbach, *The System of Nature*. Project Gutenberg E-text. http://www.gutenberg.org/dirs/etext05/8son110.txt.

Rievman, Ellen. 1976. "The Cryonics Society: A Study of Variant Behavior among the Immortalists." Unpublished dissertation, Florida Atlantic University.

Rifkin, Jeremy. 2002. "Fusion Biopolitics." *The Nation,* February 18, 2002. http://www.thenation.com/doc/20020218/rifkin.

Robbins, Joel. 2007. "Continuity Thinking and the Problem of Christian Culture: Belief, Time, and the Anthropology of Christianity." *Current Anthropology* 48 (1): 5–38.

Romain, Tiffany. 2010. "Extreme Life Extension: Investing in Cryonics for the Long, Long Term." *Medical Anthropology* 29 (2): 194–215.

Roosth, Sophia. 2014. "Life, Not Itself." *Grey Room* 57:56–81.

Roosth, Sophia, and Stefan Helmreich. 2010. "Life Forms: A Keyword Entry." *Representations* 112:27–53.

Rorty, Richard. 1989. *Contingency, Irony, and Solidarity.* New York: Cambridge University Press.

Rose, Deborah Bird, and Thom van Dooren, eds. 2011. "Introduction to 'Unloved Others: Death of the Disregarded in the Time of Extinctions.'" *Australian Humanities Review* 50:1–4.

Rose, Nikolas. 1996. *Inventing Ourselves: Psychology, Power, and Personhood.* New York: Cambridge University Press.

Rose, Nikolas. 2003. "Neurochemical Selves." *Society* 41 (1): 46–59.

Rose, Nikolas. 2007. *The Politics of Life Itself: Biomedicine, Power, and Subjectivity in the Twenty-First Century*. Princeton: Princeton University Press.

Rose, Nikolas, and Carlos Novas. 2005. "Biological Citizenship." In *Global Assemblages: Technology, Politics, and Ethics As Anthropological Problems,* ed. Aihwa Ong and Stephen Collier, 439–63. Malden, Mass.: Blackwell.

Rosenbaum, Stephen. 1993. "How to be Dead and Not Care." In *The Metaphysics of Death,* ed. John Martin Fischer, 117–34. Stanford: Stanford University Press.

Rosenberg, M., and L. I. Pearlin. 1978. "Social Class and Self-Esteem among Children and Adults." *American Journal of Sociology* 84:53–77.

Rosenblueth, Arturo, Norbert Wiener, and Julian Bigelow. 1943. "Behavior, Purpose and Teleology." *Philosophy of Science* 10 (1): 18–24.

Ross, Catherine E., and John Mirowsky. 1989. "Explaining the Social

Patterns of Depression: Control and Problem Solving—or Support and Talking." *Journal of Health and Social Behavior* 30:206–19.

Ross, Christopher. 2005. "Ufology as Anthropology." In *E.T. Culture: Anthropology in Outer Spaces,* ed. Debbora Battaglia, 38–93. Durham: Duke University Press.

Rostand, Jean. 1965. Preface. In Robert Ettinger, *The Prospect of Immortality,* 8–10. https://www.cryonics.org/images/uploads/misc/Prospect _Book.pdf.

Rothblatt, Martine. 2009. *Mindclone* blog. October 6, 2009. http://mind clones.blogspot.com/2009/10/7-how-can-we-know-consciousness-is .html.

Rothblatt, Martine. 2012. "The Terasem Mind Uploading Experiment." *International Journal of Machine Consciousness* 4 (1): 141–58.

Rottenburg, Richard, Sally E. Merry, Sung-Joon Park, and Johanna Mugler, eds. 2015. *The World of Indicators: The Making of Governmental Knowledge through Quantification*. Cambridge: Cambridge University Press.

Rouse, Carolyn. 2004. "'If She's a Vegetable, We'll Be Her Garden': Embodiment, Transcendence, and Citations of Competing Cultural Metaphors in the Case of a Dying Child." *American Ethnologist* 31 (4): 514–29.

Rousseau, Paul. 1995. "Hospice and Palliative Care." *Disease-a-Month* 41 (12): 773–842.

Russ, Ann Julienne. 2005. "Love's Labor Paid For: Gift and Commodity at the Threshold of Death." *Cultural Anthropology* 20 (1):128–55.

Russell, Bertrand. 1903. "The Free Man's Worship." http://www.positive atheism.org/hist/russell1.htm.

Ryff, C. D. 1989. "Happiness Is Everything, or Is It? Explorations on the Meaning of Psychological Well-Being." *Journal of Personality and Social Psychology* 57 (6): 1069–81.

Sancho, S., I. Casas, H. Ekwall, F. Saravia, H. Rodriguez-Martinez, J. E. Rodriguez-Gil, E. Flores, E. Pinart, M. Briz, N. Garcia-Gil, J. Bassols, A. Pruneda, E. Bussalleu, M. Yeste, and S. Bonet. 2007. "Effects of Cryopreservation on Semen Quality and the Expression of Sperm Membrane Hexose Transporters in the Spermatozoa of Iberian Pigs." *Reproduction* 134 (1): 111–21.

Sandberg, A., and N. Bostrom. 2008. "Whole Brain Emulation: A Roadmap." Technical report 2008-3, Future of Humanity Institute, Oxford University. http://www.fhi.ox.ac.uk/Reports/2008-3.pdf.

Sanders, Jason L., and Anne B. Newman. 2013. "Telomere Length in Epidemiology: A Biomarker of Aging, Age-Related Disease, Both, or Neither?" *Epidemiologic Review* 35:112–31.

Sanford, Charles. 1961. *The Quest for Paradise: Europe and the American Moral Imagination*. Urbana: University of Illinois Press.

Sartre, Jean-Paul. 1956. *Being and Nothingness*. New York: Philosophical Library.

Scheffler, Samuel. 2013. *Death and the Afterlife*. Oxford: Oxford University Press.

Scheiderman, Lawrence J., and Nancy S. Jecker. 1995. *Wrong Medicine: Doctors, Patients, and Futile Treatment*. Baltimore: Johns Hopkins University Press.

Scheier, Michael F., and Charles S. Carver. 1987. "Dispositional Optimism and Physical Well-being: The Influence of Generalized Expectancies on Health." *Journal of Personality* 55:169–210.

Scheier, Michael F., and Charles S. Carver. 1993. "On the Power of Positive Thinking: The Benefits of Being Optimistic." *Current Directions in Psychological Sciences* 2:26–30.

Scheper-Hughes, Nancy. 2000. "The Global Traffic in Human Organs." *Current Anthropology* 41 (2): 191–224.

Scheper-Hughes, Nancy, and Margaret Lock. 1987. "The Mindful Body: A Prolegomenon to Future Work in Medical Anthropology." *Medical Anthropology Quarterly* 1:6–41.

Schloendorn, John. 2006. "Making the Case for Human Life Extension: Personal Arguments." *Bioethics* 20 (4): 191–202.

Schmidt, Paul. 2006. "Basile J. Luyet and the Beginnings of Transfusion Cryobiology." *Transfusion Medicine Reviews* 20 (3): 242–46.

Schopenhauer, Arthur. 2005. "The Emptiness of Existence." In *Essays of Schopenhauer*. Ebook. Adelaide: University of Adelaide Library. https://ebooks.adelaide.edu.au/s/schopenhauer/arthur/essays/.

Schopenhauer, Arthur. 2010. "On the Vanity of Existence." In *Studies in Pessimism*, trans. T. Bailey Saunders. http://ebooks.adelaide.edu.au/s/schopenhauer/arthur/pessimism/chapter2.html.

Scott, Felicity. 2009. "Fluid Geographies: Politics and the Revolution by Design." In *New Views on Buckminster Fuller,* ed. Hsiao-Yun Chu and Roberto Trujillo, 160–76. Stanford: Stanford University Press.

Searle, John. 2002. "I Married a Computer." In *Are We Spiritual Machines? Ray Kurzweil vs. the Critics of Strong AI,* ed. Jay Richards, 56–76. Seattle: Discovery Institute.

Segal, Alan. 2004. *Life after Death: A History of the Afterlife in the Religions of the West*. New York: Doubleday.

Segal, Howard. 1985. *Technological Utopianism in American Culture*. Chicago: University of Chicago Press.

Seremetakis, C. Nadia. 1991 *The Last Word: Women, Death, and Divination in Inner Mani*. Chicago: University of Chicago Press.

Seymour, Jayne E. 2001. *Critical Moments—Death and Dying in Intensive Care*. Buckingham, UK: Open University Press.

Shannon, Claude, and Warren Weaver. 1949. *The Mathematical Theory of Communication*. Urbana: University of Illinois Press.

Shapin, Steven, and Simon Schaffer. (1985) 2005. *Leviathan and the Air-Pump*. Princeton: Princeton University Press.

Sharma, Sarah. 2015. *In the Meantime: Temporality and Cultural Politics*. Durham: Duke University Press.

Sharp, Lesley. 1995. "Organ Transplantation as a Transformative Experience: Anthropological Insights into the Restructuring of the Self." *Medical Anthropology Quarterly* 9 (3): 357-89.

Sharp, Lesley. 2001. "Commodified Kin: Death, Mourning, and Competing Claims on the Bodies of Organ Donors in the United States." *American Anthropologist* 103 (1): 112-33.

Sharp, Lesley. 2007. *Bodies, Commodities, and Biotechnologies*. New York: Columbia University Press.

Sherlock, Alexandra. 2013. "Larger than Life: Digital Resurrection and the Re-enchantment of Society." *Information Society* 29:164-76.

Sheskin, Arlene. 1979. *Cryonics: A Sociology of Death and Bereavement*. New York: Irvington.

Shewmon, D. Alan. 2004. "The ABC of PVS: Problems of Definition." In *Brain Death and Disorders of Consciousness*, ed. Calixto Machado and D. Alan Shewmon, 215-28. New York: Kluwer.

Shugarman, Lisa, Sandra Decker, and Anita Bercovitz. 2009. "Demographic and Social Characteristics and Spending at the End of Life." *Journal of Pain and Symptom Management* 38 (1): 15-26.

Shweder, Richard, and Michael Bourne. 1984. "Does the Concept of the Person Vary Cross-Culturally?" In *Culture Theory: Essays on Mind, Self, and Emotion*, ed. Richard A. Shweder and Robert A. LeVine, 158-99. New York: Cambridge University Press.

Sibley, Mulford. 1971. *Technology and Utopian Thought*. Minneapolis: Burgess.

Simpson, Audra. 2014. *Mohawk Interruptus*. Durham: Duke University Press.

Singer, Peter. 1994. *Rethinking Life and Death: The Collapse of Our Traditional Ethics*. New York: St. Martin's.

Sitkoff, Robert H., and Max M. Schanzenbach. 2008. "Perpetuities,

Taxes, and Asset Protection: An Empirical Assessment of the Jurisdictional Competition for Trust Funds." In Tina Portando, ed., *Annual Heckerling Institute on Estate Planning*. Vol. 42, Chapter 12, 2008; Harvard Law and Economics Discussion Paper No. 609. Available at SSRN: http://ssrn.com/abstract=1095421.

Skrbina, David. 2005. *Panpsychism in the West*. Cambridge: MIT Press.

Slaughter, V., and M. Lyons. 2003. "Learning about Life and Death in Early Childhood." *Cognitive Psychology* 46:1–30.

Sloane, David Charles. 1991. *The Last Great Necessity: Cemeteries in American History*. Baltimore: Johns Hopkins University Press.

Smith, Daniel. J. 2004. "Burials and Belonging in Nigeria: Rural-Urban Relations and Social Inequality in a Contemporary African Ritual." *American Anthropologist* 106 (2): 569–79.

Smith, Richard. 2000. "A Good Death: An Important Aim for Health Services and for Us All." *British Medical Journal* 320:129–30.

Snow, Robert Emerson. 1967. "The Problem of Certainty: Bacon, Descartes, and Pascal." Unpublished PhD diss., Indiana University, 1967. Ann Arbor, Mich.: University Microfilms, 1970.

Sorabji, Richard. 1999. "Soul and Self in Ancient Philosophy." In *From Soul to Self*, ed. James Crabbe, 8–32. New York: Routledge.

Speece, M. W., and S. B. Brent. 1984. "Children's Understanding of Death: A Review of Three Components of a Death Concept." *Child Development* 55:1671–86.

Steinhart, Eric. 2014. *Your Digital Afterlives*. Palgrave Macmillan UK. Kindle edition.

Steinhauser, Karen E., Nicholas A. Christakis, Elizabeth C. Clipp, Maya McNeilly, Lauren McIntyre, and James A. Tulsky. 2000. "Factors Considered Important at the End of Life by Patients, Family, Physicians, and Other Care Providers." *Journal of the American Medical Association* 284 (19): 2476–82.

Stengers, Isabelle. 2005. "The Cosmopolitical Proposal." In *Making Things Public: Atmospheres of Democracy*, ed. Bruno Latour and Peter Weibel, 994–1003. Cambridge: MIT Press

Stengers, Isabelle. 2011. *Cosmopolitics II*. Trans. Robert Bonomo. Minneapolis: University of Minnesota Press.

Stevenson, Lisa. 2014. *Life beside Itself: Imagining Care in the Canadian Arctic*. Oakland: University of California Press.

Stokes, Patrick. 2015. "Deletion as Second Death: The Moral Status of Digital Remains." *Ethics and Information Technology* 17 (4): 1–12.

Stoler, Ann Laura. 2016. *Duress: Imperial Durabilities in Our Times*. Durham: Duke University Press.

Stoler, Ann Laura. 2017. "'Interior Frontiers' as Political Concept, Di-
agnostic, and Dispositif." Hot Spots, Cultural Anthropology website,
January 18, 2017. https://culanth.org/fieldsights/1045-interior-frontiers
-as-political-concept-diagnostic-and-dispositif.

Stone, Arthur A., Joseph E. Schwartz , Joan E. Broderick, and Angus
Deaton. 2010. "A Snapshot of the Age Distribution of Psychological
Well-Being in the United States." *Proceedings of the National Academy of
Sciences USA* 107 (38): 16489–93

Strathern, Marilyn. 1988. *The Gender of the Gift: Problems with Women and
Problems with Society in Melanesia.* Berkeley: University of California
Press.

Strathern, Marilyn. 1992. *After Nature: English Kinship in the Late Twentieth Century.* Cambridge: Cambridge University Press.

Strenski, Ivan. 2002. *Contesting Sacrifice: Religion, Nationalism, and Social
Thought in France.* Chicago: University of Chicago Press.

Strozier, Charles. 1994. *Apocalypse: On the Psychology of Fundamentalism in
America.* Boston: Beacon Press.

Sudnow, David. 1967. *Passing On: The Social Organization of Dying.* Engle-
wood Cliffs, N.J.: Prentice Hall.

Sullivan, Lawrence. 2002. "The World and Its End: Cosmologies and Eschatologies of South American Indians." In *Native Religions and Cultures
of Central and South America,* ed. Lawrence Sullivan, 179–99. New York:
Continuum.

Sullivan, Winifred. 2007. *The Impossibility of Religious Freedom.* Princeton:
Princeton University Press.

Sunder Rajan, Kaushik. 2006. *Biocapital: The Constitution of Postgenomic
Life.* Durham: Duke University Press.

Sweeney, George. 1993. "Self-Immolation in Ireland: Hunger Strikes and
Political Confrontation." *Anthropology Today* 9 (5): 10–14.

Tallbear, Kim. 2017. "Beyond the Life/Not Life Binary: A Feminist-
Indigenous Reading of Cryopreservation, Interspecies Thinking and the
New Materialisms." In *Cryopolitics: Frozen Life in a Melting World,* ed.
Joanna Radin and Emma Kowal, 179–202. Cambridge: MIT Press.

Tambiah, Stanley Jeyaraja. (1990) 2000. *Magic, Science, Religion and the
Scope of Rationality.* Cambridge University Press.

Taussig, Karen-Sue, Klaus Hoeyer, and Stefan Helmreich. 2013. "The
Anthropology of Potentiality in Biomedicine." *Current Anthropology* 54
(S7): S3–S14.

Taussig, Michael. 1986. *Shamanism, Colonialism, and the Wild Man.* Chicago: University of Chicago Press.

Taylor, Anne C. 1996. "The Soul's Body and Its States: An Amazonian

Perspective on the Nature of Being Human." *Journal of the Royal Anthropological Institute* 2 (2): 201–15.

Taylor, Charles. 1991. *The Malaise of Modernity*. Toronto: Anansi.

Taylor, Charles. 2003. *Modern Social Imaginaries*. Durham: Duke University Press.

Taylor, Charles. 2007. *A Secular Age*. Cambridge: Harvard University Press.

Taylor, Timothy. 2004. *Buried Souls: How Humans Invented Death*. Boston: Beacon Press.

Tegmark, Max. 2017. *Life 3.0*. New York: Knopf.

Thacker, Eugene. 2005a. *Global Genome: Biotechnology, Politics, and Culture*. Cambridge: MIT Press.

Thacker, Eugene. 2005b. "Nomos, Nosos and Bios." *Culture Machine* 7. https://culturemachine.net/biopolitics/nomos-nosos-and-bios/.

Thomas, Yan. 2004. "Res Religiosae: On the Categories of Religion and Commerce in Roman Law." In *Law, Anthropology, and the Constitution of the Social: Making Persons and Things,* ed. Alain Pottage and Martha Mundy, 40–72. New York: Cambridge University Press.

Thompson, E. P. 1967. "Time, Work-Discipline, and Industrial Capitalism." *Past and Present,* no. 38: 56–97.

Thomson, Richard. 1964. "Current Applications of Biological Science: Part V. Cryobiology." *Bios* 35 (4): 202–7.

Ticktin, Miriam. 2011. *Casualties of Care: Immigration and the Politics of Humanitarianism in France*. Berkeley: University of California Press.

Tillich, Paul. (1952) 2000. *The Courage to Be*. New Haven: Yale University Press.

Timmermans, Stefan. 1996. "Saving Lives or Saving Multiple Identities? The Double Dynamic of Resuscitation Scripts." *Social Studies of Science* 26 (4): 767–97.

Timmermans, Stefan. 1999. *Sudden Death and the Myth of CPR*. Philadelphia: Temple University Press.

Timmermans, Stefan. 2005. "Death Brokering: Constructing Culturally Appropriate Deaths." *Sociology of Health and Illness* 27 (7): 993–1013.

Timmermans, Stefan. 2006. *Postmortem: How Medical Examiners Explain Suspicious Death*. Chicago: University of Chicago Press.

Triantaphyllou, A. C., and E. McCabe. 1989. "Efficient Preservation of Root-Knot and Cyst Nematodes in Liquid Nitrogen." *Journal of Nematology* 21 (3): 423–26.

Trummer, H., K. Tucker, C. Young, N. Kaula, and R. Meacham. 1998. "Effect of Storage Temperature on Sperm Cryopreservation." *Fertility and Sterility* 70 (6): 1162—64.

Turner, Bryan. 2009. *Can We Live Forever? A Sociological and Moral Inquiry*. New York: Anthem Press.

Turner, Fred. 2006. *From Counterculture to Cyberculture*. Chicago: University of Chicago Press.

Turner, Frederick Jackson. (1893) 1920. *The Frontier in American History*. University of Virginia. http://xroads.virginia.edu/~Hyper/TURNER/.

Tuveson, Ernest Lee. 1968. *Redeemer Nation: The Idea of America's Millennial Role*. Chicago: University of Chicago Press.

Tylor, Edward Burnett. (1871) 1958. *Primitive Culture*. Vol. 2 of *Religion in Primitive Culture*. New York: Harper and Row.

Tyndall, John. 1874. "Address Delivered before the British Association Assembled at Belfast, with Additions." *Victorian Web,* Digitized and formatted from John Tyndall's original "Address Delivered Before the British Association Assembled at Belfast, with Additions," London: Longmans, Green, and Co. http://www.victorianweb.org/science/science_texts/belfast.html.

Unamuno, Miguel de. 1972. *The Tragic Sense of Life in Men and Nations*. Trans. Anthony Kerrigan. Princeton: Princeton University Press.

Valentine, David. 2012. "Exit Strategy: Profit, Cosmology, and the Future of Humans in Space." *Anthropological Quarterly* 85 (4): 1045–67.

Valentine, David. 2016. "Atmosphere: Context, Detachment, and the Modern Subject in Outer Space." *American Ethnologist* 43 (3): 511–24.

Valentine, Stephen. 2009. *Timeship: The Architecture of Immortality*. Mulgrave, Australia: Images.

Van Nedervelde, Philippe. 2008. "Awaken the Universe—Introducing the Order of Cosmic Engineers." *Journal of Geoethical Nanotechnology* 3 (3). http://www.terasemjournals.org/GNJournal/GN0303/pvn2.html.

Veatch, Robert. 1989. *Death, Dying, and the Biological Revolution: Our Last Quest for Responsibility*. New Haven: Yale University Press.

Veatch, Robert. 1999. "The Conscience Clause: How Much Individual Choice in Defining Death Can Our Society Tolerate?" In *The Definition of Death: Contemporary Controversies,* ed. Stuart J. Youngner, Robert M. Arnold, and Renie Schapiro, 137–60. Baltimore: Johns Hopkins University Press.

Veatch, Robert. 2005. "The Death of Whole-Brain Death: The Plague of the Disaggregators, Somaticists, and Mentalists." *Journal of Medicine and Philosophy* 30:353–78.

Verducci, Tom. 2003. "What Really Happened to Ted Williams." *Sports Illustrated,* August 18, 2003, 68–73.

Vinge, Vernor. 1993. "The Coming Technological Singularity: How to

Survive in the Post-Human Era." Paper for VISION-21 Symposium, NASA Lewis Research Center and the Ohio Aerospace Institute, March 30–31, 1993. https://edoras.sdsu.edu/~vinge/misc/singularity.html.

Virno, Paolo. 2004. *A Grammar of the Multitude: For an Analysis of Contemporary Forms of Life.* Trans. Isabella Bertoletti, James Cascaito, and Andrea Casson. New York: Semiotext[e].

Visvanathan, Shiv. 2003. "Progress and Violence." In *Living with the Genie: Essays on Technology and the Quest for Human Mastery,* ed. Alan Lightman, Daniel Sarewitz, and Christina Desser, 157–80. Washington, D.C.: Island Press.

Viveiros de Castro, Eduardo. 1998. "Cosmological Deixis and Amerindian Perspectivism." *Journal of the Royal Anthropological Institute* 4 (3): 469–88.

Waldby, Catherine, and Robert Mitchell. 2006. *Tissue Economies: Blood, Organs, and Cell Lines in Late Capitalism.* Durham: Duke University Press.

Walter, Tony. 1991. "Modern Death: Taboo or Not Taboo." *Sociology* 25 (2): 293–310.

Walter, Tony. 1997. "Secularisation." In *Death and Bereavement across Cultures,* ed. Colin Murray Parkes, Pittu Laungani, and Bill Young, 133–48. New York: Routledge.

Walter, Tony. 2005. "Mediator Deathwork." *Death Studies* 29:383–412.

Warner, W. Lloyd. 1959. *The Living and the Dead.* New Haven: Yale University Press.

Wartofsky, Marx. 1988. "Beyond a Whole-Brain Definition of Death." In *Death: Beyond Whole-Brain Criteria,* ed. Richard Zaner, 219–28. Boston: Kluwer.

Washington, Harriet A. 2008. *Medical Apartheid: The Dark History of Experiments on Black Americans from Colonial Times to the Present.* New York: Anchor Books.

Weaver, Warren. 1949. "Recent Contributions to the Mathematical Theory of Communication." In *The Mathematical Theory of Communication,* ed. Claude E. Shannon and Warren Weaver, 1–28. Urbana: University of Illinois Press.

Weber, Max. (1921) 1978. *Economy and Society: An Outline of Interpretive Sociology.* Ed. Guenther Roth and Claus Wittich. Berkeley: University of California Press.

Weber, Max. 1946a. *From Max Weber: Essays in Sociology.* Trans. H. H. Gerth and C. Wright Mills. New York: Oxford University Press.

Weber, Max. 1946b. "Science as a Vocation." In *From Max Weber: Essays in Sociology,* trans. H. H. Gerth and C. Wright Mills, 129–56. New York: Oxford University Press.

Weber, Max. 1958. *The Protestant Ethic and the Spirit of Capitalism.* New York: Scribner.

Weber, Max. 1963. *The Sociology of Religion.* Trans. Ephraim Fischoff. Boston: Beacon Press.

Weheliye, Alexander G. 2014. *Habeas Viscus: Racializing Assemblages, Biopolitics, and Black Feminist Theories of the Human.* Durham: Duke University Press.

Weinberg, Albery. 1963. *Manifest Destiny: A Study of Nationalist Expansionism in American History.* Chicago: Quadrangle Books.

Weintraub, Arlene. 2010. *Selling the Fountain of Youth: How the Anti-Aging Industry Made a Disease Out of Getting Old—and Made Billions.* New York: Basic Books.

Weismann, August. 1891. *Essays upon Heredity and Kindred Biological Problems.* Trans. and ed. Edward Poulton, Selmar Shonland, and Arthur Shipley. Vol. 1. Oxford: Clarendon Press.

Welch, A. Wesley. 2004. "Law and the Making of Slavery in Colonial Virginia." *Ethnic Studies Review* 27 (1): 1–22.

Wheeler, John Archibald. 1990. "Information, Physics, Quantum: The Search for Links." In *Complexity, Entropy and the Physics of Information,* ed. Wojciech H. Zurek, 3–28. Santa Fe Institute Studies in the Science of Complexity. Redwood, Calif.: Addison-Wesley.

Whetstine, Leslie, Stephan Streat, Mike Darwin, and David Crippen. 2005. "Review: Pro/Con Ethics Debate: When Is Dead Really Dead?" *Critical Care* 9:538–42.

White, Patricia. 1988. "Should the Law Define Death?—A Genuine Question." In *Death: Beyond Whole-Brain Criteria,* ed. Richard Zaner, 101–9. Boston: Kluwer.

Whitehead, Alfred N. (1925) 1967. *Science and the Modern World.* New York: Free Press.

Whitmarsh, Ian, and Elizabeth F. S. Roberts. 2016. "Nonsecular Medical Anthropology." *Medical Anthropology* 35 (3): 203–8.

Wiener, Norbert. (1954) 1989. *The Human Use of Human Beings: Cybernetics and Society.* London: Free Association Books.

Wilkinson, R. 2004. "Why Is Violence More Common Where Inequality Is Greater?" *Annals of the New York Academy of Sciences* 1036:1–12.

Willerslev, Rane, Dorthe Reflund Christensen, and Lotte Meinert. 2013. Introduction. In *Taming Time, Timing Death: Social Technologies and*

Ritual, ed. Dorthe Reflund Christensen and Rane Willerslev, 1–16. London: Routledge.

Winter, Jay. 1995. *Sites of Memory, Sites of Mourning: The Great War in European Cultural History.* London: Cambridge University Press.

Wojcik, Daniel. 1997. *The End of the World as We Know It: Faith, Fatalism, and Apocalypse in America.* New York: NYU Press.

Wolfe, Cary. 2010. *What Is Posthumanism?* Minneapolis: University of Minnesota Press.

Woodburn, James. 1982. "Social Dimensions and Death in Four African Hunting and Gathering Societies." In *Death and the Regeneration of Life,* ed. Maurice Bloch and Jonathan Parry, 187–210. New York: Cambridge University Press.

World Transhumanist Association. 2008. *Report on the 2007 Interests and Beliefs Survey of the Members of the World Transhumanist Association.* Hartford, Conn.: World Transhumanist Association.

Wowk, Brian, Eugen Leitl, Christopher M. Rasch, Nooshin Mesbah-Karimi, Steven B. Harris, and Gregory M. Fahy. 2000. "Vitrification Enhancement by Synthetic Ice Blocking Agents." *Cryobiology* 40 (3): 228–36.

Wynter, Sylvia. 2003. "Unsettling the Coloniality of Being/Power/Truth/Freedom: Towards the Human, after Man, Its Overrepresentation—An Argument." *CR: The New Centennial Review* 3 (3): 257–337.

Young, Simon. 2006. *Designer Evolution: A Transhumanist Manifesto.* Amherst, N.Y.: Prometheus Books.

Youngner, Stuart J., Robert M. Arnold, and Renie Schapiro, eds. 1999. *The Definition of Death: Contemporary Controversies.* Baltimore: Johns Hopkins University Press.

Yudkowski, Eliezer. 2008. "Artificial Intelligence as a Positive and Negative Factor in Global Risk." In *Global Catastrophic Risks,* ed. Nick Bostrom and Milan Cirkovic, 308–43. Oxford: Oxford University Press.

Zaleski, Carol. 1996. *The Life of the World to Come: Near-Death Experience and Christian Hope.* New York: Oxford University Press.

Zaner, Richard, ed. 1988. *Death: Beyond Whole-Brain Criteria.* Boston: Kluwer.

Zee, Jacqueline. 2008. "The Revised Uniform Anatomical Gift Act: Bringing California Donation Law up to Contemporary Medical, Legal, and Bioethical Practices." *McGeorge Law Review* 39 (2): 529–43.

Zeller, Benjamin. 2010. *Prophets and Protons: New Religious Movements and Science in Late Twentieth Century America.* New York: NYU Press.

Zeller, Benjamin. 2011. "Three Physicists on Religion: When Scientists

See the Light." Metanexus Institute. http://www.metanexus.net/magazine/tabid/68/id/11001/Default.aspx.

Zhang, Baohui, Alexi Wright, Haiden A. Huskamp, Matthew E. Nilsson, Matthew L. Maciejewski, Craig C. Earle, Susan D. Block, Paul K. Maciejewski, Holly G. Prigerson. 2009. "Health Care Costs in the Last Week of Life: Associations with End-of-Life Conversations." *Archive of Internal Medicine* 169 (5): 480–88.

Ziewitz, Malte. 2016. "Governing Algorithms: Myth, Mess, and Methods." *Science, Technology and Human Values* 41:3–16.

Zigon, Jarrett. 2018. *A War on People: Drug User Politics and a New Ethics of Community*. Berkeley: University of California Press.

Zuckerman, Phil. 2016. *The Nonreligious: Understanding Secular People and Societies*. New York: Oxford University Press.

Zussman, R. 1992. *Intensive Care: Medical Ethics and the Medical Profession*. Chicago: University of Chicago Press.

Index

Abou Farman is assistant professor of anthropology at The New School for Social Research and author of *Clerks of the Passage*. As part of the artist duo caraballo-farman, he has exhibited internationally, including at the Tate Modern in the United Kingdom and MoMA PS1 in New York, and has received several grants and awards, including New York Foundation for the Arts and Guggenheim fellowships.